Lecture Notes in Control and Information Sciences 421

Qingling Zhang, Chao Liu, and Xue Zhang

Complexity, Analysis and Control of Singular Biological Systems

Authors
Qingling Zhang
Northeastern University
College of Sciences
Institute of Systems Science
Shenyang
China, People's Republic

Xue Zhang
Northeastern University
College of Sciences
Shenyang
China, People's Republic

Chao Liu
Northeastern University
College of Sciences
Shenyang
China, People's Republic

ISSN 0170-8643
e-ISSN 1610-7411
ISBN 978-1-4471-2302-6
e-ISBN 978-1-4471-2303-3
DOI 10.1007/978-1-4471-2303-3
Springer London Heidelberg New York Dordrecht

Library of Congress Control Number: 2011943813

Printed on acid-free paper

Springer is part of Springer Science+Business Media (www.springer.com)

To my Mother Fengzhi Hou.
In memory of my parents.

Qingling Zhang

Foreword

In this book, singular system theory is utilized to model the biological systems in the real world, some singular biological systems are established, and some qualitative analysis of proposed models are performed. To ensure the highest degree of reliability, accuracy, and quality in the content of this book, the collaboration of experts was enlisted, and this book presents their cooperative efforts. I believe that a serious effort has been made by the contributing authors in this book to present theory and application that will have lasting value. It is sincerely hoped that this book will be useful to the graduate students, researchers and practitioners, and others in the complexity analysis and control of biological systems.

Chinese Academy of Sciences, Beijing, P R China, *Lansun Chen*
July 25, 2011

Preface

Singular systems have been extensively investigated in the past three decades. As a valuable tool for system modeling and analysis, singular system theory has been widely utilized in different fields including nonlinear electric and electronic circuits, constrained mechanics, networks and economy. This book concentrates on the application of singular system theory and control theory into biological systems in the real world.

This book can be used in a graduate course on control theory and application. Additionally, different reader profiles might benefit from its contents. I have tried to write this book in a self-contained manner, making it accessible to as many interested readers as possible. Some background materials have been added with this aim; this includes several existence theorems of bifurcation phenomena, and detailed introductions to fundamental tools for discussing complex dynamical behavior, such as periodical, quasiperiodical, and chaotic behavior.

The work was partially supported by the National Natural Science Foundation of China under Grant Number 60974004 and 60574011.

Finally, I would like to thank in advance all the readers for choosing to read this book and be indeed very grateful if readers could possibly provide, via email: qlzhang@mail.neu.edu.cn, feedback about any problems found. Your help will certainly make any future editions of the book much better.

Northeastern University, Shenyang, P R China,

Qingling Zhang

May 13, 2011

Acknowledgements

I wish to thank Oliver, commissioning editor of Springer-Verlag, for his kind invitation to undertake this project.

Contents

Chapter 1
Introduction

Singular systems (differential-algebraic systems, descriptor variable systems, generalized state space systems, semistate systems, singular singularly perturbed systems, degenerate systems, constrained systems, etc.), which have been investigated over the past decades, are rather general kind of equations [4, 19]. They are established according to relationships among the variables. Naturally, it is usually differential or algebraic equations that form the mathematical model of the system, or the descriptor equation. The general form of a differential-algebraic system is as follows:

$$\begin{cases} A(t)\dot{x}(t) = G(x(t), u(t), t), \\ y(t) = K(x(t), u(t), t), \end{cases} \tag{1.1}$$

where $x(t)$ is the state of the system composed of state variables; $u(t)$ is the control input; $y(t)$ is the measure output; and $G(\cdot)$, $K(\cdot)$ are appropriate dimensional vector functions in $x(t)$, $u(t)$, and t. The matrix $A(t)$ may be singular. Differential-algebraic systems are suitable for describing systems that evolve over time. Especially, nonlinear differential-algebraic equations are the natural outcome of component-based modeling of complex dynamic systems. Compared with the ordinary differential systems, the advantage they offer over the more often used ordinary differential equations is that they are generally easier to formulate. The price paid is that they are more difficult to deal with (see references [4, 19] and references cited therein).

In general, differential-algebraic model systems exhibit more complicated dynamics than ordinary differential models. The differential-algebraic systems have been applied widely in power systems, aerospace engineering, chemical processes, social economic systems, biological systems, network analysis, etc. With the help of the differential-algebraic model for the power systems and bifurcation theory, complex dynamical behaviors of the power systems, especially the bifurcation phenomena, which can reveal the instability mechanism of power systems, have been extensively studied (see references [1, 13, 20] and the references therein). Furthermore, some applications of differential-algebraic models in the field of economics are given in [11, 12, 3, 22, 21, 17] in chemical processes [10] and in aerospace engineering [14]. In the past decades, ordinary differential equations system theory has

Q. Zhang et al.: Complex., Analy. and Cntrl. of Singular Biological Systems, LNCIS 421, pp. 1–5.
springerlink.com

been applied in the research field of biology [5, 7, 6, 8, 9, 15]. However, as far as the singular system theory is concerned, the related research results are few.

Example 1.1[4] Consider a class of interconnected large-scale systems with subsystems of

$$\begin{cases} \dot{x}_i(t) = A_i x_i(t) + B_i a_i(t), \\ b_i(t) = C_i x_i(t) + D_i a_i(t), \end{cases} \tag{1.2}$$

where $x_i(t)$, $a_i(t)$, and $b_i(t)$ $i = 1,2,\cdots,N$ are the substate, control input, and output of the ith subsystem, respectively. By denoting

$$x(t) = \begin{pmatrix} x_1(t) \\ x_2(t) \\ \vdots \\ x_N(t) \end{pmatrix}, a(t) = \begin{pmatrix} a_1(t) \\ a_2(t) \\ \vdots \\ a_N(t) \end{pmatrix}, b(t) = \begin{pmatrix} b_1(t) \\ b_2(t) \\ \vdots \\ b_N(t) \end{pmatrix},$$

$$A = \mathrm{diag}(A_1, A_2, \cdots, A_N), B = \mathrm{diag}(B_1, B_2, \cdots, B_N),$$
$$C = \mathrm{diag}(C_1, V_2, \cdots, C_N), D = \mathrm{diag}(D_1, D_2, \cdots, D_N),$$

(1.2) can be rewritten as

$$\begin{cases} \dot{x}(t) = Ax(t)) + Ba(t), \\ b(t) = Cx(t) + Da(t). \end{cases} \tag{1.3}$$

Assume that the subsystem interconnection is the linear interconnection:

$$\begin{aligned} a(t) &= L_{11}b(t) + L_{12}u(t) + R_{11}a(t) + R_{12}y(t), \\ y(t) &= L_{21}b(t) + L_{22}u(t) + R_{21}a(t) + R_{22}y(t), \end{aligned} \tag{1.4}$$

where $u(t)$ is the overall input of the large-scale system; $y(t)$ is its overall measure output; $L_{ij}, R_{ij}, i, j = 1,2$, are constant matrices of appropriate dimensions. Equation (1.2) and (1.3) form a large-scale system. In fact, it is easy to show that the system composed by (1.2) and (1.3) could not be equivalent to a normal system. On the other hand, if we choose the state variable $[x^T(t), a^T(t), b^T(t), y^T(t)]^T$, (1.3) and (1.4) form the system

$$\begin{pmatrix} I\,0\,0\,0 \\ 0\,0\,0\,0 \\ 0\,0\,0\,0 \\ 0\,0\,0\,0 \end{pmatrix} \begin{pmatrix} \dot{x}(t) \\ \dot{a}(t) \\ \dot{b}(t) \\ \dot{y}(t) \end{pmatrix} = \begin{pmatrix} A & B & 0 & 0 \\ C & D & -I & 0 \\ 0 & R_{11}-1 & L_{11} & R_{12} \\ 0 & R_{21} & L_{21} & R_{22}-1 \end{pmatrix} \begin{pmatrix} x(t) \\ a(t) \\ b(t) \\ y(t) \end{pmatrix} + \begin{pmatrix} 0 \\ 0 \\ L_{12} \\ L_{22} \end{pmatrix} u(t),$$

$$y(t) = \begin{pmatrix} 0\,0\,0\,I \end{pmatrix} \begin{pmatrix} x^T(t) \; a^T(t) \; b^T(t) \; y^T(t) \end{pmatrix}^T,$$

which forms the linear singular system:

$$\begin{cases} E\dot{x}(t) = Ax(t) + Bu(t), \\ \quad y(t) = Cx(t), \end{cases} \tag{1.5}$$

where $x(t) \in R^n$, $u(t) \in R^m$, $y(t) \in R^r$, $E, A \in R^{n\times n}$, $B \in R^{n\times m}$ and $C \in R^{r\times n}$ are constant matrices. Model (1.4) is the main system investigated in this book.

Example 1.2[4] The fundamental dynamic Leontief model of economic systems is a singular system. Its description model [11] is as follows:

$$x(k) = Ax(k) + B[x(k+1) - x(k)] + d(k), \tag{1.6}$$

where $x(k)$ is the n dimensional production of n vector of n sectors; $A \in R^{n\times n}$ is an input-output (or production) matrix; $Ax(k)$ stands for the fraction of production required as input for the current production, $B \in R^{n\times n}$ is the capital coefficient matrix, and $B[x(k+1) - x(k)]$ is the amount for capacity expansion, which often appears in the form of capital. $d(k)$ is the vector that includes demand or consumption. Equation (1.6) may be rewritten as

$$Bx(k+1) = (I - A + B)x(k) - d(k).$$

In multisector economic systems, production augmentation in one sector often doesn't need the investment from all other sectors, and moreover, in practical cases only a few sectors can offer investment in capital to other sectors. Thus, most of the elements in B are zero except for a few. B is often singular. In this sense, the system (1.6) is a typical discrete singular system.

Example 1.3[4] When administration is included, the oil catalytic cracking is an extremely complicated process. Some companies reportedly found a description model for this process, whose simplification is

$$\begin{cases} \dot{x}_1(t) = A_{11}x_1(t) + A_{12}x_2(t) + B_1u(t) + F_1f(t), \\ 0 = A_{21}x_1(t) + A_{22}x_2(t) + B_2u(t) + F_2f(t), \end{cases} \tag{1.7}$$

where $x_1(t)$ is a vector to be regulated, such as regenerate temperature, valve position, blower capacity, etc; $x_2(t)$ is the vector reflection value; and $f(t)$ represents extra disturbances. Model (1.7) is a standard singular system.

Example 1.4[4] Consider a simple circuit network as shown in Fig. 1.1, where voltage source $V_s(t)$ is the driver (control input); $R, L,$ and C stand for the resistor, inductor, and capacity, respectively, as well as their quantities; and their voltages are denoted by $V_R(t), V_L(t)$, and $V_C(t)$, respectively. Then from Kirchoff's laws, the following circuit equations (description equations) can be obtained:

$$\begin{pmatrix} L & 0 & 0 & 0 \\ 0 & 0 & 1 & 0 \\ 0 & 0 & 0 & 0 \\ 0 & 0 & 0 & 0 \end{pmatrix} \begin{pmatrix} \dot{I}(t) \\ \dot{V}_L(t) \\ \dot{V}_C(t) \\ \dot{V}_R(t) \end{pmatrix} = \begin{pmatrix} 0 & 1 & 0 & 0 \\ \frac{1}{C} & 0 & 0 & 0 \\ -R & 0 & 0 & 1 \\ 0 & 1 & 1 & 1 \end{pmatrix} \begin{pmatrix} I(t) \\ V_L(t) \\ V_C(t) \\ V_R(t) \end{pmatrix} + \begin{pmatrix} 0 \\ 0 \\ 0 \\ -1 \end{pmatrix} V_S(t). \tag{1.8}$$

However, as far as the biological systems are concerned, the related research results are few. Biological systems are generally complete units or entities that have many

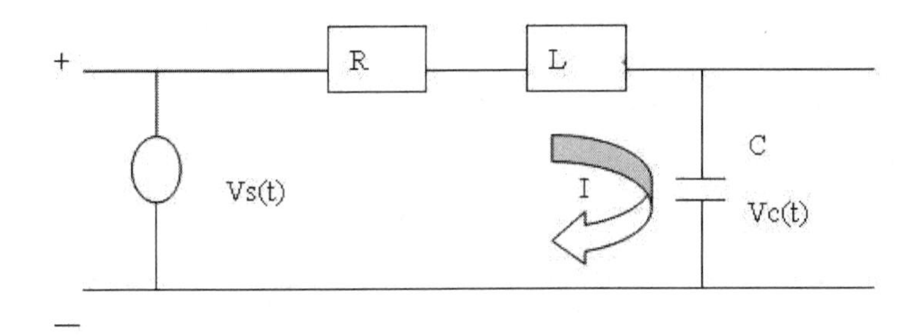

Fig. 1.1 A simple circuit network.

components interacting in a complicated way [16]. There are many species and a large amount of life on earth. Thus, it is very complicated to study the phenomenon of life. Biological complex systems are regarded as the most complex systems, with complexity higher than that of other systems [3]. From the point of view of methodology, the traditional breaking up to pieces and putting pieces together method is not suitable for most complex systems. Therefore, there is a strong need to resort to more advanced methods to study the original system as a whole. However, most of the developed theory for a complex system is still not up to a quantitative level because of the complexity of individuals interacting with each other in the system. Additionally, few reports on the quantitative study and associated control strategies of biological complex systems have been published [2, 18].

In this book, singular system theory is utilized to model the biological systems in the real world, and some singular biological systems are established. In Chapter 2 to Chapter 14, singular model systems are established to investigate the effects of harvesting on the dynamic behavior of a prey-predator system, and local stability analysis are performed to investigate the complex dynamical behavior and instability of model systems around the interior equilibrium, which are beneficial to studying coexistence and interaction mechanisms of population in biological systems. With the purpose of stabilizing a model system around the interior equilibrium, different kinds of controllers are designed that are utilized to discuss the effect of output control to restore the model system to a stable state, which are also theoretical guides to formulate related measures to maintain the sustainable development of population resources in such biological systems. Especially, Chapter 10 proposes a hybrid predator-prey economic model, which is formulated by differential-difference-algebraic equations. It shows that this hybrid model exhibits saddle-node bifurcation and singular induced bifurcation at the intersampling instants. And for the sampling instants, the system undergoes Neimark-Sacker bifurcation. In Chapter 12, a poly-chamber model of the endocrine disruptor-Diethylstibestrol-moving in a human body is established based on physiological rules. Passivity of this model is systematically analyzed, and a feedback controller is designed under a station of strict passivity. Furthermore, a singular model is constructed to express the transmission mechanics and mode of hepatitis B, and its associated control, s.t., quarantine, and isolation control is discussed. Finally, readers interested in applications of singular biological systems will find a discussion of singular harmful algal models and associated control problems in Chapter 15.

References

1. Ayasun, S., Nwankpa, C.O., Kwatny, H.G.: Computation of singular and singularity induced bifurcation points of differential-algebraic power system model. IEEE Trans. Cir. Syst. I 51(8), 1525–1537 (2004)
2. Bendoricchio, G., Jorgensen, S.E.: Fundamentals of Ecological Modeling. Elsevier, Amsterdam (2001)
3. Cheng, S.W., Feng, Z.Y.: Discovery of Complex Science. Democracy and Development Press, Beijing (1999)
4. Dai, L.: Singular Control System. Springer, New York (1989)
5. Hiroaki, K.: Foundations of Systems Biology. The MIT Press, Massachusetts (2001)
6. Hofbauer, J., Sigmund, K.: Evolutionary Games and Population Dynamics and Dynamical Systems. Cambridge University Press, Cambridge (1998)
7. Isidore, R., Gregory, S.: Systems Biology. Oxford University Press, Oxford (2006)
8. Klipp, E., Hertwig, R., Kowald, A., Wierling, C., Lehrach, H.: Systems Biology in Practice. Wiley-VCH, Berlin (2005)
9. Kot, M.: Elements of Mathematical Biology. Cambridge University Press, Cambridge (2001)
10. Kumar, A., Daoutidis, P.: Control of Nonlinear Differential-algebraic Equation Systems with Applications to Chemical Process. CRC Press, London (1999)
11. Luenberger, D.G.: Nonlinear descriptor systems. J. Eco. Dyn. Cont. 1, 219–242 (1979)
12. Luenberger, D.G., Arbel, A.: Singular dynamic Leontief systems. Econ. 45(32), 991–995 (1997)
13. Marszalek, W., Trzaska, Z.W.: Singularity-induced bifurcations in electrical power system. IEEE Trans. Pow. Syst. 20(1), 302–310 (2005)
14. Masoud, M., Masoud, S., Caro, L., Babak, N.A.: Introducing a new learning method for fuzzy descriptor systems with the aid of spectral analysis to forecast solar activity. J. Atmo. Sol.-terr. Phy. 68(18), 2061–2074 (2006)
15. May, R.M.: Stability and Complexity in Model Ecosystems. Princeton University Press, Princeton (1975)
16. Mitchell, W.: Complexity: The Emerging Science at the Edge of Order and Chaos. Touchstone, New York (1992)
17. Silva, M.S., Lima, T.P.D.: Looking for nonnegative solutions of a Leontief dynamic model. Lin. Alg. Appl. 364(1), 281–316 (2003)
18. Werner, K.: Amphibians as a model to study endocrince disruptors II: Estrogenic activity of environmental chemicals in vitro and in vivo. Sci. Tot. Envi. 225, 59–68
19. Yang, D.M., Zhang, Q.L.: Descriptor System. Science Press, Beijing (2004)
20. Yue, M., Schlueter, R.: Bifurcation subsystem and its application in power system analysis. IEEE Trans. Pow. Syst. 19(4), 1885–1893 (2004)
21. Zhang, J.S.: Singular System Economy Control Theory. Tsinghua Press, Beijing (1990)
22. Zhang, J.S.: Application of singular system theory in civil economy plan. Acta Auto. Sini. 15(5), 478–480 (1989)

Chapter 2
Preliminary Singular Systems Theory

2.1 Solution of Linear Singular System

Since the 1960s, much research has been extensively focused on system analysis and synthesis of dynamical system with the core features in modern control theory based on space system model. state space variable method not only provides us with a completely new method for system analysis and synthesis, but it also offers us more understanding of systems. Using the so-called state space variable method, state space models of systems are obtained.

Singular systems can be described as the following form:

$$\begin{cases} E(t)\dot{x}(t) = H(x(t),u(t),t), \\ y(t) = J(x(t),u(t),t), \end{cases} \tag{2.1}$$

where H, J are appropriate dimensional vector functions in $x(t)$, $u(t)$, and t. The matrix $E(t)$ may be singular. If H, J are linear functions of $x(t)$ and $u(t)$, another special form of the system (2.1) is a linear singular system

$$\begin{cases} E\dot{x}(t) = Ax(t) + Bu(t), \\ y(t) = Cx(t), \end{cases} \tag{2.2}$$

which is the main system studied in this chapter. Here, $x(t) \in R^n$, $u(t) \in R^m$, $y(t) \in R^r$. $E, A \in R^{n\times n}$, $B \in R^{n\times m}$, and $C \in R^{r\times n}$ are constant matrices.

For nonsingular matrix E, system (2.2) becomes

$$\begin{cases} \dot{x}(t) = E^{-1}Ax(t) + E^{-1}Bu(t), \\ y(t) = Cx(t). \end{cases} \tag{2.3}$$

Here, for singular systems mentioned above, we always mean the singular E, i.e., $\text{rank}E = q < n$.

In practical system analysis and control system design, many system models may be established in the form of (2.1), while they could not be described by (2.3).

Q. Zhang et al.: Complex., Analy. and Cntrl. of Singular Biological Systems, LNCIS 421, pp. 7–21.
springerlink.com

Now, we will introduce some common forms in singular systems[2].

(I) First Equivalent Form (EF1)

For any singular system (2.2), there exist two nonsingular matrices Q and P such that (2.2) is r.s.e. to

$$\begin{cases} \dot{x}_1 = A_1 x_1 + B_1 u, \\ y_1 = C_1 x_1, \end{cases} \tag{2.4}$$

$$\begin{cases} N\dot{x}_2 = x_2 + B_2 u, \\ y_2 = C_2 x_2, \end{cases} \tag{2.5}$$

$$y = C_1 x_1 + C_2 x_2 = y_1 + y_2, \tag{2.6}$$

with the coordinate transformation

$$\begin{bmatrix} x_1 \\ x_2 \end{bmatrix} = P^{-1}x, x_1 \in R^{n_1}, x_2 \in R^{n_2}, \tag{2.7}$$

and

$$\begin{aligned} QEP &= \text{diag}(I_{n_1}, N), \\ QAP &= \text{diag}(A_1, I_{n_2}), \\ QB &= \begin{pmatrix} B_1 \\ B_2 \end{pmatrix}, \\ CP &= [C_1 \quad C_2], \end{aligned} \tag{2.8}$$

where $n_1 + n_2 = n$, $N \in R^{n_2 \times n_2}$ is nilpotent.

Equation (2.4)-(2.6) is the EF1, usually called the standard decomposition. In this form, subsystems (2.4) and (2.5) are called slow and fast subsystems, respectively; x_1, x_2 are the slow and fast substates, respectively.

Generally, the matrices Q and P, which transfer a singular system into its standard EF1, are not unique, resulting in the nonuniqueness of EF1, i.e., $A_1, B_1, B_2, C_1, C_2, N$. Assume that $\bar{Q}$ and $\bar{P}$ are nonsingular and system (2.2) is transferred into its EF1, in other words, system (2.2) is r.s.e. to

$$\begin{cases} \dot{\bar{x}}_1 = \bar{A}_1 \bar{x}_1 + \bar{B}_1 u, \\ \bar{y}_1 = \bar{C}_1 \bar{x}_1, \end{cases} \tag{2.9}$$

$$\begin{cases} \bar{N}\dot{\bar{x}}_2 = \bar{x}_2 + \bar{B}_2 u, \\ \bar{y}_2 = \bar{C}_2 \bar{x}_2, \end{cases} \tag{2.10}$$

$$y = \bar{C}_1 \bar{x}_1 + \bar{C}_2 \bar{x}_2 = \bar{y}_1 + \bar{y}_2, \tag{2.11}$$

with the coordinate transformation $\bar{P}^{-1}x = [\bar{x}_1 / \bar{x}_2]$.

The following theorem shows the relationship between the matrices Q, P and $\bar{Q}$, $\bar{P}$, and the system coefficient matrices.

Theorem 2.1. *[2] Suppose that (2.4)-(2.11) are the EF1 for system (2.2). Then* $n_1 = \bar{n}_1$, $n_2 = \bar{n}_2$, *and there exist nonsingular matrices* $T_1 \in R^{n_1 \times n_1}$, $T_2 \in R^{n_2 \times n_2}$ *such that*

$$Q = diag(T_1, T_2)\bar{Q}, P = \bar{P}diag(T_1^{-1}, T_2^{-1}),$$

$$A_1 = T_1\bar{A}_1T_1^{-1}, N = T_2\bar{N}T_2^{-1},$$

$$B_i = T_i\bar{B}_i, C_i = \bar{C}_iT_i^{-1}, i = 1, 2.$$

Although different EF1s may be obtained under different coordinate transformation matrices, this theorem assures the similarity property among these EF1s. This means that EF1 is unique in the sense of similar equivalence.

(II) Second Equivalent Form (EF2)

Let $q = \text{rank}E$. From matrix theory, we know that there exist nonsingular matrices Q_1 and P_1 such that $Q_1EP_1 = diag(I_q, 0)$. By taking the coordinate transformation $P_1^{-1}x = [x_1/x_2]$, $x_1 \in R^q$, $x_2 \in R^{n-q}$, system (2.2) is r.s.e. to

$$\begin{cases} \dot{x}_1 = A_{11}x_1 + A_{12}x_2 + B_1u, \\ 0 = A_{21}x_1 + A_{22}x_2 + B_2u, \\ y = C_1x_1 + C_2x_2. \end{cases} \tag{2.12}$$

where

$$Q_1AP_1 = \begin{bmatrix} A_{11} & A_{12} \\ A_{21} & A_{22} \end{bmatrix},$$

$$Q_1B = \begin{bmatrix} B_1 \\ B_2 \end{bmatrix},$$

$$CP_1 = [C_1 \quad C_2].$$

Equation (2.12) is the second equivalent form (EF2) for system (2.2). In this transformation, matrices Q_1 and P_1 are not unique, which results in the nonuniqueness of EF2. Two EF2s may have a relationship too complicated to merit further study.

The EF2 clearly reflects the physical meaning of singular systems. In (2.12), the first equation is a differential one composed of dynamic subsystems, and the second is an algebraic equation that represents the connection between subsystems. Thus, singular systems may be viewed as a composite system formed by several interconnected subsystems. Furthermore, sub-states x_1 and x_2 reflect a layer property in some singular systems: one layer has a dynamic property (described by the differential equation); the other has interconnection, constraint, and administration properties (described by the algebraic equation).

Singular value decomposition may be used in the decomposition of E instead of rank decomposition here.

(III) Third Equivalent Form (EF3)

Under the regularity assumption, there always exists a scalar α (such as are numerous) such that $|\alpha E + A| \neq 0$. Let the transformation matrices be

$$Q_2 = (\alpha E + A)^{-1}, P_2 = I_n,$$

then $Q_2A = I_n - \alpha(\alpha E + A)^{-1}E$. Thus, system (2.2) is r.s.e. to

$$\begin{cases} \hat{E}\dot{x} = (I - \alpha\hat{E})x + \hat{B}u, \\ y = Cx, \end{cases} \tag{2.13}$$

where $\hat{E} = Q_2E$ and $\hat{B} = Q_2B$, which is the third equivalent form (EF3) for (2.2). Obviously, for a fixed α, the EF3 is unique in the sense of algebraic equivalence (similarity).

For simplicity, the bar will be omitted in the second equivalent form (2.9)-(2.11), where $x_1 \in R^{n_1}$, $x_2 \in R^{n_2}$, $n_1 + n_2 = n$, and N are nilpotent whose nilpotent index is denoted by h, and

$$QEP = \mathrm{diag}(I,N), QAP = \mathrm{diag}(A_1,I), CP = [C_1,C_2],$$

$$P^{-1}x = [x_1/x_2], QB = [B_1/B_2].$$

The state and output responses of singular system (2.2) are given by

$$\begin{aligned} x(t) &= P[I/0]x_1(t) + P[0/I]x_2(t) \\ &= P[I/0](e^{A_1t}x_1(0) + \int_0^t e^{A_1(t-\tau)}B_1u(\tau)\mathrm{d}\tau) - P[0/I]\sum_{i=0}^{h-1} N^iB_2u^{(i)}(t). \end{aligned}$$

$$\begin{aligned} y(t) &= Cx(t) \\ &= CP\begin{bmatrix} I \\ 0 \end{bmatrix}(e^{A_1t}x_1(0) + \int_0^t e^{A_1(t-\tau)}B_1u(\tau)\mathrm{d}\tau) - CP\begin{bmatrix} 0 \\ I \end{bmatrix}\sum_{i=0}^{h-1} N^iB_2u^{(i)}(t). \end{aligned}$$

Particularly by setting $t > 0$, $t \to 0^+$, we obtain

$$x(0^+) = P\begin{bmatrix} I \\ 0 \end{bmatrix}x_1(0) - P\begin{bmatrix} 0 \\ I \end{bmatrix}\sum_{i=0}^{h-1} N^iB_2u^{(i)}(0^+), \tag{2.14}$$

which is the so-called consistent initial condition imposed on initial value $x(0)$.

Compared with results in linear system theory, the normal system always has one unique solution if the input function is piecewise continuous, or weaker. However, equation (2.14) shows that the singular system (2.2) has a unique solution only for the consistent initial vector $x(0)$, and for the h times piecewise continuously differentiable input function $u(t)$. The latter is stronger than that in the normal system case. Such characters stand for the special feature of singular systems.

2.2 Time Domain Analysis

Practical systems can be established by selecting the proper state variables. Time domain analysis is the method of analyzing the system based on this description equation. By using time domain analysis, a fair understanding of the system's structural features and its internal properties may be obtained. In this section, we introduce the fundamentals in system theory such as reachability, controllability, observability, system decomposition, and transfer matrix.

Consider the regular singular system

$$\begin{cases} E\dot{x} = Ax + Bu, \\ y = Cx, \end{cases} \tag{2.15}$$

where $x \in R^n$, $u \in R^m$, $y \in R^r$ are its state, control input, and measure output, respectively. $E, A \in R^{n\times n}$, $B \in R^{n\times m}$, $C \in R^{r\times n}$ are constant matrices. Without loss of generality, $q = \text{rank}E < n$ is supposed. For the system, as pointed out in the previous section, two nonsingular matrices Q and P exist such that it is r.s.e. to (EF1):

$$\begin{cases} \dot{x}_1 = A_1 x_1 + B_1 u, \\ N\dot{x}_2 = x_2 + B_2 u, \\ y = C_1 x_1 + C_2 x_2, \end{cases} \tag{2.16}$$

where $x_1 \in R^{n_1}$, $x_2 \in R^{n_2}$, $n_1 + n_2 = n$, $N \in R^{n_2 \times n_2}$ is nilpotent, the nilpotent index is denoted by h, and

$$QEP = \text{diag}(I,N), QAP = \text{diag}(A_1,I), CP = [C_1 \quad C_2],$$

$$P^{-1}x = [x_1/x_2], QB = [B_1/B_2].$$

Definition 2.1. [4] Any vector $w \in R^n$ in $n-$dimensional vector space is said to be reachable. If there exists an initial condition $x_1(0)$, admissible control input $u(t) \in C_p^{h-1}$, and $t_1 > 0$ such that

$$x(t_1) = [x_1(t_1), x_2(t_2)] = w,$$

let $R(0)$ represent the state reachable set from the zero initial condition $(x_1(0) = 0)$, then

$$R(0) = < A_1|B_1 > \oplus < N|B_2 >,$$

where $\oplus$ is the direct sum in vector space.

Denote

$$H(x_1(0)) = \{x = [x_1/x_2] | x_1 = e^{A_1 t} x_1(0) \in R^{n_1}, x_2 = 0 \in R^{n_2}\},$$

which represents the free state reachable set from starting point $x_1(0)$.

If R is the reachable set for system (2.16) defined as the union of all reachable sets from all possible initial conditions $x_1(0) \in R^{n_1}$, i.e.,

$$R = \cup_{x_1(0)}(R(0) + H(x_1(0))) = R^{n_1} \oplus < N|B_2 > .$$

Definition 2.2. [2] System (2.16) is called controllable, if there exists a control input $u(t) \in C_p^{h-1}$ such that $x(t_1) = w$ for any $t_1 > 0$, $x(0) \in R^n$ and $w \in R^n$.

This definition states that under the controllability assumption, for any initial condition $x(0)$, we may always choose a control input such that the state response starting from $x(0)$ may arrive at any prescribed position in R^n in any given time period. It is easy to see the definitions natural generalization of controllability concepts in the normal case.

Rewriting system (2.16) in the slow-fast subsystem form , we get

$$\begin{cases} \dot{x}_1 = A_1 x_1 + B_1 u, \\ y_1 = C_1 x_1. \end{cases} \tag{2.17}$$

$$\begin{cases} N\dot{x}_2 = x_2 + B_2 u, \\ y_2 = C_2 x_2. \end{cases} \tag{2.18}$$

$$y = C_1 x_1 + C_2 x_2 = y_1 + y_2. \tag{2.19}$$

Theorem 2.2. *[2]*

(1) Slow subsystem (2.17) is controllable

$$rank[sE - A, B] = n, \forall s \in C,$$

where C represents the complex plane.

(2) The following statements are equivalent:

(a) The fast subsystem (2.18) is controllable,
(b) $rank[B_2, NB_2, \cdots, N^{h-1}B_2] = n_2$,
(c) $rank[N \quad B_2] = n_2$,
(d) $rank[E \quad B] = n$,
(e) For any nonsingular matrices Q_1 and P_1 satisfying $E = Q_1 diag(I, 0) P_1$, Let $Q_1 B = [\bar{B}_1 / \bar{B}_2]$. Then $\bar{B}_2$ is of full row rank, $rank\bar{B}_2 = n - rankE$,

(3) The following statements are equivalent:

(a) System (2.16) is controllable.
(b) Both its slow and fast subsystems are controllable.
(c) $rank[B_1, A_1 B_1, \cdots, A_1^{n_1-1} B_1] = n_1$, $rank[B_2, NB_2, \cdots, N^{h-1}B_2] = n_2$.
(d) $rank[sE - A \quad B] = n$, $\forall s \in C$, *and* $rank[E \quad B] = n$.
(e) The matrix

$$D_1 = \begin{bmatrix} -A & & & & B & & & \\ E & -A & & & & B & & \\ 0 & E & & & & & B & \\ & & \cdots & -A & & & & \cdots \\ & & & E & & & & B \end{bmatrix}_{n_2 \times (n+m-1)n}$$

is of full row rank n_2.

Definition 2.3. [1] Singular system (2.16) is called R-controllable, if it is controllable in the reachable set, or more precisely, for any prescribed $t_1 > 0$, $x_1(0) \in R$

and $w \in R$, there always exists an admissible control input $u(t) \in C_p^{h-1}$ such that $x(t_1) = w$.

The R-controllability guarantees our controllability for the system from any admissible initial condition $x_1(0)$ to any reachable state and this process will be finished in any given time period if the control $u(t)$ is suitably chosen.

Theorem 2.3. *[3] The following statements are equivalent:*
(a) Singular system (2.16) is R-controllable.
(b) The slow subsystem (2.17) is controllable.
(c) $rank[sE - A, B] = n$, $\forall s \in C$ *and* s *is finite.*
(d) $rank[B_1, A_1B_1, \cdots, A_1^{n_1-1}B_1] = n_1$.
(e) Denote

$$D_2 = \begin{bmatrix} -A & & & & B & & & \\ E & -A & & & & B & & \\ 0 & E & & & & & B & \\ & & \cdots & -A & & & & \cdots \\ & & & E & & & & B \end{bmatrix}_{kn\times(n+m)k}$$

where $k > n_1$ *may be any positive number; for example,* $k = rankE$ *or* n*. The matrix* D_2 *is of full row rank* kn*.*

Definition 2.4. [2] System (2.16) is termed impulse controllable, if for any initial condition $x(0)$, $\tau \in R$ and $w \in R^{n_2}$, there exists an admissible control input $u \in C_p^{h-1}$ such that

$$x_\tau(t) = I_\tau(w, t).$$

Theorem 2.4. *[2] The following statements are equivalent:*
(a) System (2.16) is impulse controllable.
(b) Its fast subsystem (2.18) is impulse controllable.
(c) $KerN + Im[B_2, NB_2, \cdots, N^{h-1}B_2] = R^{n_2}$.
(d) $ImN = Im[NB_2, N^2B_2, \cdots, N^{h-1}B_2]$.
(e) $ImN + ImB_2 + KerN = R^{n_2}$.
(f) Let the controllability decomposition of (N, B_2) *be*

$$\left(\begin{bmatrix} N_{11} & N_{12} \\ 0 & N_{22} \end{bmatrix}, \begin{bmatrix} B_{21} \\ 0 \end{bmatrix} \right),$$

then either N_{22} *disappears or a matrix* M *exists such that* $[N_{12}/N_{22}] = [N_{11}/0]M$.
(g) $rank \begin{bmatrix} E & 0 & 0 \\ A & E & B \end{bmatrix} = n + rankE$.

Clearly, from Theorems 2.4 and 2.3, a system is impulse controllable if it is controllable. Its inverse is false.

Definition 2.5. [4] System (2.16) is observable if the initial condition $x(0)$ may be uniquely determined by $u(t)$, $y(t)$, $0 \leq t < \infty$.

The observability states that the state of the observable system may be determined by observing the initial condition $x(0)$, followed by constructing the state response at any time t.

Let

$$R_s = \ker[C_1/C_1A_1/\cdots/C_1A_1^{n_1-1}],$$
$$R_f = \ker[C_2/C_2N/\cdots/C_2N^{h-1}],$$

then the following is true.

Theorem 2.5. *[1] Consider system (2.16)*

(1) *Let $u(t) = 0$, then $y(t) = 0, t \geq 0$ if $x(0) \in R_s \bigoplus R_f$.*
(2) *Its slow subsystem (2.17) is observable if*

$$rank[sE - A/C] = n, \forall s \in C.$$

(3) *The following statements are equivalent:*

(a) *Its fast subsystem (2.18) is observable.*
(b) *$rank[C_2/C_2N/\cdots/C_2N^{h-1}] = n_2$.*
(c) *$Ker[N/C_2] = \{0\}$.*
(d) *$rank[N/C_2] = n_2$.*
(e) *$rank[E/C] = n$.*
(f) *For any two nonsingular matrices Q_1, P_1 satisfying $Q_1EP_1 = diag(I_q, 0)$, $CP = [\bar{C}_1, \bar{C}_2]$, $\bar{C}_2$ is of full column rank, $rank\bar{C}_2 = n - rankE$.*

(4) *The following statements are equivalent:*

(a) *System (2.16) is observable.*
(b) *Both its slow and fast subsystems are observable.*
(c) *$rank[sE - A/C] = n$, $\forall s \in C$, s is finite, and $rank[E/C] = n$.*
(d) *The matrix*

$$\theta_1 = \begin{bmatrix} -A & E & & & & \\ & -A & E & & & \\ & & & \cdots & & \\ & & & & -A & E \\ C & & & & & \\ & C & & & & \\ & & C & & & \\ & & & \cdots & & \\ & & & & C & \\ & & & & & C \end{bmatrix}_{(n+r-1)n \times n^2}$$

is of full column rank n^2.

Definition 2.6. [2] System (2.16) is R-observable if it is observable in the reachable set, or, in other words, any state in the reachable set may be uniquely determined by $y(\tau)$ and $u(\tau)$, $0 \le \tau \le t$.

Definition 2.7. [4] If $x_\tau(t)$ may be uniquely determined by $y_\tau(t)$ and $\Delta_\tau(t)$ for any $\tau \ge 0$, system (2.16) will be called impulse observable.

For a given system, its observability, R-observability, and impulse observability characterize its state reconstruction ability from different aspects. Apparently, system is impulse observable if it is observable, but its inverse is not true.

A careful examination of controllabilities and observabilities would reveal the important fact that they are very similar. For example, the necessary and sufficient conditions of controllability for system (2.16) are

$$\text{rank}[B_1, A_1B_1, \cdots, A_1^{n_1-1}B_1] = n_1,$$

$$\text{rank}[B_2, NB_2, \cdots, N^{h-1}B_2] = n_2.$$

Transposing these matrices, we know the equation above is equivalent to

$$\text{rank}[B_1^\tau, B_1^\tau A_1^\tau, \cdots, B_1^\tau (A_1^\tau)^{n_1-1}] = n_1,$$

$$\text{rank}[B_2^\tau, B_2^\tau N^\tau, \cdots, B_2^\tau (N^\tau)^{n_1-1}] = n_2,$$

which are the necessary and sufficient conditions of observability for the system $(E^\tau, A^\tau, C^\tau, B^\tau)$.

Summing up these arguments, system (E,A,B,C) is controllable if and only if $(E^\tau, A^\tau, B^\tau, C^\tau)$ is observable; conversely, system (E,A,B,C) is observable if $(E^\tau, A^\tau, B^\tau, C^\tau)$ is controllable. This is the so-called dual principle.

The system

$$\begin{cases} E^\tau \dot{z} = A^\tau z + C^\tau v, \\ y = B^\tau z, \end{cases} \tag{2.20}$$

is called its dual system.

Theorem 2.6. *[2] (Dual Principle) Consider system (2.15).*
(1) System (2.15) is controllable (observable) if its dual system (2.20) is observable (controllable).
(2) System (2.15) is R-controllable (R-observable) if its dual system (2.20) is R-observable (R-controllable).
(3) System (2.15) is impulse controllable (impulse observable) if its dual system (2.20) is impulse observable (impulse controllable).

The dual principle plays an important role in system theory. For this law, there is no need to study the controllabilities and observabilities separately. We may study the system itself or its dual system at our convenience.

Based on the state variable concept, the state space description method for system analysis and synthesis obtains its state space model (or description equation) from the physical sense of variables and their relationships. Thus, this method allows us to understand the inner structure of systems. On the other hand, the transfer matrix reflects the outer structure: input-output relationship, or transfer relationship from input to output. To compare, the state space method may characterize the inner properties but is sometimes too complicated for practical interests; while the transfer matrix is usually unable to characterize inner properties, it provides a dependence relationship between input and output.

The Laplace transformation $L[f]$ of a function $f(t)$ is defined as

$$L[f] = \int_0^{\infty} e^{-st} f(t)\mathrm{d}t,$$

which is assumed to exist. It has the basic property

$$L[\alpha f_1 + \beta f_2] = \alpha L[f_1] + \beta L[f_2],$$

$$L[\dot{f}] = sL[f] - f(0).$$

Taking Laplace transformation on both sides of the system equation (2.15), we have

$$\begin{aligned} sEx(s) - Ex(0) &= Ax(s) + Bu(s), \\ y(s) &= Cx(s). \end{aligned}$$

Under the assumption of regularity $(sE-A)^{-1}$ exists, thus

$$y(s) = C(sE-A)^{-1}(Ex(0) + Bu(s)).$$

When $x(0) = 0$, the preceding equation yields the input-output relationship $y(s) = G(s)u(s)$, in which $G(s) = C(sE-A)^{-1}B$ is called the transfer matrix for system (2.15).

For a given singular system, we may obtain its transfer matrix. Conversely, we will be interested in that for a given transfer matrix $G(s)$, a certain singular system may be found whose transfer matrix is $G(s)$. This is the so-called realization theory. To be precise, we define the following.

Definition 2.8. [2] Assume that $G(s) \in R^{r \times m}$ is a rational matrix. If there exist matrices E, A, B, C such that

$$G(s) = C(sE-A)^{-1}B,$$

where $E, A \in R^{n \times n}$, $B \in R^{n \times m}$, and $C \in R^{r \times n}$ are constant matrices, the system (2.15) will be called a singular system realization of $G(s)$, simply, a realization of $G(s)$.

Furthermore, if any other realization has an order greater than n, system (2.15) will be called the minimal order realization or, simply, minimal realization.

Theorem 2.7. *[4] Any $r \times m$ rational matrix $G(s)$ may be represented as*

$$G(s) = G_1(s) + G_2(s),$$

where $G_1(s)$ is a strictly proper rational matrix and $G_2(s)$ is a polynomial matrix.

Theorem 2.8. *[2] Any rational matrix may have a realization of (2.15), which satisfies $G(s) = C(sE - A)^{-1}B$. Furthermore, the realization is minimal if and only if the system is both controllable and observable.*

2.3 Feedback Control

The most commonly used methods to change the system's dynamic or static properties are state feedback and static output feedback control. In practical system analysis and design, the evaluation of a system's properties is based on some features that characterize the system, such as stability, static behavior, and response speed. However, we can not obtain all of the characteristics at the same time. This need to investigate for adjustment. In this section, the improvement of the properties are mentioned, which refers to system design and many topics of control engineering and control theory.

Consider the system (2.15). Let the state feedback control have the form of

$$u = K_1 x - K_2 \dot{x} + v, \tag{2.21}$$

where the first term is proportional feedback and the second is derivative feedback, which may be viewed as the speed feedback. $v(t)$ is the new input control, and $K_1, K_2 \in R^{m\times n}$ are constant matrices. Control (2.21) is the proportional and derivative (P-D) state feedback.

In singular systems, control input in the form of (2.21) is often used.

The closed-loop system forming (2.15) and (2.21) is

$$\begin{cases} (E + BK_2)\dot{x} = (A + BK_1)x + Bu, \\ y = Cx. \end{cases} \tag{2.22}$$

Particularly, when $K_2 = 0$, the feedback control (2.21) becomes

$$u = K_1 x + v, \tag{2.23}$$

which is a pure proportional (P-) state feedback, as in the normal system case. The corresponding closed-loop system is

$$\begin{cases} E\dot{x} = (A + BK_1)x + Bu, \\ y = Cx. \end{cases} \tag{2.24}$$

To guarantee the existence and uniqueness of a solution for any control input, we will hereafter assume that feedback control is confined to the set such that both (2.22) and (2.24) are regular.

In what follows,

$$\sigma(E,A) = \{s|s \in C, |sE - A| = 0\}$$

is used to denote the finite pole set for the system, and $\sigma(I,A)$ will be specified as $\sigma(A)$.

Theorem 2.9. *[4] System (2.15) is stable if and only if*

$$\sigma(E,A) \subset C^-$$

where $\subset= \{s|s \in C, Re(s) < 0\}$ *represents the open left half complex plane.*

Definition 2.9. [4] System (2.15) is called stabilizable if there exists a state feedback (2.23) such that the closed-loop system (2.24) is stable.

From Definition 2.9 and Theorem 2.9, the following theorem can be obtained.

Theorem 2.10. *[1] System (2.15) is stabilizable if and only if*

$$rank[sE - A, B] = n, \forall s \in \bar{C}^+ \tag{2.25}$$

where $\bar{C}^+$ *represents the closed right half complex plane;* $\bar{C}^+ = \{s|s \in C, Re(s) \geq 0\}$, *which, in turn, is equivalent to the stabilizability of the slow subsystem.*

Corollary 2.1. *System (2.15) is stabilizable if it is R-controllable, but the inverse is not true.*

For regular system (2.15), there exist nonsingular matrices Q_1 and P_1 such that (2.15) is r.s.e. to the controllability canonical form

$$\begin{bmatrix} E_{11} & E_{12} \\ 0 & E_{22} \end{bmatrix} \begin{bmatrix} \dot{x}_1 \\ \dot{x}_2 \end{bmatrix} = \begin{bmatrix} A_{11} & A_{12} \\ 0 & A_{22} \end{bmatrix} \begin{bmatrix} x_1 \\ x_2 \end{bmatrix} + \begin{bmatrix} B_1 \\ 0 \end{bmatrix} u$$
$$y = [C_1, C_2][x_1/x_2] \tag{2.26}$$

where $Q_1EP_1 = \begin{bmatrix} E_{11} & E_{12} \\ 0 & E_{22} \end{bmatrix}$, $Q_1AP_1 = \begin{bmatrix} A_{11} & A_{12} \\ 0 & A_{22} \end{bmatrix}$, $Q_1B = \begin{bmatrix} B_1 \\ 0 \end{bmatrix}$, $CP_1 = [C_1, C_2]$, $x = P_1[x_1/x_2]$, and (E_{11}, A_{11}, B_1) is controllable.

It is easy to verify from (2.25) that the following corollary holds.

Corollary 2.2. *System (2.15) is stabilizable if and only if in its canonical form (2.26)* A_{22} *is a stable matrix,* $\sigma(E_{22}, A_{22}) \subset C^-$.

This corollary shows that the stabilizable system has only stable uncontrollable poles. It also shows that there exists a feedback control

$$u(t) = f(x(t),t), \tag{2.27}$$

where $f(x(t),t)$ is a vector function of $x(t)$ and t, such that the closed-loop system formed by (2.15) and (2.27) is stable if and only if it is stabilizable. Thus, feedback in the form of (2.27) can't stabilize system (2.15) if it is not stabilizable. This indicates the equivalence of the existence of (2.21) stabilizing (2.22) and that of (2.23) stabilizing (2.24).

The stabilizability characterizes the controllability of the system's stability. The dual concept of stabilizability is detectability, which is defined as follows.

Definition 2.10. [2] System (2.15) is detectable if its dual system $(E^{\tau},A^{\tau},C^{\tau})$ is stabilizable.

Theorem 2.11. *[2] System (2.15) is detectable if and only if*

$$rank[sE - A/C] = n, \forall s \in \bar{C}^{+},$$

which is equivalent to the detectability of its slow subsystem.

Thus stabilizability and detectability are determined by the slow subsystem and alone. This indicates the fact that they really only characterize properties of the slow subsystem although they are defined for the whole system.

The R-observability, along with Theorem 2.11, shows that detectability may be deduced from R-observability, but its inverse is not true.

Corollary 2.3. *Let*

$$\left(\begin{bmatrix} E_{11} & 0 \\ E_{21} & E_{22} \end{bmatrix}, \begin{bmatrix} A_{11} & 0 \\ A_{21} & A_{22} \end{bmatrix}, \begin{bmatrix} B_1 \\ B_2 \end{bmatrix}, \begin{bmatrix} C_1 & 0 \end{bmatrix}\right)$$

be an observability canonical form for system (2.15) under r.s.e. equivalence, where (E_{11},A_{11},B_1,C_1) is observable. Then, the necessary and sufficient condition of detectability is that the system has only stable unobservable poles, i.e., $\sigma(E_{22},A_{22}) \subset C^{-}$.

Example: It is easy to verify from Corollary 2.3 that

$$\begin{bmatrix} 1 & 0 \\ 0 & 0 \end{bmatrix} \dot{x} = \begin{bmatrix} 1 & 0 \\ 0 & 1 \end{bmatrix} x + \begin{bmatrix} 1 \\ 0 \end{bmatrix} u, y = [01]x.$$

Definition 2.11. [1] System (2.15) is called normalizable if a feedback control (2.21) may be chosen such that its closed-loop (2.22) is normal, i.e.,

$$|E + BK_2| \neq 0, \tag{2.28}$$

while for any matrix $K_1 \in R^{m\times n}$, the closed-loop system under (2.23) is always singular. Under the assumption of normalizability for system (2.15), we may choose appropriate feedback control (2.21) such that its closed-loop system is normal, enabling the application of results in normal system theory to singular ones. As long as condition (2.30) is satisfied, the closed-loop system (2.22) would become

$$\begin{cases} \dot{x} = (E+BK_2)^{-1}(A+BK_1)x+(E+BK_2)^{-1}Bv, \\ y = Cx, \end{cases} \tag{2.29}$$

and its plain feature is its n finite poles, without any infinite poles. Under the normalizability assumption, a singular system's infinite poles may be set to finite via feedback (2.21). Thus, no impulse terms appear in the closed-loop state response.

Theorem 2.12. *[2] The following statements are equivalent:*
(a) System (2.15) is normalizable.
(b) $rank[EB] = n$.
(c) The fast subsystem (2.18) is controllable.
(d) In subsystem (2.18), $rank[N, B_2] = n_2$.
(e) For any nonsingular matrices Q_1 *and* P_1 *satisfying* $Q_1EP_1 = diag(I, 0)$, $Q_1B = [\bar{B}_1/\bar{B}_2]$, $\bar{B}_2$ *is of full row rank n-rankE.*

As previously pointed out, for a given singular system, stabilizability and detectability characterize its slow subsystem properties, while normalizability is the opposite, reflecting only its fast subsystem properties. As for the whole system, normalizability clarifies the difference between singular and normal systems — a normalizable singular system may be changed into a normal one via suitable selection of feedback (2.21).

From Theorem 2.11 and the controllability criterion, we see that a system is normalizable if it is controllable. Thus, if (2.26) is the controllability canonical form under r.s.e., the necessary and sufficient condition for the normalizability of system (2.15) is that E_{22} is nonsingular (2.26). A system is impulse controllable if it is normalizable; the inverse is false.

Definition 2.12. [2] System (2.15) is called normalizable if a feedback control (2.21) may be chosen such that its closed-loop (2.22) is normal, i.e.,

$$|E+BK_2| \neq 0. \tag{2.30}$$

Definition 2.13. [2] System (2.15) is dual normalizable if its dual system is normalizable.

Theorem 2.13. *[2] System (2.15) is dual normalizable if and only if* $rank[E/C] = n$.

It has been shown above that any regular system (2.15) is r.s.e. to EF3:

$$\begin{cases} \hat{E}\dot{x} = (I-\alpha\hat{E})x+\hat{B}u, \\ y = Cx, \end{cases} \tag{2.31}$$

where α satisfies $|\alpha E+A| \neq 0$. Under this equivalent form, we can obtain the following theorem.

Theorem 2.14. *[1] Consider system (2.31)*
(1) It is stabilizable if and only if

$$rank[sI - \hat{E}, \hat{B}] = n, \forall s \neq 0, s \in \bar{C}^+, s \text{ is finite}.$$

Or equivalently, the uncontrollable poles of $(\hat{E}, \hat{B})$ must be stable or zero.
(2) It is normalizable if and only if $rank[\hat{E}, \hat{B}] = n$.
(3) It is detectable if and only if $rank[sI - \hat{E}/C] = n, \forall s \neq 0, s \in \bar{C}^+, s$ is finite.
(4) It is dual normalizable if and only if $rank[\hat{E}/C] = n$.

Above all, some fundamental concepts in singular system theory have been introduced, including controllability, R-controllability, impulse controllability, observability, R-observability, impulse observability, normalizability, dual normalizability, stabilizability, and detectability. Similarities and differences of these introduced concepts are clear. Their relationships are complicated, but they characterize the singular system from different points of view. As far as their intrinsic properties are concerned, controllability and observability characterize the properties for the whole system, R-controllability, R-observability, stabilizability, and detectability characterize properties for the slow subsystem, and impulse controllability, impulse observability, normalizability, and dual normalizability for the fast subsystem. Furthermore, these concepts can't be substituted for each other[2].

References

1. Campbell, S.L.: Singular Systems of Differential Equations. Priman, London (1980)
2. Dai, L.: Singular Control System. Springer, New York (1989)
3. Mueller, P.C.: Linear mechanical descriptor systems: Identification, analysis and design. In: IFAC Control of Industrial Systems, Belfort, France, pp. 501-506 (May 1997)
4. Yang, D.M., Zhang, Q.L.: Descriptor System. Science Press, Beijing (2004)

Chapter 3
Bifurcations

3.1 Differential Equations and Dynamical Systems

Differential equation can be described as the following form:

$$\frac{\mathrm{d}x}{\mathrm{d}t} = \dot{x} = f(x), \tag{3.1}$$

where $x = x(t) \in R^n$ is a vector valued function of an independent variable (usually time) and $f : U \to R^n$ is a smooth function defined on some subset $U \subseteq R^n$. We say that the vector field f generates a flow $\phi_t : U \to R^n$, where $\phi_t(x) = \phi(x,t)$ is a smooth function defined for all x in U and t in some interval $I = (a,b) \subseteq R$, and ϕ satisfies (3.1) in the sense that

$$\frac{\mathrm{d}}{\mathrm{d}t}(\phi(x,t))|_{t=\tau} = f(\phi(x,\tau)), \tag{3.2}$$

for all $x \in U$ and $\tau \in I$. We note that (in its domain of definition) ϕ_t satisfies the group properties (i) $\phi_0 = id$, and (ii) $\phi_{t+s} = \phi_t \circ \phi_s$. Systems of the form (3.1), in which the vector field does not contain time explicitly, are called autonomous.

Usually, the initial condition is

$$x(0) = x_0 \in U, \tag{3.3}$$

in which case we seek a solution $\phi(x_0,t)$ such that

$$\phi(x_0,0) = x_0. \tag{3.4}$$

Sometimes we also write such a solution as $x(x_0,t)$, or simply $x(t)$.

Theorem 3.1. *[1] Let $U \subset R^n$ be an open subset of real Euclidean space (or of a differentiable manifold M), let $f : U \to R^n$ be a continuously differentiable (C^1) map, and let $x_0 \in U$. Then, there is some constant $c > 0$ and a unique solution*

Q. Zhang et al.: Complex., Analy. and Cntrl. of Singular Biological Systems, LNCIS 421, pp. 23–42.
springerlink.com

$\phi(x_0,\cdot):(-c,c)\to U$ satisfying the differential equation $\dot{x}=f(x)$ with initial condition $x(0)=x_0$.

Here, f need satisfy Lipschitz, i.e., $|f(y)-f(x)|\le K|x-y|$ for some $K<\infty$, where K is called Lipschitz constant for f. Thus, we can deal with piecewise linear functions, such as one gets in stich-slip friction problems and in clock problems.

In fact, any solution may leave U after sufficient time. Hence, this theorem is only local. Vector fields is constructed $f:U\to R^n$ such that $x(t)$ leaves any subset $U\subset R^n$ in a finite time.

Equilibrium, also called fixed point, is a class of solutions of a differential equation. Equilibrium is defined by the vanishing of the vector field $f(x): f(\bar{x})=0$. A equilibrium $\bar{x}$ is said to be stable if a solution $x(t)$ remains close to $\bar{x}$ for all time, i.e., if for every neighborhood V of $\bar{x}$ in U there is a neighborhood $V_1\subset V$ such that every solution $x(x_0,t)$ with $x_0\in V_1$ is defined and lies in V for all $t>0$. If, in addition, V_1 can be chosen so that $x(t)\to\bar{x}$ as $t\to\infty$, then $\bar{x}$ is said to be asymptotically stable.

We can use Lyapunov function approach suffices to show that an energy-like quantity decreases for $|x|$ sufficiently large, in order that $x(t)$ remains bounded for all t and all (bounded) initial conditions $x(0)$. In what follows, we will give the method here. The important step is finding a positive definite function $V:U\to R$, called the Lyapunov function, which can decrease along solution curves of the differential equation:

Theorem 3.2. *[3] Let $\bar{x}$ be a fixed point for (3.1) and $V:W\to R$ be a differentiable function defined on some neighborhood $W\subseteq U$ of $\bar{x}$ such that*
(i) $V(\bar{x})=0$ and $V(x)>0$ if $x\neq\bar{x}$;
(ii) $\dot{V}(x)\le 0$ in $W-\{\bar{x}\}$.
Then $\bar{x}$ is stable. Moreover, if
(iii) $\dot{V}(x)<0$ in $W-\{\bar{x}\}$,
then $\bar{x}$ is asymptotically stable.

Here
$$\dot{V}=\sum_{j=1}^{n}\frac{\partial V}{\partial x_j}\dot{x}_j=\sum_{j=1}^{n}\frac{\partial V}{\partial x_j}f_j(x)$$
is the derivative of V along solution curves of (3.1).

If $W(=U)=R^n$ can be chosen in case (iii), then we say that x is globally asymptotically stable. Hence, it shows that all solutions remain bounded and approach $\bar{x}$ as $t\to\infty$. However, there is no general method for finding suitable Lyapunov functions.

For problems with multiple equilibria, local Lyapunov functions can be sought or one can attempt to find a compact hypersurface $S\subset R^n$ such that the vector field is directed everywhere inward on S. If such a surface exists, then any solution starting on or inside S can never leave the interior of S and thus must remain bounded for all time.

The local existence theorem becomes global in all cases when we work on compact manifolds M instead of open spaces like R^n.

Theorem 3.3. *[4] The differential equation* $\dot{x} = f(x), x \in M$*, with* M *compact, and* $f \in C^1$*, has solution curves defined for all* $t \in R$*.*

Since there is no way in which solutions can escape from such manifolds, flows on spheres and tori are globally defined. The local theorem can be extended to show that solutions depend on initial conditions:

Theorem 3.4. *[6] Let* $U \subseteq R^n$ *be open and suppose* $f : U \to R^n$ *has a Lipschitz constant* K*. Let* $y(t)$*,* $z(t)$ *be solutions to* $\dot{x} = f(x)$ *on the closed interval* $[t_0, t_1]$*. Then, for all* $t \in [t_0, t_1]$

$$|y(t) - z(t)| \leq |y(t_0) - z(t_0)| e^{K(t-t_0)}.$$

3.2 Nonlinear System

It is admitted that almost nothing beyond general statements can be made about most nonlinear systems. A method to start the study of the nonlinear system $\dot{x} = f(x)$ is to find the equilibrium of f or the fixed points of (3.1).

We suppose that $\bar{x}$ is a fixed point, satisfying $f(\bar{x}) = 0$. The behavior of solutions near $\bar{x}$ will be characterized by linearizing (3.1) at $\bar{x}$, i.e., by studying the linear system

$$\dot{\eta} = Df(\bar{x})\eta, \quad \eta \in R^n, \tag{3.5}$$

where $Df = [\partial f_i / \partial x_j]$ is the Jacobian matrix of first partial derivatives of the function $f = (f_1(x_1, \cdots, x_n), f_2(x_1, \cdots, x_n), \cdots, f_n(x_1, \cdots, x_n))^T$ (T denotes transpose), and $x = \bar{x} + \eta$, $|\eta| \ll 1$. Since (3.5) is just a linear system of the form $\dot{x} = Ax$, we can do this easily. In particular, the linearized flow map $D\phi_t(\bar{x})\eta$ arising from (3.1) at a fixed point $\bar{x}$ is obtained from (3.5) by integration:

$$D\phi_t(\bar{x})\eta = e^{tDf(\bar{x})}\eta. \tag{3.6}$$

The important question is, what can we say about the solutions of (3.1) based on our knowledge of (3.5). The answer is provided by two fundamental results of dynamical systems theory, which we give below, and may be summed up by saying that local behavior does carry over in certain nice cases.

Theorem 3.5. *[2] If* $Df(\bar{x})$ *has no zero or purely imaginary eigenvalues, then there is a homeomorphism* h *defined on some neighborhood* U *of* $\bar{x}$ *in* R^n *locally taking orbits of the nonlinear flow* ϕ_t *of (3.1), to those of the linear flow* $e^{tDf(\bar{x})}$ *of (3.5). The homeomorphism preserves the sense of orbits and can also be chosen to preserve parameterization by time.*

A more delicate situation in which the nonlinear and linear flows are related via diffeomorphisms requires certain non-resonance conditions among the eigenvalues of $Df(\bar{x})$.

When $Df(\bar{x})$ has no eigenvalues with a zero real part, $\bar{x}$ is called a hyperbolic or nondegenerate fixed point and the asymptotic behavior of solutions near it (and hence its stability type) is determined by the linearization. If any one of the eigenvalues has zero real part, then stability cannot be determined by linearization.

Before the next result, we need a couple of definitions. We define the local stable and unstable manifolds of $\bar{x}$, $W^s_{loc}(\bar{x})$, $W^u_{loc}(\bar{x})$ as follows:

$$\begin{aligned} W^s_{loc}(\bar{x}) &= \{x \in U \,|\, \phi_t(x) \to \bar{x}, t \to \infty, \phi_t(x) \in U, for \quad t \geq 0\}, \\ W^u_{loc}(\bar{x}) &= \{x \in U \,|\, \phi_t(x) \to \bar{x}, t \to -\infty, \phi_t(x) \in U, for \quad t \leq 0\}, \end{aligned} \tag{3.7}$$

where $U \subset R^n$ is a neighborhood of the fixed point $\bar{x}$. The invariant manifolds W^s_{loc} and W^u_{loc} provide nonlinear analogues of the flat stable and unstable eigenspaces E^s, E^u of the linear problem (3.5). The next result tells us that W^s_{loc} and W^u_{loc} are in fact tangent to E^s, E^u at $\bar{x}$.

Theorem 3.6. *[8] (Stable Manifold Theorem for a Fixed Point) Suppose that $\dot{x} = f(x)$ has a hyperbolic fixed point $\bar{x}$. Then, there exist local stable and unstable manifolds W^s_{loc}, W^u_{loc}, of the same dimensions n_s, n_u as those of the eigenspaces E^s, E^u of the linearized system (3.5), and tangent to E^s, E^u at $\bar{x}$. W^s_{loc} and W^u_{loc} are as smooth as the function f.*

We have seen how the linear system $\dot{x} = Ax$ gives rise to flow map $e^{tA} : R^n \to R^n$, when e^{tA} is an $n \times n$ matrix. For fixed $t = \tau$, let $e^{\tau A} = B$; then B is a constant coefficient matrix and the difference equation

$$x_{n+1} = Bx_n, \text{or} \quad x \mapsto Bx$$

is a discrete dynamical system. Similarly, a nonlinear system and its flow ϕ_t give rise to a nonlinear map

$$x_{n+1} = G(x_n), \text{or} \quad x \mapsto G(x),$$

where $G = \phi_\tau$ is a nonlinear vector valued function. If the flow ϕ_t is smooth (say r-times continuously differentiable), then G is a smooth map with a smooth inverse: i.e., a diffeomorphism.

An orbit of a linear map $x \to Bx$ is a sequence of points $\{x_i\}_{i=-\infty}^{\infty}$ defined by $x_{i+1} = Bx_i$. Any initial point generates a unique orbit provided that B has no zero eigenvalues.

In spite of problems caused by multiplicities, if B has no eigenvalues of unit modulus, the eigenvalues alone serve to determine stability. In this case $x = 0$ is called a hyperbolic fixed point and, in general, if $\bar{x}$ is a fixed point for G ($G(\bar{x}) = \bar{x}$) and $DG(\bar{x})$ has no eigenvalues of unit modulus, then $\bar{x}$ is called a hyperbolic fixed point.

Theorem 3.7. *[5] Let $G : R^n \to R^n$ be a (C^1) diffeomorphism with a hyperbolic fixed point $\bar{x}$. Then, there exists a homeomorphism h defined on some neighborhood U on $\bar{x}$ such that $h(G(\xi)) = DG(\bar{x})h(\xi)$ for all $\xi \in U$.*

Theorem 3.8. *[4] Let $G : R^n \to R^n$ be a (C^1) diffeomorphism with a hyperbolic fixed point $\bar{x}$. Then, there are local stable and unstable manifolds $W^s_{loc}(\bar{x}), W^u_{loc}(\bar{x})$, tangent to the eigenspaces $E^s_{\bar{x}}$, $E^u_{\bar{x}}$ of $DG(\bar{x})$ at $\bar{x}$ and of corresponding dimensions. $W^s_{loc}(\bar{x})$, $W^u_{loc}(\bar{x})$ are smooth as the map G.*

Global stable and unstable manifolds are defined as for flows, by taking unions of backward and forward iterates of the local manifolds. We have

$$W^s_{loc}(\bar{x}) = \{x \in U | G^n(x) \to \bar{x}, n \to +\infty, G^n(x) \in U, \forall n \geq 0\},$$
$$W^u_{loc}(\bar{x}) = \{x \in U | G^n(x) \to \bar{x}, n \to +\infty, G^n(x) \in U, \forall n \leq 0.\}$$

and

$$W^s(\bar{x}) = \bigcup_{n\geq 0} G^{-n}(W^s_{loc}(\bar{x})),$$
$$W^u(\bar{x}) = \bigcup_{n\geq 0} G^{n}(W^u_{loc}(\bar{x})).$$

Suppose that we are given a differential equation

$$\begin{aligned} \dot{x} &= f(x,y), \\ \dot{y} &= g(x,y), \end{aligned} \quad (x,y) \in U \subseteq R^2 \tag{3.8}$$

where f and g are (sufficiently smooth) functions specified by some physical model. In approaching Eqn.(3.8), we normally first seek fixed points, at which $f(x,y) = g(x,y) = 0$. Linearizing (3.8) at such a point $(\bar{x},\bar{y})$, we obtain

$$\begin{pmatrix} \dot{\xi}_1 \\ \dot{\xi}_2 \end{pmatrix} = \begin{pmatrix} \frac{\partial f}{\partial x}(\bar{x},\bar{y}) & \frac{\partial f}{\partial y}(\bar{x},\bar{y}) \\ \frac{\partial g}{\partial x}(\bar{x},\bar{y}) & \frac{\partial g}{\partial y}(\bar{x},\bar{y}) \end{pmatrix} \begin{pmatrix} \xi_1 \\ \xi_2 \end{pmatrix} \tag{3.9}$$

or

$$\dot{\xi} = Df(\bar{x},\bar{y})\xi.$$

If the eigenvalues of the matrix $Df(\bar{x},\bar{y})$ have nonzero real parts, then the solution $\xi(t) = e^{tDf(\bar{x},\bar{y})}\xi(0)$ of (3.9) not only yields local asymptotic behavior, but also provides the local topological structure of the phase portrait.

Theorem 3.9. *[2] (The Poincaré-Bendixson theorem) A nonempty compact ω- or $\alpha-$ limit set of a planar flow, which contains no fixed points, is a closed orbit.*

Theorem 3.10. *[3](Bendixson's criterion) If on a simply connected region $D \subseteq R^2$ the expression $\partial f/\partial x + \partial g/\partial y$ is not identically zero and does not change sign, then Eqn. (3.8) has no closed orbits lying entirely in D.*

In fact, for planar flows, all the possible nonwandering sets fall into three classes:
(i) Fixed points;
(ii) closed orbits; and
(iii) the unions of fixed points and the trajectories connecting them.
The latter are referred to as heteroclinic orbits when they connect distinct points and homoclinic orbits when they connect a point to itself. Closed paths formed

of heteroclinic orbits are called homoclinic cycles. We note that the fixed points contained in such cycles must all be saddle points (if they are hyperbolic), since sinks and sources necessarily have wandering points in their neighborhoods.

If C is chosen to encircle a single, isolated fixed point, $\bar{x}$, then k is called the index of $\bar{x}$. The following statements can be verified by direct examination of the vector fields or by evaluation of the curvilinear integral

$$\begin{aligned} k &= \frac{1}{2\pi}\int_C \mathrm{d}\left\{\arctan\left(\frac{\mathrm{d}y}{\mathrm{d}x}\right)\right\} \\ &= \frac{1}{2\pi}\int_C \mathrm{d}\left\{\arctan\left(\frac{g(x,y)}{f(x,y)}\right)\right\} = \frac{1}{2\pi}\int_C \frac{f\mathrm{d}g - g\mathrm{d}f}{f^2+g^2}. \end{aligned}$$

Proposition 3.1. *[1] (i) The index of a sink, a source, or a center is* $+1$.
(ii) The index of a hyperbolic saddle point is -1.
(iii) The index of a closed orbit is $+1$.
(iv) The index of a closed curve not containing any fixed points is 0.
(v) The index of a closed curve is equal to the sum of the indices of the fixed points within it.

As a direct corollary to these statements, we find the following:

Corollary 3.1. *[1] Inside any closed orbit* γ *there must be at least one fixed point. If there is only one, then it must be a sink or a source. If all the fixed points within* γ *are hyperbolic, then there must be an odd number,* $2n+1$*, of which* n *are saddle and* $n+1$ *either sinks or sources.*

A further simple but useful technique for the global approximation of solution curves is provided by the method of isoclines. Eliminating explicit time dependence from (3.8), we obtain the first-order system

$$\frac{\mathrm{d}y}{\mathrm{d}x} = \frac{g(x,y)}{f(x,y)}. \tag{3.10}$$

Neglecting for the moment the fact that (3.10) might not be well defined on $f(x,y) = 0$, we seek curves $y = h(x)$ or $x = h(y)$ on which the slope of the vector field $\mathrm{d}y/\mathrm{d}x = c$ is constant. Such curves are given (perhaps implicitly) by solving the equation

$$g(x,y) = cf(x,y)$$

and are called isoclines.

Theorem 3.11. *[1] (Peixoto). A* C^r *vector field on a compact two-dimensional manifold* M^2 *is structurally stable if and only if*
(i) the number of fixed points and closed orbits is finite and each is hyperbolic;
(ii) there are no orbits connecting saddle points;
(iii) the nonwandering set consists of fixed points and periodic orbits alone.
Moreover, if M^2 *is orientable, the set of structurally stable vector fields is open-dense in* $\aleph^r(M^2)$.

3.3 Example of Lorenz System

In this section, the Lorenz system will be introduced, which possesses fascinating properties.

In 1963, E. N. Lorenz attempted to set up a system of differential equations that would explain some of the unpredictable behavior of the weather [7]. Most viable models for weather involve partial differential equations; Lorenz sought a much simpler and easier-to-analyze system, which can be written by

$$\begin{cases} \dot{x} = \sigma(y-x), \\ \dot{y} = rx - y - xz, \\ \dot{z} = xy - bz, \end{cases} \tag{3.11}$$

where three parameters are involved: the Prandtl number σ, the Rayleigh number r, and another parameter b that is related to the physical size of the system. In this system, all three parameters are assumed to be positive and, moreover, $\sigma > b+1$. We denote this system by

$$X' = \ell(X).$$

In Fig. 3.1, we have displayed the solution curves through two different initial conditions $P = (0,2,0)$ when the parameters are $\sigma = 10$, $b = 8/3$, and $r = 28$. These are the original parameters that led to Lorenz's discovery. Note how both solutions start out very differently but eventually have more or less the same fate: They both seem to wind around a pair of points, alternating at times which point they encircle. This is the first important fact about the Lorenz system: All nonequilibrium solutions tend eventually to the same complicated set, the so-called Lorenz attractor.

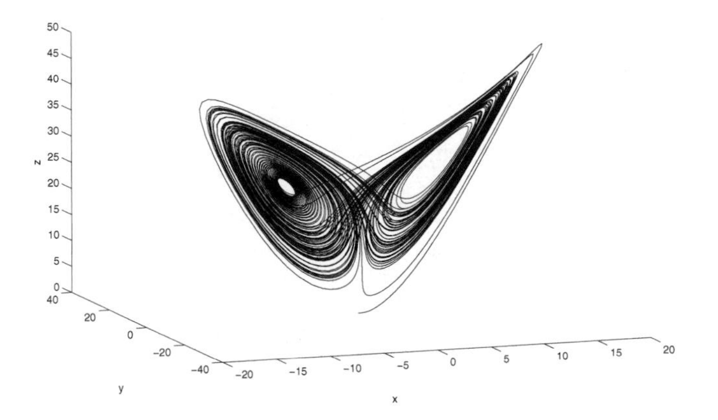

Fig. 3.1 The Lorenz attractor with initial condition $(0,2,0)$.

There is another important ingredient lurking in the background here. In Fig. 3.1, we started with two relatively far apart initial conditions. Had we started with two very close initial conditions, we would not have observed the transient behavior apparent in Fig. 3.1. Rather, more or less the same picture would have resulted for each solution. This, however, is misleading. When we plot the actual coordinates of the solutions, we see that these two solutions actually move quite far apart during their journey around the Lorenz attractor. This is illustrated in Fig. 3.2, where we have graphed the x coordinates of two solutions that start out nearby, one at $(0,2,0)$, the other (in dot) at $(0,2.01,0)$. These graphs are nearly identical for a certain time period, but then they differ considerably as one solution travels around one of the lobes of the attractor while the other solution travels around the other. No matter how close two solutions start, they always move apart in this manner when they are close to the attractor. This is sensitive dependence on initial conditions, one of the main features of a chaotic system.

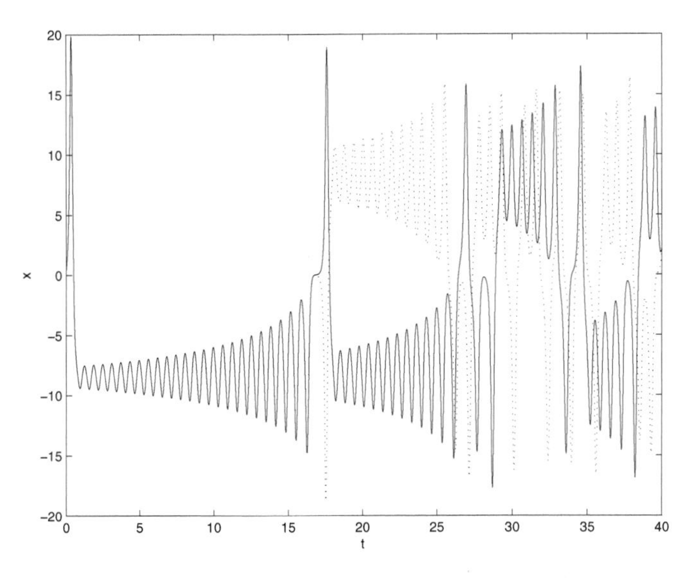

Fig. 3.2 The $x(t)$ graphs for two nearby initial conditions $P_1 = (0,2,0)$ and $P_2 = (0,2.01,0)$.

As usual, to analyze this system, we begin by finding the equilibria. Some simple algebra yields three equilibrium points, the origin, and

$$Q_\pm = (\pm\sqrt{b(r-1)}, \pm\sqrt{b(r-1)}, r-1).$$

The latter two equilibria exist only when $r > 1$, so already we see that we have a bifurcation when $r = 1$.

Linearizing, we find the system

$$Y' = \begin{pmatrix} -\sigma & \sigma & 0 \\ r-z & -1 & -x \\ y & x & -b \end{pmatrix} Y.$$

At the origin, the eigenvalues of this matrix are $-b$ and

$$\lambda_{\pm} = \frac{1}{2}(-(\sigma+1) \pm \sqrt{(\sigma+1)^2 - 4\sigma(1-r)}).$$

Note that both $\lambda_{\pm}$ are negative when $0 \leq r < 1$. Hence, the origin is a sink in this case.

The Lorenz vector field $\ell(x)$ possesses a symmetry. If $S(x,y,z) = (-x,-y,z)$, then we have $S(\ell(x)) = \ell(S(x))$. That is, reflection through the z-axis preserves the vector field. In particular, if $(x(t),y(t),z(t))$ is a solution of the Lorenz equations, then so is $(-x(t),-y(t),z(t))$.

When $x = y = 0$, we have $x' = y' = 0$, so the z-axis is invariant. On this axis, we have simply $z' = -bz$, so all solutions tend to the origin on this axis. In fact, the solution through any point in R^3 tends to the origin when $r < 1$, for we have the following:

Proposition 3.2. *[1] Suppose $r < 1$. Then, all solutions of the Lorenz system tend to the equilibrium point at the origin.*

Proof. We construct a strict Lyapunov function on all of R^3. Let

$$L(x,y,z) = x^2 + \sigma y^2 + \sigma z^2.$$

Then, we have

$$\dot{L}(x,y,z) = -2\sigma(x^2 + y^2 - (1+r)xy) - 2\sigma b z^2.$$

We therefore have $\dot{L} < 0$ away from the origin provided that

$$g(x,y) = x^2 + y^2 - (1+r)xy > 0,$$

for $(x,y) \neq (0,0)$. This is clearly true along the y-axis. Along any other straight line $y = mx$ in the xy-plane, we have

$$g(x,mx) = x^2(m^2 - (1+r)m + 1).$$

But the quadratic term $m^2 - (1+r)m + 1$ is positive for all m if $r < 1$, as is easily checked. Hence, $g(x,y) > 0$ for $(x,y) \neq (0,0)$.

When r increases through 1, two things happen. First, the eigenvalue λ_+ at the origin becomes positive, so the origin is now a saddle with a two-dimensional stable surface and an unstable curve. Second, the two equilibria $Q_{\pm}$ are born at the origin when $r = 1$ and move away as r increases.

Proposition 3.3. *[1] The equilibrium points $Q_{\pm}$ are sinks provided*

$$1 < r < r^* = \sigma(\frac{\sigma+b+3}{\sigma-b-1}).$$

Proof. From the linearization, we calculate that the eigenvalues at $Q_{\pm}$ satisfy the cubic polynomial

$$f_r(\lambda) = \lambda^3 + (1+b+\sigma)\lambda^2 + b(\sigma+r)\lambda + 2b\sigma(r-1) = 0.$$

When $r = 1$, the polynomial f_1 has distinct roots at 0, $-b$, and $-\sigma - 1$. These roots are distinct since $\sigma > b+1$ so that

$$-\sigma - 1 < \sigma + 1 < -b < 0.$$

Hence, for r close to but greater than 1, f_r has three real roots close to these values. Note that $f_r(\lambda) > 0$ for $\lambda \geq 0$ and $r > 1$. Looking at the graph of f_r, it follows that, at lease for r close to 1, the three roots of f_r must be real and negative.

We now let r increase and ask what is the lowest value of r for which f_r has an eigenvalue with a zero real part. Note that this eigenvalue must in fact be of the form $\pm i\omega$ with $\omega \neq 0$, since f_r is a real polynomial that has no roots equal to 0 when $r > 1$. Solving $f_r(i\omega) = 0$ by equating both real and imaginary parts to zero then yields the result (recall that we have assumed $\sigma > b+1$).

We remark that a Hopf bifurcation is known to occur at r^*, but proving this is beyond the scope of this book.

When $r > 1$, it is no longer true that all solutions tend to the origin. However, we can say that solutions that start far from the origin do at least move closer in. To be precise, let

$$V(x,y,z) = rx^2 + \sigma y^2 + \sigma(z-2r)^2.$$

Note that $V(x,y,z) = \nu > 0$ defines an ellipsoid in R^3 centered at $(0,0,2r)$. We will show the following:

Proposition 3.4. *[1] There exists ν^* such that any solution that starts outside the ellipsoid $V = \nu^*$ eventually enters this ellipsoid and then remains trapped therein for all future time.*

Proof. We compute

$$\begin{aligned} \dot{V} &= -2\sigma(rx^2 + y^2 + b(z^2 - 2rz)) \\ &= -2\sigma(rx^2 + y^2 + b(z-r)^2 - br^2), \end{aligned}$$

and the equation

$$rx^2 + y^2 + b(z-r)^2 = \mu$$

also defines an ellipsoid when $\mu > 0$. When $\mu > br^2$, we have $\dot{V} < 0$. Thus, we may choose ν^* large enough so that the ellipsoid $V = \nu^*$ strictly contains the ellipsoid

$$rx^2 + y^2 + b(z-r)^2 = br^2$$

in its interior. Then $\dot{V} < 0$ for all $\nu \geq \nu^*$.

As a consequence, all solutions starting far from the origin are attracted to a set that sits inside the ellipsoid $V = \nu^*$. Let Λ denote the set of all points whose solutions remain for all time (forward and backward) in this ellipsoid. Then, the ω-limit set of any solution of the Lorenz system must lie in Λ. Theoretically, Λ could be a large set, perhaps bounding an open region in R^3. However, for the Lorenz system, this is not the case.

To see this, recall from calculus that the divergence of a vector field $F(X)$ on R^3 is given by

$$\mathrm{div}F = \sum_{i=1}^{3} \frac{\partial F_i}{\partial x_i}(X).$$

The divergence of F measures how fast volumes change under the flow ϕ_t of F. Suppose D is a region in R^3 with a smooth boundary, and let $D(t) = \phi_t(D)$, the image of D under the time t map of the flow. Let $V(t)$ be the volume of $D(t)$. Then, Liouville's theorem asserts that

$$\frac{\mathrm{d}V}{\mathrm{d}t} = \int_{D(t)} \mathrm{div}F \, \mathrm{d}x\mathrm{d}y\mathrm{d}z.$$

For the Lorenz system, we compute immediately that the divergence is the constant $-(\sigma + 1 + b)$ so that volume decreases at a constant rate

$$\frac{\mathrm{d}V}{\mathrm{d}t} = -(\sigma + 1 + b)V.$$

Solving this simple differential equation yields

$$V(t) = e^{-(\sigma+1+b)t} V(0),$$

so that any volume must shrink exponentially fast to 0. In particular,

Proposition 3.5. *[1] The volume of Λ is zero.*

The natural question is what more can we say about the structure of the attractor Λ? In dimension two, such a set would consist of a collection of limit cycles, equilibrium points, and solutions connecting them.

Definition 3.1. [1] Let $X' = F(X)$ be a system of differential equations in R^n with flow ϕ_t. A set Λ is called an attractor if
(1) Λ is compact and invariant;
(2) there is an open set U containing Λ such that for each $X \in U$, $\phi_t(X) \in U$ for all $t \geq 0$ and $\bigcap_{t \geq 0} \phi_t(U) = \Lambda$;
(3) (transitivity) given any points $Y_1, Y_2 \in \Lambda$ and any open neighborhoods U_j about Y_j in U, there is a solution curve that begins in U_1 and later passes through U_2.

The transitivity condition in this definition may seem a little strange. Basically, we include it to guarantee that we are looking at a single attractor rather than a collection of dynamically different attractors. For example, the transitivity condition rules out situations such as that given by the planar system

$$\begin{aligned} x' &= x - x^3, \\ y' &= -y. \end{aligned}$$

Note that any solution of this system enters an open set U and then tends to one of the three equilibrium points: either to one of the sinks at $(\pm 1,0)$ or to the saddle $(0,0)$. The forward intersection of the flow ϕ_t applied to U is the interval $-1 \leq x \leq 1$. This interval meets conditions (1) and (2) in the definition, but condition (3) is violated, because none of the solution curves passes close to points in both the left and right half of this interval. We choose not to consider this set an attractor since most solutions tend to one of the two sinks. We really have two distinct attractors in this case.

As a remark, there is no universally accepted definition of an attractor in mathematics; some people choose to say that a set Λ that meets only conditions (1) and (2) is an attractor. For the planar system, condition (3) is usually easily verified; in higher dimensions, however, this can be much more difficult, as we shall see.

For the rest of this section, we restrict attention to the very special case of the Lorenz system where the parameters are given by $\sigma = 10$, $b = 8/3$, and $r = 28$. Historically, these are the values Lorenz used when he first encountered chaotic phenomena in this system. Thus, the specific Lorenz system we consider is as follows:

$$X' = \ell(X) = \begin{pmatrix} 10(y-x) \\ 28x - y - xz \\ xy - \frac{8}{3}z \end{pmatrix}.$$

As in the previous section, we have three equilibria: the origin and $Q_{\pm} = (\pm 6\sqrt{2}, \pm 6\sqrt{2}, 27)$. At the origin, we find eigenvalues $\lambda_1 = -8/3$ and

$$\lambda_{\pm} = -\frac{11}{2} \pm \frac{\sqrt{1201}}{2}.$$

For later use, note that these eigenvalues satisfy

$$\lambda_- < -\lambda_+ < \lambda_1 < 0 < \lambda_+.$$

The linearized system at the origin is then

$$Y' = \begin{pmatrix} \lambda_- & 0 & 0 \\ 0 & \lambda_+ & 0 \\ 0 & 0 & \lambda_1 \end{pmatrix} Y.$$

Note that all solutions in the stable plane of this system tend to the origin to the z-axis.

As $Q_{\pm}$ a computation shows that there is a single negative real eigenvalue and a pair of complex conjugate eigenvalues with positive real parts. Note that the symmetry in the system forces the rotations about Q_+ and Q_- to have opposite orientations.

If a numerical computation of a portion of the left- and right-hand branches of the unstable curve at the origin is displayed, we note that the right-hand portion of this curve comes close to Q_- and then spirals away. The left portion behaves symmetrically under reflection through the z-axis.

These unstable curves appear to circulate around the two equilibria curves continually intersecting the portion of the plane $z = 27$ containing $Q_{\pm}$ in which the vector field points downward.

3.4 Different Bifurcations

In this section, we give some theory for the local bifurcations of vector fields and maps. The discussion in brief can be found in [1]. As we have seen, systems of physical interest typically have parameters that appear in the defining systems of equations.

As these parameters are varied, changes may occur in the qualitative structure of the solutions for certain parameters' values.

These changes are called bifurcations and the parameter values are called bifurcation values.

In what follows, we describe four codimension one bifurcations of equilibria: saddle-node bifurcation, transcritical bifurcation, pitchfork bifurcation, and Hopf bifurcation [1, 2, 3, 4, 5, 6, 8]. These are represented by the following four differential equations, which depend on a single parameter μ:

$$\dot{x} = \mu - x^2 \text{(saddle-node)}$$

$$\dot{x} = \mu x - x^2 \text{(transcritical)}$$

$$\dot{x} = \mu x - x^3 \text{(pitchfork)}$$

and

$$\begin{cases} \dot{x} = -y + x(\mu - (x^2 + y^2)), \\ \dot{y} = x + y(\mu - (x^2 + y^2)). \end{cases} \text{(Hopf)}$$

The bifurcation diagrams for these four equations are depicted in Fig. 3.4-3.6. Each of the equations above arises naturally in a suitable context as determining the local qualitative behavior of the generic bifurcation of an equilibrium. In what follows, we will describe how, and under what conditions, one can reduce the study of the general equation to one of these four specific examples.

3.4.1 The Saddle-Node Bifurcation

Consider a system of equations

$$\dot{x} = f_{\mu}(x), \tag{3.12}$$

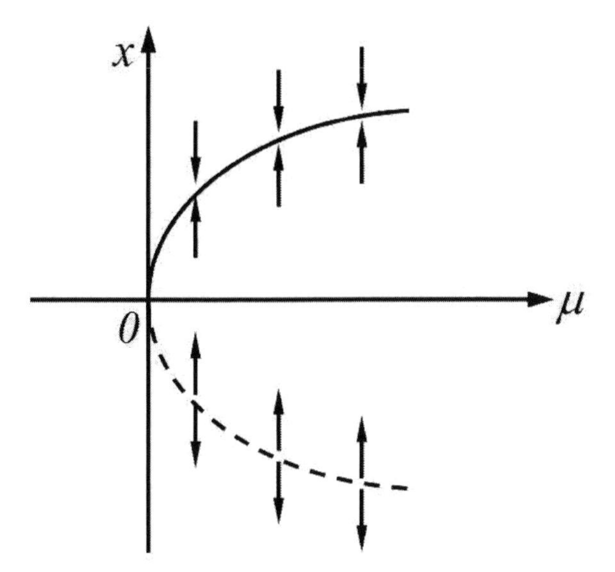

Fig. 3.3 Saddle-node bifurcation.

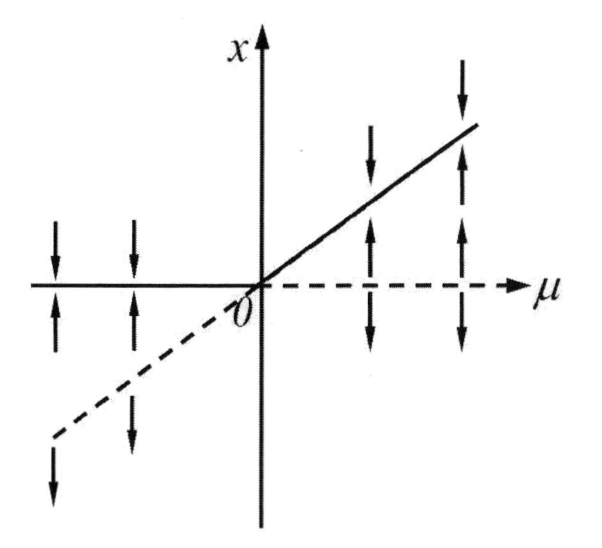

Fig. 3.4 Transcritical bifurcation.

with $x \in R^n$, $\mu \in R$ and f_μ smooth. Assume that at $\mu = \mu_0$, $x = x_0$, the system (3.12) has an equilibrium at which there is a zero eigenvalue (for the linearization). Usually, this zero eigenvalue will be simple, and the center manifold theorem allows us to reduce the study of this kind of bifurcation problem to one in which x is one

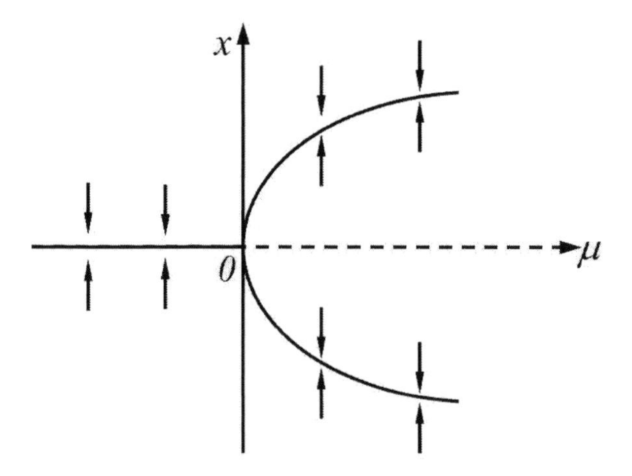

Fig. 3.5 Pitchfork bifurcation (supercritical).

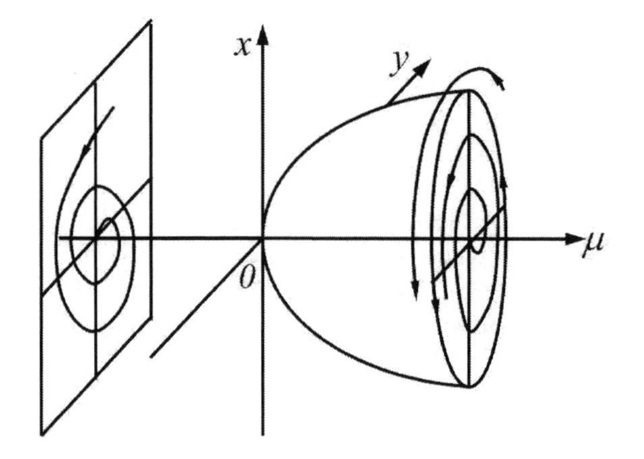

Fig. 3.6 Hopf bifurcation.

dimensional [2]. More precisely, we can find a two-dimensional center manifold $\sum \subset R^n \times R$ passing through (x_0, μ_0) such that
(1) the tangent space of $\sum$ at (x_0, μ_0) is spanned by an eigenvector of 0 for $Df_{\mu_0}(x_0)$ and a vector parallel to the μ-axis.
(2) for any finite r, $\sum$ is C^r if restricted to a small enough neighborhood of (x_0, μ_0).
(3) the vector field of the system (3.12) is tangent to $\sum$.
(4) there is a neighborhood U of (x_0, μ_0) in $R^n \times R$ such that all trajectories contained entirely in U for all time lie in $\sum$.

Restricting the system (3.12) to $\sum$, we obtain a one-parameter family of equations on the one-dimensional curves $\sum_\mu$ in $\sum$ obtained by fixing μ. This one-parameter family is our reduction of the bifurcation problem.

Let us now formulate transversality conditions for the system (3.12) with $n=1$, which yield the saddle-node bifurcation.

We have $(\mathrm{d}f_{\mu_0}/\mathrm{d}x)(x_0)=0$, but we take $(\partial f_{\mu_0}/\partial\mu)(x_0)\neq 0$ as a transversality condition. The implicit function theorem then implies that the equilibria of the system (3.12) form a curve that will be tangent to the line $\mu=\mu_0$. An additional transversality condition $(\mathrm{d}^2 f_{\mu_0}/\mathrm{d}x^2)(x_0)\neq 0$ implies that the curve of equilibria has a quadratic tangency with $\mu=\mu_0$ and locally lies to one side of this line. This information is already sufficient to imply that the local phase portraits of this system are topologically equivalent to those of a family $\dot{x}=\pm(\mu-\mu_0)\pm(x-x_0)^2$. However, we can also formulate these transversality conditions for an n-dimensional system without recourse to the center manifold reduction. The following theorem states the necessary conditions.

Theorem 3.12. *[1] Let $\dot{x}=f_\mu(x)$ be a system of differential equations in R^n depending on the single parameter μ. When $\mu=\mu_0$, assume that there is an equilibrium for which the following hypotheses are satisfied:*
(SN1) $D_x f_{\mu_0}(p)$ has a simple eigenvalue 0 with right eigenvector v and left eigenvalue w. $D_x f_{\mu_0}(p)$ has k eigenvalues with negative real parts and $(n-k-1)$ eigenvalues with positive real parts (counting multiplicity).
(SN2) $w((\partial f_\mu/\partial\mu)(p,\mu_0))\neq 0$.
(SN3) $w(D_x^2 f_{\mu_0}(p)(v,v))\neq 0$.
Then there is a smooth curve of equilibria in $R^n\times R$ passing through (p,μ_0), tangent to the hyperplane $R^n\times\{\mu_0\}$. Depending on the signs of the expressions in (SN2) and (SN3), there are no equilibria near (p,μ_0) when $\mu<\mu_0$ ($\mu>\mu_0$) and two equilibria near (p,μ_0) for each parameter value $\mu>\mu_0$ ($\mu<\mu_0$). The two equilibria for $\dot{x}=f_\mu(x)$ near (p,μ_0) are hyperbolic and have stable manifolds of dimensions k and $k+1$, respectively. The set of equations $\dot{x}=f_\mu(x)$ that satisfy (SN1)-(SN3) is open and dense in the space of C^∞ one-parameter families of vector fields with an equilibrium at (p,μ_0) with a zero eigenvalue.

The formal theorem merely expresses the fact that the genetic saddle-node bifurcation is qualitatively like the family of equations $\dot{x}=\mu-x^2$ in the direction of the zero eigenvector, with hyperbolic behavior in the complementary directions. Hypotheses (SN2) and (SN3) are the transversality conditions that control the nondegeneracy of the behavior with respect to the parameter and the dominant effect of the quadratic nonlinear term.

The results obtained from Theorem 3.12 are limited in two different ways. On the one hand, it is possible that more quantitative information about the flows near bifurcation can be extracted. For example, one can use the system $\dot{x}=\mu-x^2$ to give estimates of how rapid the convergence to the various equilibria are. Higher-order terms in the Taylor expansion of an equation can be used to refine the estimates. The second limitation of Theorem 3.12 is that there may be global changes in a phase portrait associated with a saddle-node bifurcation.

3.4.2 Transcritical and Pitchfork Bifurcation

The importance of the saddle-node bifurcation is that all bifurcations of one-parameter families at an equilibrium with a zero eigenvalue can be perturbed to saddle-node bifurcations. Thus, one expects that the zero eigenvalue bifurcations encountered in applications will be saddle-nodes. If they are not, then there is probably something special about the formulation of the problem that restricts the context so as to prevent the saddle-node from occurring. The transcritical bifurcation is one example that illustrates how the setting of the problem can rule out the saddle-node bifurcation [3, 4].

In classical bifurcation theory, it is often assumed that there is a trivial solution from which bifurcation occurs. Thus, the system (3.12) is assumed to satisfy $f_\mu(0) = 0$ for all μ, so that $x = 0$ is an equilibrium for all parameter values. Since the saddle-node families contain parameter values for which there are no equilibria near the point of bifurcation, this situation is qualitatively different. To formulate the appropriate transversality conditions, we look at the one-parameter families that satisfy the constraint that $f_\mu(0) = 0$ for all μ. This prevents hypothesis (SN2) of Theorem 3.1 from being satisfied. If we replace this condition with the requirement that $w((\partial^2 f/\partial\mu\partial x)(v)) \neq 0$ at $(0,\mu)$, then the phase portraits of the family near the bifurcation will be topologically equivalent to those of Fig. 3.4 and we have a transcritical bifurcation or exchange of stability.

A second setting in which the saddle-node does not occur involves systems that have a symmetry. Many physical problems are formulated so that the equations defining the system do have symmetries of some kind. For example, the Duffing equation is invariant under the transformation $(x,y) \to (-x,-y)$ and the Lorenz equation is symmetric under the transformation $(x,y,z) \to (-x,-y,z)$. In one dimension, a differential equation (3.12) is symmetric or equivariant with respect to the symmetry $x \to -x$ if $f_\mu(-x) = -f_\mu(x)$. Thus, the equivariant vector fields are ones for which f_μ is an odd function of x. In particular, all such equations have an equilibrium at 0. The transcritical bifurcation can't occur in these systems, however, because an odd function f_μ can't satisfy the condition $\partial^2 f_\mu/\partial x^2 \neq 0$ required by the transcritical bifurcation (SN3). If this condition is replaced by the transversality hypothesis $\partial^3 f_\mu/\partial x^3 \neq 0$, then one obtains the pitchfork bifurcation. At the point of bifurcation, the stability of the trivial equilibrium changes, and a new pair of equilibria (related by the symmetry) appear to one side of the point of bifurcation in parameter space, as in Fig. 3.5.

We note that the direction of the bifurcation and stability of the branches in these examples is determined by the sign of $\partial^2 f_\mu/\partial x^2$ or $\partial^3 f_\mu/\partial x^3$. In the last case, if $\partial^3 f_\mu/\partial x^3$ is negative, then the branches occur above the bifurcation value and we have a supercritical pitchfork bifurcation, whereas we have a subcritical bifurcation if it is positive.

3.4.3 Hopf Bifurcation

Consider now a system (3.12) with a parameter value μ_0 and equilibrium $p(\mu_0)$ at which Df_{μ_0} has a simple pair of pure imaginary eigenvalues, $\pm i\omega$, $\omega > 0$, and no other eigenvalues with a zero real part. The implicit function theorem guarantees (since Df_{μ_0} is invertible) that for each μ near μ_0 there will be an equilibrium $p(\mu)$ near $p(\mu_0)$ that varies smoothly with μ. Nonetheless, the dimensions of stable and unstable manifolds of $p(\mu)$ do change if the eigenvalues of $Df(p(\mu))$ cross the imaginary axis at μ_0. This qualitative change in the local flow near $p(\mu)$ must be marked by some other local changes in the phase portraits not involving fixed points [2].

A clue to what happens in the generic bifurcation problem involving an equilibrium with pure imaginary eigenvalues can be gained from examining linear systems in which there is a change of this type. For example, consider the system

$$\begin{cases} \dot{x} = \mu x - \omega y, \\ \dot{y} = \omega x + \mu y, \end{cases} \tag{3.13}$$

whose solutions have the form

$$\begin{pmatrix} x(t) \\ y(t) \end{pmatrix} = e^{\mu t} \begin{pmatrix} \cos\omega t & -\sin\omega t \\ \sin\omega t & \cos\omega t \end{pmatrix} \begin{pmatrix} x_0 \\ y_0 \end{pmatrix}. \tag{3.14}$$

When $\mu < 0$, solutions spiral into the origin, and when $\mu > 0$, solutions spiral away from the origin. When $\mu = 0$, all solutions are periodic. Even in a one-parameter family of equations, it is highly special to find a parameter value at which there is a whole family of periodic orbits, but there is still a surface of periodic orbits that appears in the general problem.

The normal form theorem gives us the required information about how the generic problem differs from the system (3.13). By smooth changes of coordinates, the Taylor series of degree 3 for the general problem can be brought to the following form

$$\begin{cases} \dot{x} = (d\mu + a(x^2+y^2))x - (\omega + c\mu + b(x^2+y^2))y, \\ \dot{y} = (\omega + c\mu + b(x^2+y^2))x + (d\mu + a(x^2+y^2))y, \end{cases} \tag{3.15}$$

which is expressed in polar coordinates as

$$\begin{cases} \dot{r} = (d\mu + ar^2)r, \\ \dot{\theta} = (\omega + c\mu + br^2). \end{cases} \tag{3.16}$$

Since the $\dot{r}$ equation in (3.16) separates from θ, we see that there are periodic orbits of (3.15) that are circles $r = const.$, obtained from the nonzero solutions of $\dot{r} = 0$ in (3.16). If $a \neq 0$ and $d \neq 0$, these solutions lie along the parabola $\mu = -ar^2/d$. This implies that the surface of periodic orbits has a quadratic tangency with its tangent plane $\mu = 0$ in $R^2 \times R$. The content of the Hopf bifurcation theorem is that the qualitative properties of (3.15) near the origin remain unchanged if higher-order terms are added to the system:

Theorem 3.13. *[1] Suppose that the system $\dot{x} = f_\mu(x)$, $x \in R^n$, $\mu \in R$ has an equilibrium (x_0, μ_0) at which the following properties are satisfied:*
(H1) $D_x f_{\mu_0}(x_0)$ has a simple pair of pure imaginary eigenvalues and no other eigenvalues with zero real parts.

Then, (H1) implies that there is a smooth curve of equilibria $(x(\mu), \mu)$ with $x(\mu_0) = x_0$. The eigenvalues $\lambda(\mu)$, $\bar{\lambda}(\mu)$ of $D_x f_{\mu_0}(x(\mu))$ which are imaginary at $\mu = \mu_0$ vary smoothly with μ. If, moreover,
(H2)

$$\frac{d}{d\mu}(Re\lambda(\mu))|_{\mu=mu_0} = d \neq 0,$$

then there is a unique three-dimensional center manifold passing through (x_0, μ_0) in $R^n \times R$ and a smooth system of coordinates (preserving the planes μ = const.) for which the Taylor expansion of degree 3 on the center manifold is given by the system (3.15). If $a \neq 0$, there is a surface of periodic solutions in the center manifold that has quadratic tangency with the eigenspace of $\lambda(\mu_0)$, $\bar{\lambda}(\mu_0)$ agreeing to second order with the paraboloid $\mu = -(a/d)(x^2 + y^2)$. If $a < 0$, then these periodic solutions are stable limit cycles, while if $a > 0$, the periodic solutions are repelling.

For large systems of equations, computation of the normal form (3.15) and the cubic coefficient a, which determines the stability, can be a substantial undertaking.

In a two-dimensional system of the form

$$\begin{pmatrix} \dot{x} \\ \dot{y} \end{pmatrix} = \begin{pmatrix} 0 & -\omega \\ \omega & 0 \end{pmatrix} \begin{pmatrix} x \\ y \end{pmatrix} \begin{pmatrix} f(x,y) \\ g(x,y) \end{pmatrix}, \tag{3.17}$$

with $f(0) = g(0) = 0$ and $Df(0) = Dg(0) = 0$, the normal form calculation, which we sketch in the appendix to this section, yields

$$\begin{aligned} a = \frac{1}{16}[f_{xxx} + f_{xyy} + g_{xxy} + g(yyy)] + \frac{1}{16\omega}[f_{xy}(f_{xx} + f_{yy}) \\ -g_{xy}(g_{xx} + g_{yy}) - f_{xx}g_{xx} + f_{yy}g_{yy}], \end{aligned} \tag{3.18}$$

where f_{xy} denotes $(\partial^2 f/\partial x \partial y)(0,0)$, etc. In applying this formula to systems of dimensions greater than two, however, the quadratic terms that play a role in the center manifold calculations can affect the value of a.

References

1. Guckenheimer, J., Holmes, P.: Nonlinear Oscillations, Dynamical Systems and Bifurcations of Vector Field. Springer, New York (1983)
2. Hale, J.K.: Methods of Bifurcation Theory. Springer, New York (1982)
3. Hartman, P.: Ordinary Differential Equations. Cambridge University Press, Cambridge (2002)
4. Kielhoefer, H.: Bifurcation Theory: An Introduction with Applications to PDEs. Springer, New York (2004)

5. Kuznetsov, Y.: Elements of Applied Bifurcation Theory. Springer, New York (2004)
6. Ma, T., Wang, S.: Bifurcation Theory and Applications. World Scientific, Singapore (2005)
7. Parker, T.S., Chua, L.O.: Practical Numerical Algorithms for Chaotic Systems. Springer, New York (1989)
8. Seydel, R.: Practical Bifurcation and Stability Analysis. Springer, Berlin (2010)

Chapter 4
Bifurcations and Control in a Singular Biological Economic Model

4.1 Introduction

In the natural world, there are many species whose individuals have a life history that takes them through two stages, juvenile stage and adult stage . Individuals in each stage are identical in biological characteristics, and some vital rates (rates of survival, development, and reproduction) of individuals in a population almost always depend on stage structure. Furthermore, there is a strong interaction relationship between the mature population and the immature population , which is to some extent relevant to the persistence and extinction of the related population. Consequently, it is constructive to investigate the dynamics of such ecosystem without ignorance of stage structure for population.

Before 1990, the stage structure has attracted some attention from several authors [6, 20, 21, 23, 24, 31, 48, 53, 29]. The well-known biological system of single species with stage structure is established by O. Aiello and H. Freedman [1], which represents a milestone in this research field. In the following decades, much research efforts are put into the investigation of biological dynamical systems with stage structure for population. Some related research results can be found in [1, 2, 3, 10, 14, 26, 8, 39, 34, 35, 36, 37, 38, 40, 44, 41, 42, 43, 45, 47, 51, 52, 54, 25, 16, 9] and the references therein.

Recently, the dynamics of a class of stage-structured prey-predator models with discrete time delay have been studied by several authors [4, 55, 18, 58, 5] and the references therein.

Arino et al. [4] suggested that the time delay to adulthood should be state dependent and careful formulation of such state dependent time delays can lead to models that produce periodic solutions. Xu et al. [55] studied the persistence and stability of a delayed prey-predator model with stage structure for predator. Gourley et al. [18] formulated a general and robust prey predator model with stage-structure with constant maturation time delay and performed a systematic mathematical and computational

Q. Zhang et al.: Complex., Analy. and Cntrl. of Singular Biological Systems, LNCIS 421, pp. 43–66.
springerlink.com

study. They have shown that there is a window in the maturation time delay parameter that generates sustainable oscillatory dynamics. Zhang et al. [58] established a prey-predator model with stage structure for prey species to obtain the necessary and sufficient condition for the permanence of prey-predator and the extinction of one species or two species. By assuming that the reproduction of predator after predating the prey is not instantaneous but mediates by some discrete time lag required for gestation of predator, Bandyopadhyay et al. [5] extended the work done by Zhang et al. [58] by incorporating a delay in the model proposed in [58], which made the model more realistic. The model proposed in [5] can be expressed as follows:

$$\begin{cases} \dot{x}_1(t) = r_1 x_2(t) - d_1 x_1(t) - \alpha x_1(t) - s_1 x_1^2(t) - \beta x_1(t) y(t), \\ \dot{x}_2(t) = \alpha x_1(t) - d_2 x_2(t), \\ \dot{y}(t) = \beta x_1(t-\tau) y(t-\tau) - d_3 y(t) - s_2 y^2(t), \end{cases} \tag{4.1}$$

where $x_1(t)$ and $x_2(t)$ represent the densities of the juveniles and adults of the prey population, respectively. $y(t)$ represents the density of the predator population. All these populations are growing in a closed homogeneous environment . At any time $(t > 0)$, the birth rate of the juvenile prey is proportional to the density of existing adults with proportionality constant r_1, the rate of transformation of the adult prey is proportional to the density of existing juveniles with proportionality constant α. The death rates of the juvenile prey, adult prey, and predator are proportional to the density of existing juvenile prey, adult prey, and predator with respective proportionality constants d_1, d_2, and d_3. The juvenile prey are density restricted and the predators compete among themselves for food; s_1 and s_2 are the intra-specific competition coefficients of the juvenile prey and predator population, respectively. The predator consumes the juvenile prey at the rate β, and the reproduction of predator after predating the prey is not instantaneous but will be mediated by some discrete time lag required for gestation of predator. Let τ represent the discrete time delay, which is the time interval between the moments when an individual prey is killed and when the corresponding biomass is added to the predator population. The constants mentioned above are all positive .

4.2 Model Formulation

It is well known that harvesting has a strong impact on the dynamic evolution of a population. Nowadays, the biological resources in the prey-predator ecosystem are commercially harvested and sold with the aim of achieving economic interest. Furthermore, the harvest effort is usually influenced by the variation of economic interest of harvesting. By considering the economic interest of the harvest effort on the predator, a differential-algebraic model is proposed in this chapter. It is utilized to investigate the dynamical behavior of the prey-predator ecosystem due to the variation of economic interest of harvesting. Humans are facing the dual problems of resource shortages and environmental degradation. Concerning conservation for the long-term benefits of humanity, there is a wide range of interest in analysis and modeling of bio-economic systems. We aim to obtain some results that are theoretically beneficial to maintaining the sustainable development of the ecosystem as well as keeping the persistent prosperity of harvesting.

In 1954, Gordon proposed the economic theory of a common-property resource in [17], which studies the effect of harvest effort on the ecosystem from an economic perspective. In [17], an equation is proposed to investigate the economic interest of the yield of harvest effort, which takes form as follows:

$$\text{Net Economic Revenue (NER)} = \text{Total Revenue (TR)} - \text{Total Cost (TC)}, \tag{4.2}$$

and let $E(t)$ and $Y(t)$ represent the harvest effort and the density of the harvested population, respectively. $\text{TR} = wE(t)Y(t)$ and $\text{TC} = cY(t)$, where w represents the unit price of the harvested population, and c represents the cost of the harvest effort. Associated with model system (4.1), an algebraic equation, which considers the economic interest v of the harvest effort on the predator, is established as follows:

$$E(t)(wy(t) - c) = v. \tag{4.3}$$

Recently, Zhang et al. [59, 60, 61, 33] established a class of singular biological economic models, which are established by several differential equations and an algebraic equation. The differential-algebraic biological economic models proposed in [59, 60, 61, 33] are used to investigate the dynamical behavior of the ecosystem. Especially, based on the economic theory proposed by [17], an algebraic equation is established to investigate the effect of harvest effort on some population in the ecosystem. Compared with the models proposed in the references [46, 32, 57, 13], the advantages of the models proposed in [59, 60, 61, 33] are that these models not only investigate the interaction mechanism of the prey-predator with stage structure for predator but also offer a simpler way to study the effect of harvest effort on the ecosystem from an economic perspective.

As mentioned above, [60, 61, 59] have investigated some biological-economic systems by using the differential-algebraic model. However, they investigate only the dynamical behavior of the model system of a single population and prey-predator model without stage structure. To authors' best knowledge, the differential-algebraic model system describing two more species with stage structure has not been investigated. In this chapter, we will establish a differential-algebraic model of two species (prey-predator) to study a harvested prey-predator system with stage structure for prey. Based on (4.1) and (4.3), a differential-algebraic model that consists of three differential equations and an algebraic equation can be established as follows:

$$\begin{cases} \dot{x}_1(t) = r_1 x_2(t) - d_1 x_1(t) - \alpha x_1(t) - s_1 x_1^2(t) - \beta x_1(t) y(t), \\ \dot{x}_2(t) = \alpha x_1(t) - d_2 x_2(t), \\ \dot{y}(t) = \beta x_1(t-\tau) y(t-\tau) - d_3 y(t) - s_2 y^2(t) - E(t) y(t), \\ 0 = E(t)(wy(t) - c) - v, \end{cases} \tag{4.4}$$

where $x_1(t)$, $x_2(t)$, $y(t)$, r_1, d_1, d_2, d_3, s_1, s_2, α, β, w, c, and v have the same interpretations as mentioned in (4.1) and (4.3).

The differential-algebraic model system (4.4) can be expressed in the following form:

$$A(t)\dot{X}(t) = G(x_1(t), x_2(t), y(t), E(t)), \tag{4.5}$$

where $X(t) = (x_1(t), x_2(t), y(t), E(t))^T$, $A(t) = \begin{bmatrix} 1 & 0 & 0 & 0 \\ 0 & 1 & 0 & 0 \\ 0 & 0 & 1 & 0 \\ 0 & 0 & 0 & 0 \end{bmatrix}$,

$$G(x(t), y_1(t), y_2(t), E(t)) = \begin{bmatrix} r_1 x_2(t) - d_1 x_1(t) - \alpha x_1(t) - s_1 x_1^2(t) - \beta x_1(t) y(t) \\ \alpha x_1(t) - d_2 x_2(t) \\ \beta x_1(t-\tau) y(t-\tau) - d_3 y(t) - s_2 y^2(t) - E(t) y(t) \\ E(t)(w y(t) - c) - v \end{bmatrix}.$$

Remark 4.1. The algebraic equation in differential-algebraic model system (4.5) contains no differentiated variables, hence the leading matrix $A(t)$ in model system (4.5) has a corresponding zero row.

Remark 4.2. The models proposed in [5], which consist of differential equations, discuss only the interaction and coexistence mechanism of prey-predator. Compared with the models proposed in [5], an algebraic equation is introduced into the model established in this chapter, which focuses on the economic interest of harvesting. Hence, the established model investigates not only the interaction and coexistence mechanism of the harvested prey-predator system, but also the dynamical behavior due to the variation of the economic interest of harvesting. The theorem in the following section investigates dynamical behavior of differential-algebraic model system (4.4) around the interior equilibrium due to variation of the economic interest of harvesting.

Remark 4.3. It should be noted that the singular models presented in [59, 33, 60, 61] investigate only some biological-economic systems with a single population or population with stage structure. Compared with these works, the model proposed in this chapter investigates dynamical behavior of three more interacting species (immature prey, mature predator, and harvested predator), which means the work done here has some new and positive features.

4.3 Local Bifurcation Analysis

Consider the effects of harvest effort on the predator and discrete time delay for the gestation of the predator, the dynamical behavior of the proposed model system is investigated by using the differential-algebraic system theory and bifurcation theory. By performing the local stability analysis of the model system, the bifurcation phenomena, which can reveal the instability mechanism of the model system, are investigated, and the biological interpretations of the bifurcation phenomena are also discussed. It is of inspiration for people to maintain the sustainable development of the prey-predator ecosystem as well as keep the economic interest of harvesting at an ideal level. In this chapter, we concentrate only on the interior equilibrium of model system (4.4), since the biological interpretation of the interior equilibrium implies that juvenile prey, adult prey, predator, and harvest effort on predator all exist, which are relevant to our study.

4.3.1 Case I: Discrete Time Delay $\tau = 0$

In this subsection, the local stability of singular model system (4.4) without discrete time delay at the interior equilibrium will be investigated. A stability switch caused by the variation of economic interest of the harvest effort on the predator is also studied. Furthermore, a state feedback controller is designed to stabilize the model system at the interior equilibrium.

4.3.1.1 Singularity Induced Bifurcation

Theorem 4.1. *The singular model system (4.4) without discrete time delay has a singularity induced bifurcation at the interior equilibrium, and $v = 0$ is a bifurcation value. Furthermore, a stability switch occurs as v increases through* 0.

Proof. According to the economic theory of a common-property resource [17], there is a phenomenon of bio-economic equilibrium when the economic interest of harvesting is zero, i.e., $v = 0$. For model system (4.4) without discrete time delay, an interior equilibrium $P^*(x_1^*, x_2^*, y^*, E^*)$ can be obtained in the case of a phenomenon of bio-economic equilibrium, where $x_1^* = \dfrac{w\alpha r_1 - (d_1 + \alpha)d_2 w - c\beta d_2}{s_1 d_2 w}$, $x_2^* = \dfrac{\alpha}{d_2} x_1^*$, $y^* = \dfrac{c}{w}$, $E^* = \beta x_1^* - d_3 - \dfrac{cs_2}{w}$. According to the biological interpretation of the interior equilibrium, it follows that $x_1^* > 0, x_2^* > 0, y^* > 0$, and $E^* > 0$. In order to guarantee the existence of interior equilibrium, some inequalities are satisfied:

$$\begin{cases} w\alpha r_1 - (d_1 + \alpha)d_2 w - c\beta d_2 > 0, \\ (\beta x_1^* - d_3)w - cs_2 > 0. \end{cases} \tag{4.6}$$

Let v be a bifurcation parameter, and D be a differential operator (i.e., $D_E h$ represents the partial differential operator of $h(H(t), E(t), v)$ with respect to $E(t)$), $H(t) = (x_1(t), x_2(t), y(t))^T$,

$$h(H(t),E(t),v) = \begin{bmatrix} r_1x_2(t) - d_1x_1(t) - \alpha x_1(t) - s_1x_1^2(t) - \beta x_1(t)y(t) \\ \alpha x_1(t) - d_2x_2(t) \\ \beta x_1(t)y(t) - d_3y(t) - s_2y^2(t) - E(t)y(t) \end{bmatrix} \text{ and}$$

$$g(H(t),E(t),v) = E(t)(wy(t) - c) - v.$$

By simple computing, $g(H(t),E(t),v)|_{P^*} = \begin{cases} 0, & \text{v=0;} \\ \text{nonzero}, & \text{v} \neq 0. \end{cases}$,which implies that

$$\text{dimker}(D_E g(H(t),E(t),v)|_{P^*}) = 1. \tag{4.7}$$

It can be calculated that

$$\text{trace}(D_E h \text{adj}(D_E g)(D_H g, D_E g))|_{P^*} = y^*[cs_2 - w(\beta x_1^* - d_3)],$$

and it follows from (4.6) that

$$\text{trace}(D_E h \text{adj}(D_E g)(D_H g, D_E g))|_{P^*} \neq 0. \tag{4.8}$$

Furthermore, it can also be calculated that

$$\begin{vmatrix} D_H h & D_E h \\ D_H g & D_E g \end{vmatrix}_{P^*} = \frac{cd_2s_1x_1^*[(\beta x_1^* - d_3)w - cs_2]}{w},$$

and by virtue of (4.6), it is easy to show that

$$\begin{vmatrix} D_H h & D_E h \\ D_H g & D_E g \end{vmatrix}_{P^*} \neq 0. \tag{4.9}$$

According to Part A of Section IV in reference [49], $\Delta(H(t),E(t),v)$ can be defined as follows:

$$\Delta(H(t),E(t),v) = \det(D_E g) = wy(t) - c.$$

By simple computing,

$$\begin{vmatrix} D_H h & D_E h & D_v h \\ D_H g & D_E g & D_v g \\ D_H \Delta & D_E \Delta & D_v \Delta \end{vmatrix}_{P^*} = cs_1d_2x_1^*,$$

it follows from (4.6) that

$$\begin{vmatrix} D_H h & D_E h & D_v h \\ D_H g & D_E g & D_v g \\ D_H \Delta & D_E \Delta & D_v \Delta \end{vmatrix}_{P^*} \neq 0. \tag{4.10}$$

Based on the above analysis, five items (i-iv) can be obtained as follows:

(i) According to (4.8), $\text{trace}(D_E h \text{adj}(D_E g)(D_H g, D_E g))|_{P^*} \neq 0$;

(ii) it follows from (4.9) that $\begin{bmatrix} D_H h & D_E h \\ D_H g & D_E g \end{bmatrix}$ is nonsingular at P^*;

(iii) by virtue of (4.10), it can be shown that $\begin{bmatrix} D_H h & D_E h & D_v h \\ D_H g & D_E g & D_v g \\ D_H \Delta & D_E \Delta & D_v \Delta \end{bmatrix}$ is nonsingular at P^*, hence rank $\begin{bmatrix} D_H h & D_E h & D_v h \\ D_H g & D_E g & D_v g \\ D_H \Delta & D_E \Delta & D_v \Delta \end{bmatrix} = 5$;

(iv) it is easy to show that $\mathrm{rank}(h(H(t),E(t),v)) = 3$ and $\mathrm{rank}(g(H(t),E(t),v)) = 1$. It is easy to show that

$$\begin{aligned} &\mathrm{rank}\begin{bmatrix} D_H h & D_E h & D_v h \\ D_H g & D_E g & D_v g \\ D_H \Delta & D_E \Delta & D_v \Delta \end{bmatrix} \\ &= \mathrm{rank}(h(H(t),E(t),v)) + \mathrm{rank}(g(H(t),E(t),v)) + 1. \end{aligned}$$

The conditions for singularity induced bifurcation, which is introduced in Section III (A) of reference [49], consists of three conditions: i.e., SI1, SI2, and SI3. According to the above items (i-iv), SI1, SI2 and SI3 are all satisfied, hence model system (4.4) without discrete time delay has a singularity induced bifurcation at the interior equilibrium P^* and the bifurcation value is $v = 0$.

Remark 4.4. Some preliminaries of bifurcations in singular systems are introduced below. Parameter dependent differential-algebraic system of the form

$$\begin{cases} \dot{x}(t) = f(x(t),y(t),\lambda), & h: R^n \times R^m \times R^p \to R^n, \\ 0 = g(x(t),y(t),\lambda), & g: R^n \times R^m \times R^p \to R^m. \end{cases} \tag{4.11}$$

is widely used to model the dynamics of differential-algebraic system, where assuming that $x(t)$, $y(t)$ and λ have appropriate dimensions.

Definition 4.1. [49] $EQ = \{(x,y,\mu) \in X \times Y \times \Lambda | f(x,y,\lambda) = 0, g(x,y,\lambda) = 0\}$ is called equlibria set of model system (4.11).

Definition 4.2. [49] Let $S = \{(x,y,\mu) | g(x,y,\mu) = 0, \Lambda(x,y,\mu) = \det[D_y g(x,y,\mu)] = 0\}$, then S is called singular surface.

It has been shown recently that there are generically three types of codimension one local bifurcation associated with the singular system (4.11), namely saddle-node bifurcation, Hopf bifurcation, and singularity induced bifurcation (see [49]).

The singularity induced bifurcation is firstly introduced and analyzed in references [49, 50]. It is a new type of bifurcation and does not occur in usual ordinary differential equation system, which has been characterized for differential-algebraic system, and later improved in [56]. Roughly speaking, the singularity induced bifurcation refers to a stability change of the singular system (4.11) owing to some eigenvalues of a related linearization $f_x - f_y g_y^{-1} g_x$ diverging to infinity when the Jacobian g_y is singular. One of the important consequences of the singularity induced bifurcation is that it leads to an impulse phenomenon of the differential-algebraic system, which may result in the collapse of the differential-algebraic system (see reference [7]).

Lemma 4.1. *(Singularity Induced Bifurcation Theorem) [49] Consider the system (4.11) with a one-dimensional parameter space. Suppose the following conditions are satisfied at* $(0,0,\mu_0)$, *where D is a differential operator:*

SI1 $f(0,0,\mu_0)=0$, $g(0,0,\mu_0)=0$, *D_yg has a simple zero eigenvalue, and* $trace[D_yf adj(D_yg)D_xg]$ *is nonzero.*

SI2 $\begin{pmatrix} D_xf & D_yf \\ D_xg & D_yg \end{pmatrix}$ *is nonsingular.*

SI3 $\begin{pmatrix} D_xf & D_yf & D_\mu f \\ D_xg & D_yg & D_\mu g \\ D_x\Delta & D_y\Delta & D_\mu\Delta \end{pmatrix}$ *is nonsingular;*

$$rank\begin{pmatrix} D_xf & D_yf & D_\mu f \\ D_xg & D_yg & D_\mu g \\ D_x\Delta & D_y\Delta & D_\mu\Delta \end{pmatrix} = rankf(x,y,\lambda)+rankg(x,y,\lambda).$$

Then there exists a smooth curve of equilibria in R^{n+m+1} *that passes through* $(0,0,\mu_0)$ *and is transversal to the singular surface at* $(0,0,\mu_0)$. *When* μ *increases through* μ_0, *one eigenvalue of the system moves from* C^- *to* C^+ *if* $\frac{M}{N}>0$ *(respectively, from* C^+ *to* C^- *if* $\frac{M}{N}<0$*) along the real axis by diverging through* ∞. *The other* $(n-1)$ *eigenvalues remain bounded and stay away from the origin. The constants M and N can be computed by evaluating*

$$M = -trace[D_yf adj(D_yg)D_xg],$$

$$N = D_\mu\Delta - (D_x\Delta, D_y\Delta)\begin{pmatrix} D_xf & D_xg \\ D_yf & D_yg \end{pmatrix}^{-1}\begin{pmatrix} D_\mu f \\ D_\mu g \end{pmatrix}$$

at $(0,0,\mu_0)$.

More detailed introductions of the singularity induced bifurcation can be found in the references [49, 50, 7, 56].

Along with the line of the above proof, for differential-algebraic model system (4.4) without discrete time delay, by simple computing,

$$M = -\text{trace}(D_Eh\text{adj}(D_Eg)(D_Hg,D_Eg))|_{P^*} = y^*[w(\beta x_1^* - d_3) - cs_2],$$

$$N = [D_v\Delta - [D_H\Delta, D_E\Delta]\begin{bmatrix} D_Hh & D_Eh \\ D_Hg & D_Eg \end{bmatrix}^{-1}\begin{bmatrix} D_vh \\ D_vg \end{bmatrix}]|_{P^*} = \frac{wy^*}{c},$$

it follows from (4.6) that

$$\frac{M}{N} = \frac{c(\beta x_1^* - d_3)w - c^2 s_2}{w} > 0. \tag{4.12}$$

The inequality (4.12) satisfies Lemma 4.1. According to Lemma 4.1, when v increases through 0, one eigenvalue (denoted by λ_1) of differential-algebraic model system (4.4) without discrete time delay moves from C^- to C^+ along the real axis by diverging through infinity; the movement behavior of this eigenvalue influences the stability of model system (4.4) without discrete time delay.

Since the Jacobian of model system (4.4) without discrete time delay evaluated at P^* takes the form as follows:

$$J_{P^*} = \begin{bmatrix} -d_1 - \alpha - 2s_1x_1^* - \beta y^* & r_1 & -\beta x_1^* & 0 \\ \alpha & -d_2 & 0 & 0 \\ \beta y^* & 0 & \beta x_1^* - d_3 - 2s_2y^* - E^* & -y^* \\ 0 & 0 & w^*E^* & 0 \end{bmatrix}, \tag{4.13}$$

according to the leading matrix $A(t)$ in model system (4.5) and J_{P^*}, the characteristic equation of model system (4.4) without discrete time delay at P^* is $\det(\lambda A - J_{P^*}) = 0$. By virtue of simple computation, the characteristic equation can be expressed as follows:

$$\lambda^2 + (d_2 + \frac{r_1\alpha}{d_2} + s_1x_1^*)\lambda + s_1d_2x_1^* = 0.$$

Since $(d_2 + \frac{r_1\alpha}{d_2} + s_1x_1^*) > 0$ and $s_1d_2x_1^* > 0$, it can be concluded that the remaining eigenvalues (denoted by λ_2 and λ_3) have negative real parts by using the Routh-Hurwitz criteria [30].

Furthermore, the expressions (4.9), (4.10), and (4.11) satisfy Lemma 4.1. It follows from Lemma 4.1 that there is only one eigenvalue diverging to infinity as v increases through 0, and the remaining eigenvalues are continuous, are nonzero, and cannot jump from one half open complex plane to another one as v increases through 0. It has been shown that λ_1 moves from C^- to C^+ along the real axis by diverging through infinity. Therefore, λ_2 and λ_3 are continuous and bounded in the C^- half plane as v increases through 0 and their movement behaviors have no influence on the stability of differential-algebraic model system (4.4) without discrete time delay at the interior equilibrium P^*.

Table 4.1 shows the change in the signs of the real parts of eigenvalues (λ_1, λ_2 and λ_3) due to the variation of economic interest of the harvest effort.

Table 4.1 Signs of eigenvalues' real parts of model system (4.4) without discrete time delay at P^*

	$\text{Re}\lambda_1$	$\text{Re}\lambda_2$	$\text{Re}\lambda_3$
$v < 0$	$-$	$-$	$-$
$v > 0$	$+$	$-$	$-$

According to Table 4.1 and the stability theory, it can be concluded that differential-algebraic model system (4.4) without discrete time delay is stable at P^* as $v < 0$ and differential-algebraic model system (4.4) without discrete time delay is unstable at P^* as $v > 0$. Consequently, a stability switch occurs as v increases through 0.

Remark 4.5. The research results obtained above show that when the economic interest becomes positive, there is a phenomenon of singularity induced bifurcation at the interior equilibrium, which can cause a stability switch of the differential-algebraic model system. From the biological and economic perspectives, the prey-predator ecosystem can be influenced in the following two aspects.

- As stated in Remark 4.4 of this chapter, the singularity induced bifurcation can result in an impulse phenomenon, which may lead to the collapse of the proposed model system. In the prey-predator ecosystem, the impulse phenomenon of the ecosystem is usually connected with the rapid growth of the species population. If this phenomenon lasts a period of time, the species population will be beyond the carrying capacity of the environment and the prey-predator ecosystem will be out of balance, which is disastrous for the prey-predator ecosystem.
- People always try to make a profit of harvesting, so they are usually interested in the case of positive economic interest of harvesting. However, Theorem 4.1 of this chapter shows that differential-algebraic model system (4.4) without discrete time delay becomes unstable when the economic interest is positive, which hampers the sustainable development of harvesting on the ecosystem.

With the purpose of maintaining the sustainable development of the biological resource as well as economic interest of the harvest effort at an ideal level, some related measures should be taken to eliminate the impulse phenomenon caused by singularity induced bifurcation and stabilize differential-algebraic model system (4.4) without discrete time delay when the economic interest of harvesting is positive.

4.3.1.2 State Feedback Control for Singularity Induced Bifurcation

A state feedback controller is designed to stabilize model system (4.4) without discrete time delay at the interior equilibrium P^*.

According to the leading matrix $A(t)$ in model system (4.5) and J_{P^*} in (4.13), it can be calculated that $\mathrm{rank}(J_{P^*}, AJ_{P^*}, A^2J_{P^*}, A^3J_{P^*}) = 4$. By using Theorem 2-2.1 in reference [12], it is easy to show that differential-algebraic model system (4.4) without discrete time delay is locally controllable at P^*. Consequently, a state feedback controller can be applied to stabilize differential-algebraic model system (4.4) without discrete time delay at P^*. By using Theorem 3-1.2 in reference [12], a state feedback controller $u(t) = k(E(t) - E^*)$ (k is a feedback gain and E^* is the component of the interior equilibrium P^*) can be applied to stabilize model system (4.4) without discrete time delay at P^*, and a controlled differential-algebraic model system is as follows:

$$\begin{cases} \dot{x}_1(t) = r_1x_2(t) - d_1x_1(t) - \alpha x_1(t) - s_1x_1^2(t) - \beta x_1(t)y(t), \\ \dot{x}_2(t) = \alpha x_1(t) - d_2x_2(t), \\ \dot{y}(t) = \beta x_1(t)y(t) - d_3y(t) - s_2y^2(t) - E(t)y(t), \\ 0 = E(t)(wy(t) - c) + k(E(t) - E^*) - v, \end{cases} \tag{4.14}$$

where $x_1(t)$, $x_2(t)$, $y(t)$, r_1, d_1, d_2, d_3, s_1, s_2, α, β, w, c, and v have the same interpretations as mentioned in differential-algebraic model system (4.4). The feedback controller $u(t) = k(E(t) - E^*)$ is designed in the following Theorem 4.2.

Theorem 4.2. *If the feedback gain k satisfies the following inequality*
$k > \max\{\frac{s_1[(\beta x_1^* - d_3)w - cs_2]}{s_1 s_2 + \beta^2}, \frac{c[(\beta x_1^* - d_3)w - cs_2]}{w[d_2 + s_2 y^* + \frac{r_1\alpha}{d_2} + s_1 x_1^*]}, \frac{c(\frac{r_1\alpha}{d_2} + s_1 x_1^* + d_2)[(\beta x_1^* - d_3)w - cs_2]}{w[s_2 d_2 y^* + s_1 d_2 x_1^* + \frac{r_1 \alpha s_2 y^*}{d_2} + s_1 s_2 x_1^* y^* + \beta^2 x_1^* y^*]}\}$,
then differential-algebraic model system (4.14) is stable at P^*.

Proof. The Jacobian of model system (4.14) evaluated at the interior equilibrium P^* takes the form,

$$\tilde{J}_{P^*} = \begin{bmatrix} -d_1 - \alpha - 2s_1 x_1^* - \beta y^* & r_1 & -\beta x_1^* & 0 \\ \alpha & -d_2 & 0 & 0 \\ \beta y^* & 0 & \beta x_1^* - d_3 - 2s_2 y^* - E^* & -y^* \\ 0 & 0 & wE^* & k \end{bmatrix}.$$

According to the leading matrix $A(t)$ in model system (4.5) and $\tilde{J}_{P^*}$ in (4.13), the characteristic equation of differential-algebraic model system at P^* is $\det(\tilde{\lambda} A - \tilde{J}_{P^*}) = 0$, which can be expressed as follows:

$$\tilde{\lambda}^3 + \tilde{\Delta}_1 \tilde{\lambda}^2 + \tilde{\Delta}_2 \tilde{\lambda} + \tilde{\Delta}_3 = 0,$$

where

$$\tilde{\Delta}_1 = d_2 + s_2 y^* + \tfrac{r_1\alpha}{d_2} + s_1 x_1^* - \tfrac{c[(\beta x_1^* - d_3)w - cs_2]}{wk},$$

$$\tilde{\Delta}_2 = s_2 d_2 y^* + s_1 d_2 x_1^* + \tfrac{r_1 \alpha s_2 y^*}{d_2} + s_1 s_2 x_1^* y^* + \beta^2 x_1^* y^* - \tfrac{c(\frac{r_1\alpha}{d_2} + s_1 x_1^* + d_2)[(\beta x_1^* - d_3)w - cs_2]}{wk},$$

$$\tilde{\Delta}_3 = s_1 s_2 d_2 x_1^* y^* + \beta^2 d_2 x_1^* y^* - \tfrac{cs_1 d_2 x_1^*[(\beta x_1^* - d_3)w - cs_2]}{wk}.$$

By using the Routh-Hurwitz criteria [30], the sufficient and necessary condition for the stability of differential-algebraic model system (4.14) at P^* is that the feedback gain k satisfies
$k > \max\{\frac{cs_1[(\beta x_1^* - d_3)w - cs_2]}{wy^*(s_1 s_2 + \beta^2)}, \frac{c[(\beta x_1^* - d_3)w - cs_2]}{w[d_2 + s_2 y^* + \frac{r_1\alpha}{d_2} + s_1 x_1^*]}, \frac{c(\frac{r_1\alpha}{d_2} + s_1 x_1^* + d_2)[(\beta x_1^* - d_3)w - cs_2]}{w[s_2 d_2 y^* + s_1 d_2 x_1^* + \frac{r_1 \alpha s_2 y^*}{d_2} + s_1 s_2 x_1^* y^* + \beta^2 x_1^* y^*]}\}$.

Consequently, if the feedback gain satisfies the above inequality, differential-algebraic model system (4.14) is stable at P^*.

Remark 4.6. After applying the feedback controller into model system (4.4) without discrete time delay, the model system can be stabilized at the interior equilibrium. The elimination of the singularity induced bifurcation means that the prey-predator ecosystem restores ecological balance. Furthermore, according to the design of the feedback controller, the biological interpretation of the feedback controller is also obvious. It means that by enhancing the harvest effort on the predator, both sustainable development of the prey-predator ecosystem and the ideal economic interest of harvesting can be obtained, as is in line with intuition.

4.3.2 Case II: Discrete Time Delay $\tau > 0$

In this subsection, a stability switch in model system (4.4) due to variation of the discrete time delay τ will be investigated in the case of positive economic interest of harvesting.

By the analysis of the roots in Eqn. (4.15), it can be obtained that if there exists

$$v_{sn} = \frac{[s_1 d_2 d_3 w - w\beta(r_1\alpha - d_1 d_2 - \alpha d_2) + c d_2(s_1 s_2 + \beta^2)]^2}{4 w s_1 d_2^2 (s_1 s_2 + \beta^2)} > 0,$$

(1) if $v > v_{sn}$, there are no interior equilibria;

(2) if $v = v_{sn}$, there is only one interior equilibrium $\tilde{P}^*(\tilde{x}_1^*, \tilde{x}_2^*, \tilde{y}^*, \tilde{E}^*)$, where $\tilde{x}_1^* = \frac{1}{s_1}(\frac{r_1\alpha}{d_2} - d_1 - \alpha - \beta\tilde{y}^*)$, $\tilde{x}_2^* = \frac{\alpha}{d_2}\tilde{x}_1^*$, $\tilde{E}^* = \frac{v_{sn}}{w\tilde{y} - c}$, and $\tilde{y}^* = \frac{w\beta(r_1\alpha - d_1 d_2 - \alpha d_2) - w s_1 d_2 d_3 + c d_2(s_1 s_2 + \beta^2)}{2 w d_2 (s_1 s_2 + \beta^2)}$;

(3) if $v < v_{sn}$, there're two interior equilibria $\tilde{P}_1^*(\tilde{x}_{11}^*, \tilde{x}_{12}^*, \tilde{y}_1^*, \tilde{E}_1^*)$, $\tilde{P}_2^*(\tilde{x}_{21}^*, \tilde{x}_{22}^*, \tilde{y}_2^*, \tilde{E}_2^*)$, and the coordinate of $\tilde{y}$ satisfies the following equation:

$$B_1 \tilde{y}^2 + B_2 \tilde{y} + B_3 = 0, \tag{4.15}$$

where

$B_1 = w(s_1 s_2 + \beta^2)$,

$B_2 = -w\beta(\frac{r_1\alpha}{d_2} - d_1 - \alpha) + s_1 d_2 w - c(s_1 s_2 + \beta^2)$,

$B_3 = c\beta(\frac{r_1\alpha}{d_2} - d_1 - \alpha) - s_1 d_3 c + s_1 v$.

4.3.2.1 Saddle-Node Bifurcation

Let v denote a bifurcation parameter, and then the following result is obtained:

Theorem 4.3. *If* $\frac{c}{w} < \frac{w\beta(r_1\alpha - d_1 d_2 - \alpha d_2) - w s_1 d_2 d_3 + c d_2(s_1 s_2 + \beta^2)}{2 w d_2 (s_1 s_2 + \beta^2)} < \frac{1}{\beta}(\frac{r_1\alpha}{d_2} - d_1 - \alpha)$, *then model system (4.4) undergoes a saddle-node bifurcation around* $v = v_{sn}$.

Proof. The characteristic polynomial of model system (4.4) around the interior equilibrium $\tilde{P}^*$ is given as follows:

$$\lambda^3 + l_1\lambda^2 + l_2\lambda + l_3 = 0, \tag{4.16}$$

where

$$l_1 = s_1\tilde{x}_1^* + \frac{r_1\alpha}{d_2} + d_2 + s_2\tilde{y}^* - \frac{w\tilde{E}^*\tilde{y}^*}{w\tilde{y}^* - c},$$

$$l_2 = (s_2\tilde{y}^* - \frac{w\tilde{E}^*\tilde{y}^*}{w\tilde{y}^* - c})(s_1\tilde{x}^* + \frac{r_1\alpha}{d_2} + d_2) + d_2 s_1\tilde{x}_1^* + \beta^2\tilde{x}_1^*,$$

$l_3 = s_1 d_2 \tilde{x}_1^*(s_2\tilde{y}^* - -\frac{w\tilde{E}^*\tilde{y}^*}{w\tilde{y}^* - c}) + d_2\beta^2\tilde{x}_1^*\tilde{y}^*.$

It follows from simple calculation that there is a geometrically simple zero eigenvalue with left eigenvector $u_l = \begin{pmatrix} d_2 & r_1 & \frac{s_1 d_2 \tilde{x}_1^*}{\beta\tilde{y}^*} \end{pmatrix}$ and right eigenvector $u_r = \begin{pmatrix} d_2 & \alpha & -\frac{s_1 d_2}{\beta} \end{pmatrix}$.

Furthermore,

$$u_l D_v h|_{\tilde{P}^*} = u_l (D_v f - D_E f (D_E g)^{-1} D_v g)|_{\tilde{P}^*},$$

$$u_r D_H h(u_r, u_r, u_r)|_{\tilde{P}^*} = u_l \sum_{i=1}^{3} (e_i u_r^T D_H (D_H f_i)^T u_r)|_{\tilde{P}^*} = -\frac{2 s_1^2 d_2^3 \tilde{x}_1^*}{\beta^3 \tilde{y}^*}(s_1 s_2 + \beta^2),$$

where $e_i (i = 1,2,3)$ represents the unit vector.

According to [19], model system (4.4) undergoes a saddle-node bifurcation around the interior equilibrium $\tilde{P}^*$ when $v = v_{sn}$.

4.3.2.2 Hopf Bifurcation

In this subsection, qualitative analysis about model system (4.4) is made in the case of positive economic interest of harvesting, i.e., $0 < v < v_{sn}$. From the discussion of Section 4.2.2, there exist two interior equilibria, $\tilde{P}_1^*, \tilde{P}_2^*$.

As mentioned above, there are two interior equilibria (denoted by $\tilde{P}_1^*$ and $\tilde{P}_2^*$) when $0 < v < \tilde{v}_1$. In the following part, we investigate only the dynamical behavior of model system (4.4) with discrete time delay at the interior equilibrium $\tilde{P}_1^*$. Some symmetric results about the interior equilibrium $\tilde{P}_2^*$ can also be obtained.

In the case of positive discrete time delay and $0 < v < \tilde{v}_1$, according to the Jacobian evaluated at the interior equilibrium $\tilde{P}_1^*$ and the leading matrix $A(t)$ in model system (4.5), we can obtain the characteristic equation of differential-algebraic model system (4.4) at $\tilde{P}_1^*$, which can be expressed as follows:

$$\begin{vmatrix} \lambda + d_1 + \alpha + 2s_1\tilde{x}_1^* + \beta\tilde{y}^* & -r_1 & \beta\tilde{x}_1^* & 0 \\ -\alpha & \lambda + d_2 & 0 & 0 \\ -\beta\tilde{y}^* e^{-\lambda\tau} & 0 & \lambda - \beta\tilde{x}_1^* e^{-\lambda\tau} + d_3 + 2s_2\tilde{y}^* + E^* & \tilde{y}^* \\ 0 & 0 & -w\tilde{E}^* & -\frac{v}{\tilde{E}^*} \end{vmatrix} = 0,$$

$\Rightarrow$

$$P(\lambda) + Q(\lambda)e^{-\lambda\tau} = 0 \tag{4.17}$$

where

$P(\lambda) = \lambda^3 + p_1\lambda^2 + p_2\lambda + p_3,$
$Q(\lambda) = q_1\lambda^2 + q_2\lambda + q_3,$
$p_1 = d_2 + \frac{r_1\alpha}{d_1} + (s_1 + \beta)\tilde{x}_1^* + s_2\tilde{y}^* + \tilde{E}^*(1 + \frac{c\tilde{E}^*}{v}),$

$p_2 = d_2 s_1 \tilde{x}_1^* + (\beta \tilde{x}_1^* + s_2 \tilde{y}^*)(d_2 + s_1 \tilde{x}_1^* + \frac{r_1 \alpha}{d_2}) + \tilde{E}^*(1 + \frac{c\tilde{E}^*}{v})(s_1 \tilde{x}_1^* + d_2 + \frac{\alpha r_1}{d_2})$,
$p_3 = d_2 s_1 \tilde{x}_1^* (\beta \tilde{x}_1^* + s_2 \tilde{y}^*) + \tilde{E}^*(1 + \frac{c\tilde{E}^*}{v}) s_1 d_2 \tilde{x}_1^*$,
$q_1 = -\beta \tilde{x}_1^*$,
$q_2 = -\beta \tilde{x}_1^* (d_2 + s_1 \tilde{x}_1^* - \beta \tilde{y}^* + \frac{r_1 \alpha}{d_2})$,
$q_3 = -\beta d_2 \tilde{x}_1^* (s_1 \tilde{x}_1^* - \beta \tilde{y}^*)$.

Now substituting $\lambda = i\sigma$ (σ is a positive real number) into Eq. (4.17) and separating the real and imaginary parts, two transcendental equations can be obtained as follows:

$$\sigma^3 - p_2 \sigma = q_2 \sigma \cos(\sigma\tau) - (q_3 - q_1 \sigma^2) \sin(\sigma\tau), \tag{4.18}$$

$$p_1 \sigma^2 - p_3 = (q_3 - q_1 \sigma^2) \cos(\sigma\tau) + q_2 \sigma \sin(\sigma\tau). \tag{4.19}$$

By squaring and adding (4.18) and (4.19), it can be calculated that

$$(q_3 - q_1 \sigma^2)^2 + q_2^2 \sigma^2 = (p_1 \sigma^2 - p_3)^2 + (\sigma^3 - p_2 \sigma)^2, \tag{4.20}$$

$\Rightarrow$

$$\sigma^6 + C_1 \sigma^4 + C_2 \sigma^2 + C_3 = 0, \tag{4.21}$$

where $C_i (i = 1, 2, 3)$ can be expressed as follows:

$$\begin{aligned} C_1 &= d_2^2 + (s_1 \tilde{x}_1^*)^2 + (\frac{r_1 \alpha}{d_2})^2 + 2\frac{r_1 \alpha}{d_2}(d_2 + s_1 \tilde{x}_1^*) + (E^*(1 + \frac{c\tilde{E}^*}{v}))^2 \\ &+ (s_2 \tilde{y}^*)^2 + 2 s_2 \beta \tilde{x}_1^* \tilde{y}^* + 2\tilde{E}^*(\beta \tilde{x}_1^* + s_2 \tilde{y}^*)(1 + \frac{c\tilde{E}^*}{v}) > 0, \end{aligned}$$

$$\begin{aligned} C_2 &= (d_2 s_1 \tilde{x}_1^*)^2 + [d_2^2 + (s_1 \tilde{x}_1^*)^2 + (\frac{\alpha r_1}{d_2})^2 + 2\alpha r_1 + \frac{2\alpha r_1 s_1 \tilde{x}_1^*}{d_2}] \\ &[\beta \tilde{x}_1^* + s_2 \tilde{y}^* + \tilde{E}^*(1 + \frac{c\tilde{E}^*}{v})] - (\beta \tilde{x}_1^*)^2 [d_2^2 + (\frac{\alpha r_1}{d_2})^2 + (s_1 \tilde{x}_1^* - \beta \tilde{y}^*)^2 \\ &+ 2\alpha r_1 + \frac{2\alpha r_1 (s_1 \tilde{x}_1^* - \beta \tilde{y}^*)}{d_2}], \end{aligned}$$

$$C_3 = d_2^2 (\tilde{x}_1^*)^2 (2\beta s_1 \tilde{x}_1^* + (s_1 s_2 - \beta^2)\tilde{y}^* + s_1 \tilde{E}^*(1 + \frac{c\tilde{E}^*}{v}))[(\beta^2 + s_1 s_2)\tilde{y}^* + s_1 \tilde{E}^*(1 + \frac{c\tilde{E}^*}{v})].$$

According to the values of $C_i, (i = 1, 2, 3)$ and the Routh-Hurwitz criteria [30], a simple assumption that Eq. (4.21) has a positive real root σ_0 is $C_3 < 0$. Hence, under this assumption, the characteristic equation (4.21) will have a pair of purely imaginary roots of the form $\pm i\sigma_0$, eliminating $\sin(\sigma\tau)$ from (4.18) and (4.19), and it can be calculated that the τ_n^* corresponding to σ_0 is as follows:

$$\tau_n^* = \frac{1}{\sigma_0}\arccos[\frac{(p_1\sigma_0^2 - p_3)(q_3 - q_1\sigma_0^2) + q_2\sigma_0(\sigma_0^3 - p_2\sigma_0)}{(q_3 - q_1\sigma_0^2)^2 + (q_2\sigma_0)^2}] + \frac{2n\pi}{\sigma_0}, \tag{4.22}$$

where $n = 0, 1, 2, \cdots$. By using Butler's lemma [15], the interior equilibrium $\tilde{P}_1^*$ remains stable for $\tau < \tau_0$, where $\tau_0 = \tau_0^*$ as $n = 0$.

Subsequently, the conditions for Hopf bifurcation in reference [22] are utilized to investigate whether there is a phenomenon of Hopf bifurcation as τ increases through τ_0.

As mentioned above, let $\lambda = i\sigma$ represent the purely imaginary root of Eq. (4.20). It follows from Eq. (4.17) that $|P(i\sigma_0)| = |Q(i\sigma_0)|$, which determines a set of possible values of σ_0 [19]. In the following part, we determine the direction of motion of $\lambda = i\sigma$ as τ is varied, namely, we determine

$$\Theta = \text{sign}[\frac{\mathrm{d}(\text{Re}\lambda)}{\mathrm{d}\tau}]_{\lambda=i\sigma_0} = \text{sign}[\text{Re}(\frac{\mathrm{d}\lambda}{\mathrm{d}\tau})^{-1}]_{\lambda=i\sigma_0}.$$

By differentiating (4.17) with respect to τ, it can be obtained that

$$\begin{aligned}(\frac{\mathrm{d}\lambda}{\mathrm{d}\tau})^{-1} &= \frac{3\lambda^2 + 2p_1\lambda + p_2}{\lambda e^{-\lambda\tau}(q_1\lambda^2 + q_2\lambda + q_3)} + \frac{2q_1\lambda + q_2}{\lambda(q_1\lambda^2 + q_2\lambda + q_3)} - \frac{\tau}{\lambda} \\ &= \frac{2\lambda^3 + p_1\lambda^2 - p_3}{-\lambda^2(\lambda^3 + p_1\lambda^2 + p_2\lambda + p_3)} + \frac{q_1\lambda^2 - q_3}{\lambda^2(q_1\lambda^2 + q_2\lambda + q_3)} - \frac{\tau}{\lambda}.\end{aligned}$$

By virtue of (4.20) and the above equation, it can be obtained that

$$\begin{aligned}\Theta &= \text{sign}[\text{Re}(\frac{\mathrm{d}\lambda}{\mathrm{d}\tau})^{-1}]_{\lambda=i\sigma_0} \\ &= \frac{1}{\sigma_0^2}\text{sign}[\frac{2\sigma_0^6 + C_1\sigma_0^4 - C_3}{(q_3 - q_1\sigma_0^2)^2 + (q_2\sigma_0)^2}].\end{aligned}$$

According to the values of $C_i, (i = 1, 2, 3)$ given in the previous part of this chapter, it can be computed that $C_1 > 0$ for $v \in (0, v_1)$. Furthermore, based on the assumption that Eq. (4.21) has a positive real root, it follows that $C_3 < 0$. Hence, it can be concluded that $2\sigma_0^6 + C_1\sigma_0^4 - C_3 > 0$ for $v \in (0, v_2)$, and then $\text{sign}[\frac{\mathrm{d}(\text{Re}\lambda)}{\mathrm{d}\tau}]_{\tau=\tau_0, \sigma=\sigma_0} > 0$. Therefore, the transversality condition holds and Hopf bifurcation occurs at $\sigma = \sigma_0$, $\tau = \tau_0$.

Remark 4.7. $\text{Sign}[\frac{\mathrm{d}(\text{Re}\lambda)}{\mathrm{d}\tau}]_{\tau=\tau_0} > 0$ signifies that there exists at least one eigenvalue with a positive real part for $\tau > \tau_0$, and the conditions for Hopf bifurcation in reference [22, 19] are also satisfied yielding the required periodic solution.

Based on the above analysis, the following theorem is obviously obtained.

Theorem 4.4. *If the economic interest of harvesting* $v \in (0, \min\{v_1, v_2, v_{sn}\})$, *the system (4.4) undergoes Hopf bifurcation around the equilibrium* $\tilde{P}_1^*$ *when* $\tau > \tau_0$

as $n = 0, 1, 2, \cdots$. *Furthermore, an attracting invariant closed curve bifurcates from the interior equilibrium* $\tilde{P}_1^*$ *for* $\tau > \tau_0$ *and* $|\tau - \tau_0| \ll 1$.

4.4 Numerical Simulation

With the help of MATLAB, numerical simulations are provided to substantiate the theoretical results that have been established in the previous sections of this chapter.

4.4.1 Numerical Simulation of Control for Singularity Induced Bifurcation

The values of hypothetic parameters are set in appropriate units. $r_1 = 7.889$, $d_1 = 1.225$, $\alpha = 2.657$, $s_1 = 0.7$, $\beta = 1.5$, $d_2 = 0.9$, $d_3 = 0.7$, $s_2 = 0.8$, $w = 0.2$, and $c = 1$. A numerical simulation is provided to illustrate the effectiveness of the state feedback controller designed in Section 4.2.1. In the absence of discrete time delay, differential-algebraic model system (4.4) with the given values of parameters takes the following form:

$$\begin{cases} \dot{x}_1(t) = 7.889x_2(t) - 1.225x_1(t) - 2.657x_1(t) - 0.7x_1^2(t) - 1.5x_1(t)y(t), \\ \dot{x}_2(t) = 2.657x_1(t) - 0.9x_2(t), \\ \dot{y}(t) = 1.5x_1(t)y(t) - 0.7y(t) - 0.8y^2(t) - E(t)y(t), \\ 0 = E(t)(0.2y(t) - 1) - v. \end{cases} \tag{4.23}$$

By using Theorem 4.1 of this chapter, it can be shown that differential-algebraic model system (4.23) has a singularity induced bifurcation at the interior equilibrium $P^*(17.0116, 50.222, 5, 20.8174)$, and a stability switch occurs as v increases through 0. Based on the analysis in Section 4.1.2 of this chapter, a state feedback controller $u(t)$ can be applied to stabilize differential-algebraic model system (4.23) at P^*, where $u(t) = k(E(t) - 20.8174)$, and then model system (4.23) with the state feedback controller takes the form as follows:

$$\begin{cases} \dot{x}_1(t) = 7.889x_2(t) - 1.225x_1(t) - 2.657x_1(t) - 0.7x_1^2(t) - 1.5x_1(t)y(t), \\ \dot{x}_2(t) = 2.657x_1(t) - 0.9x_2(t), \\ \dot{y}(t) = 1.5x_1(t)y(t) - 0.7y(t) - 0.8y^2(t) - E(t)y(t), \\ 0 = E(t)(0.2y(t) - 1) - v + k(E(t) - 20.8174). \end{cases} \tag{4.24}$$

By using Theorem 4.2 of this chapter, if the feedback gain k satisfies $k > 0.2729$, then differential-algebraic model system (4.24) is stable at P^* and the singularity induced bifurcation of differential-algebraic model system (4.23) is also eliminated. The dynamical responses of the differential-algebraic model system (4.24) can be shown in the following Fig. 4.1.

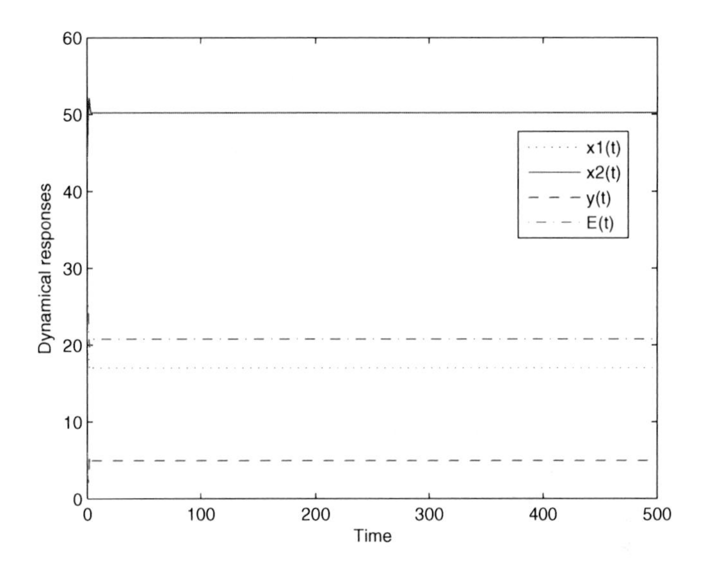

Fig. 4.1 Dynamical responses of differential-algebraic model system (4.24).

4.4.2 Numerical Simulation for Saddle-Node Bifurcation

The values of hypothetic parameters are set in appropriate units, which are given as follows: $r_1 = 8$, $d_1 = 1.2$, $\alpha = 2.6$, $s_1 = 0.2$, $\beta = 1.5$, $d_2 = 0.9$, $d_3 = 0.7$, $s_2 = 0.8$, $w = 2$, and $c = 0.5$. By simple calculation, it can be obtained that there is a unique interior equilibrium $\tilde{P}^*(50.763, 146.65, 6.106, 70.561)$ when $v = v_s n = 826.352$; there are two interior equilibria $\tilde{P}_1^*(82.309, 237.781, 1.899, 121.244)$ and $\tilde{P}_2^*(19.218, 55.518, 10.312, 19.877)$ when $v = 400 < v_{sn}$. Eigenvalues at equilibrium $\tilde{P}_1^*$ $\tilde{P}_2^*$ are $(-0.33, -38.141, 136.074)$ and $(-7.319, -4.297 + i5.646, -4.297 - i5.646)$, respectively. It implies that model system (4.4) undergoes saddle-node bifurcation when $v = v_{sn}$.

4.4.3 Numerical Simulation for Hopf Bifurcation

The values of hypothetic parameters are set in appropriate units. $r_1 = 7.889$, $d_1 = 1.225$, $\alpha = 2.657$, $s_1 = 0.7$, $\beta = 1.5$, $d_2 = 0.9$, $d_3 = 0.7$, $s_2 = 0.8$, $w = 0.2$, and $c = 1$. A numerical simulation is carried out to show the stability switch due to the change of discrete time delay. According to the given values of parameters $v_{sn} = 2.5189$. Based on the analysis in Section 4.2.2, there are two interior equilibria (denoted by $\tilde{P}_1^*$ and $\tilde{P}_2^*$) when $0 < v < v_{sn}$. In the following part, we focus only on the case of $0 < v < v_{sn}$, and the economic interest is set as $v = 2 < v_{sn}$

in the appropriate unit. By virtue of the given values of parameters, two interior equilibria can be obtained as follows: $\tilde{P}_1^* = (5.3497, 15.7935, 10.3571, 1.8667)$ and $\tilde{P}_2^* = (11.6714, 34.4566, 5.1578, 12.6809)$. Only the dynamical responses and corresponding phase portrait of model system (4.4) around $\tilde{P}_1^*$ are plotted, and some symmetric results about $\tilde{P}_2^*$ can also be obtained. Furthermore, it can be computed that $C_3 = -6.4076 \times 10^3 < 0$. Based on the analysis in Section 4.2.2, it satisfies the assumption that Eq. (4.21) has a positive root, and then the corresponding $\tau_0 = 12.37$ can be calculated by solving Eq. (4.22). By using Butler's lemmas [15], the interior equilibrium $\tilde{P}_1^*$ remains stable for $\tau < \tau_0$, which can be shown in Fig. 4.2. It should be noted that $\tau = 6.8$ in Fig. 4.2 is randomly selected in the interval $(0, 12.37)$, which is enough to merit the above mathematical study. As τ increases through τ_0, a periodic solution caused by the phenomenon of Hopf bifurcation occurs, which can be shown in Fig. 4.3. The Fig. 4.4 shows a limit cycle corresponding to the periodic solution in Fig. 4.3, which forms around the fixed point. Furthermore, Fig. 4.5 and Fig. 4.6 show that model system (4.4) remains unstable for sufficiently large τ, but show complex structures with increasing oscillations.

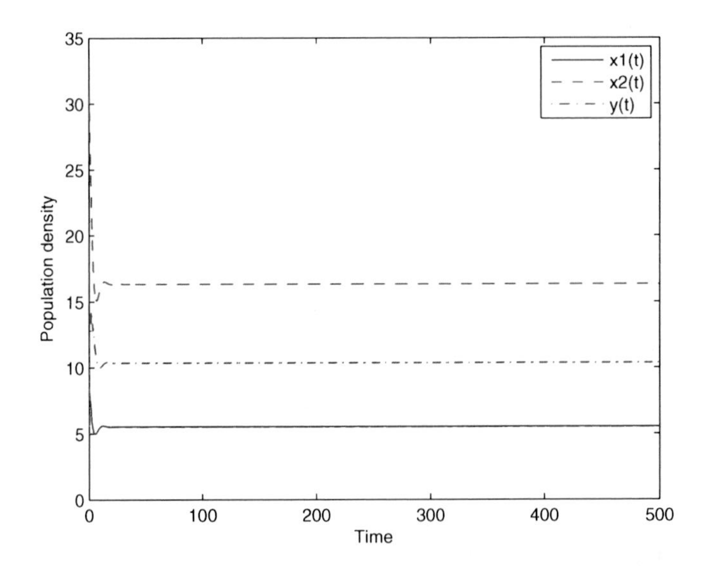

Fig. 4.2 The population density of model system (4.4) with $\tau = 6.8$ and initial values $x_1(0) = 17, x_2(0) = 29, y(0) = 30, E(0) = 5$.

Here, we take two practical examples from the real world: fur seals and cod, baleen whales and the krill ecosystem. When there is less trade and economic interest than usual, the bio-economic measures based on the theoretical analysis in this chapter can be taken to deal with the impact of the current recession on fishery, s.t., harvesting on the economic marine population (cod and krill). From a long view of

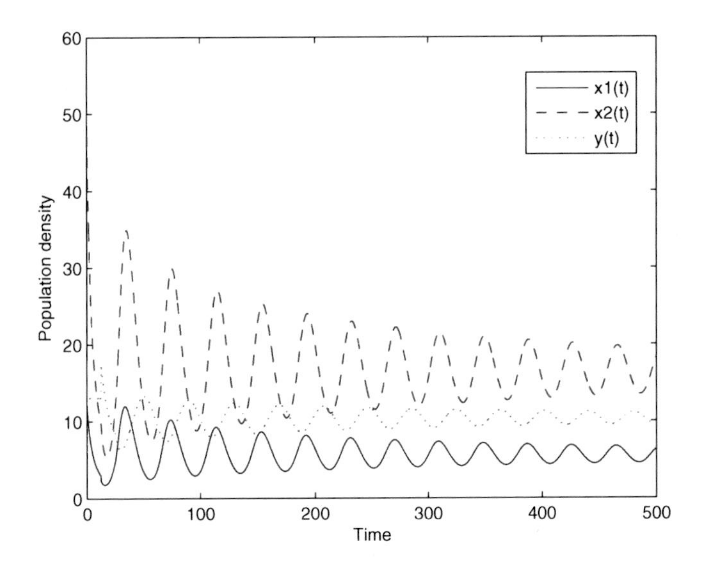

Fig. 4.3 The population density of model system (4.4) with $\tau = 12.37$ and initial values $x_1(0) = 17, x_2(0) = 29, y(0) = 30, E(0) = 5$.

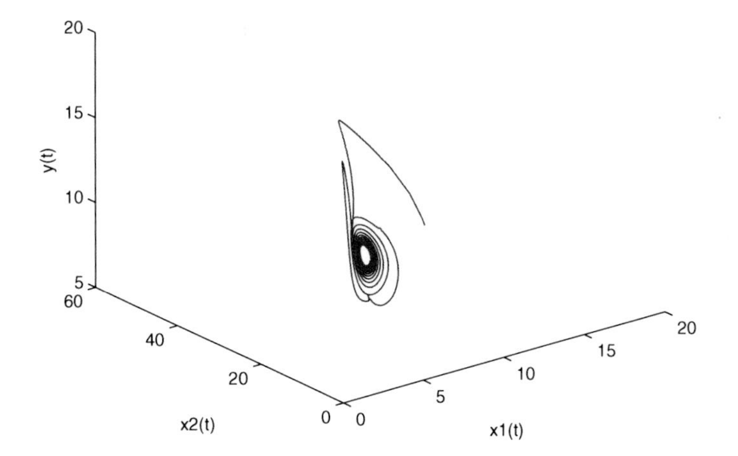

Fig. 4.4 A limit cycle for model system (4.4) with $\tau = 12.37$ corresponding to the periodic solution in Fig. 4.3.

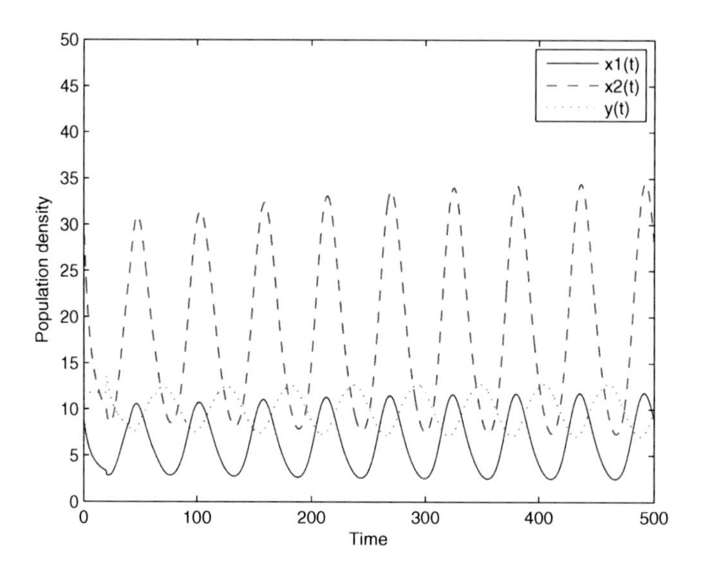

Fig. 4.5 The population density of model system (4.4) with $\tau = 21.63$ and initial values $x_1(0) = 17, x_2(0) = 29, y(0) = 30, E(0) = 5$.

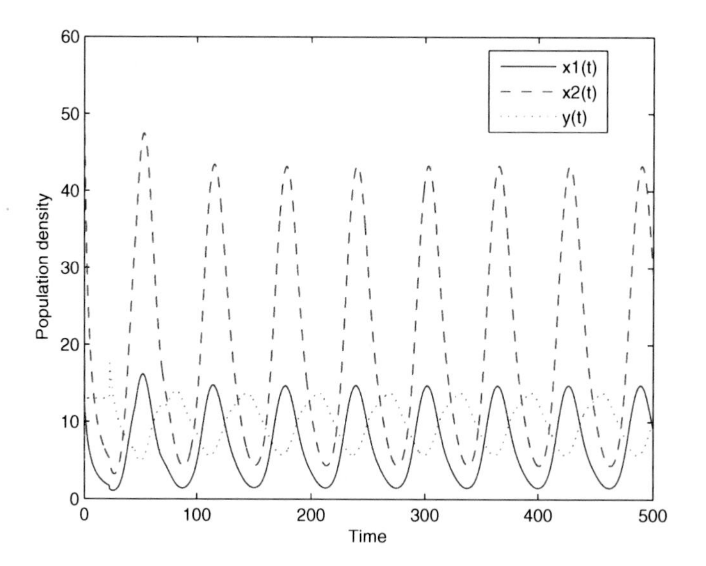

Fig. 4.6 The population density of model system (4.4) with $\tau = 29.87$ and initial values $x_1(0) = 17, x_2(0) = 29, y(0) = 30, E(0) = 5$.

maintaining the ecological balance and the sustainable development for fishery, it is desirable for the government to provide subsidies and formulate special policies for fishermen to encourage them to maintain the amount of capture effort on fishery. Under these measures, the above mentioned economic marine population resource (cod and krill), which used to face extinction, has significantly recovered.

Further attempts are made to understand the effect of gestation delay on the stability of the model system. Local stability analysis reveals that the discrete time delay is responsible for the stability switch of the model system, and a phenomenon of Hopf bifurcation occurs as the discrete time delay increases through a certain threshold.

The model system proposed in [5] does not consider the harvest effort on predators. For the model system proposed in [5], by using the values of parameters given in this chapter, the threshold value of discrete time delay where Hopf bifurcation occurs can be calculated, which is $\tilde{\tau}_0 = 13.92$. However, the harvest effort on predators is considered in model system (4.4) of this chapter. As calculated in Section 4.3.3 of this chapter, in the case of positive economic interest, the Hopf bifurcation occurs at $\tau_0 = 12.37$. It is obvious that $\tau_0 < \tilde{\tau}_0$, which implies that the harvesting has an impact on the dynamical behavior of a model system and the length of time delay to preserve stability of the model system is shortened with the introduction of harvest effort on the predator. Based on the above analysis, since there is a world wide trend of commercial harvesting on biological resources, it is more challenging for the population with a relatively long gestation delay to survive because their population may more easily fluctuate in a harvested prey-predator ecosystem.

It should be noted that the differential-algebraic models presented in [61, 59, 60] investigate only some biological-economic systems with a single population, and the gestation delay in the stage-structured prey-predator ecosystem is not discussed in [33]. Compared with these works, the model proposed in this chapter investigates dynamical behavior of two species (prey-predator) as well as stage structure for prey, which gives the work studied in this chapter some new and positive features.

References

1. Aiello, W.G., Freedman, H.I.: A time-delay model of single-species growth with stage structure. Math. Bios. 101, 139–153 (1990)
2. Aiello, W.G., Freedman, H.J., Wu, J.: Analysis of a model representing stage-structured population growth with state-dependent time delay. SIAM J. Appl. Math. 52, 855–869 (1992)
3. Al-Omari, J., Gourley, S.: Stability and traveling fronts in Lotka-Volterra competition models with stage structure. SIAM J. Appl. Math. 63, 2063–2086 (2003)
4. Arino, O., Sanchez, E., Fathallah, A.: State-dependent delay differential equations in population dynamics: Modeling and analysis. Fiel. Inst. Commu. Amer. Math. Soci. 29, 19–36 (2001)
5. Bandyopadhyay, M., Banerjee, S.: A stage-structured prey-predator model with discrete time delay. Appl. Math. Compu. 182, 1385–1398 (2006)

6. Barclay, H.J., Driessche, V.D.: A model for a single species with two life history stages and added mortality. Ecol. Model 11, 157–166 (1980)
7. Beardmore, R.E.: The singularity-induced bifurcation and its kronecker normal form. SIAM J. Matr. Ana. Appl. 23(1), 126–137 (2001)
8. Chen, L.S., Ruan, S.G., Zhu, J.: Adv. Top. Biomath. Oscillations in a stage structured predator-prey system with time dependent coefficients. World Scientific, Singapore (1997)
9. Clark, C.W.: Mathematical Bioeconomics: The Optimal Management of Renewable Resource, 2nd edn. John Wiley and Sons, New York (1990)
10. Cui, J., Chen, L.S., Wang, W.D.: The effect of dispersal on population growth with stage-structure. Compu. Math. Appl. 39, 91–102 (2000)
11. Dai, G.R., Tang, M.X.: Coexistence region and global dynamics of a harvested predator-prey system. SIAM J. Appl. Math. 58(1), 193–210 (1998)
12. Dai, L.: Singular Control System. Springer, New York (1989)
13. Edwards, H.J., Dytham, C., Pitchford, J.W., Righton, D.: Prey selection, vertical migrations and the impacts of harvesting upon the population dynamics of a predator-prey system. Bull. Math. Biol. 69, 1827–1846 (2007)
14. Freedman, H.I., Joseph, W.H.S., Wu, J.H.: A model for the growth of a population exhibiting stage structure: Cannibalism and cooperation. J. Compu. Appl. Math. 52, 177–198 (1994)
15. Freedman, H., Rao, V.S.H.: The trade-off between mutual interference and time lags in predator-prey systems. Bull. Math. Biol. 45, 991–1004 (1983)
16. Gakkhar, S., Singh, B.: The dynamics of a food web consisting of two preys and a harvesting predator. Chaos Soli. Frac. 34(4), 1346–1356 (2007)
17. Gordon, H.S.: The economic theory of a common property resource: The fishery. J. Poli. Eco. 62(2), 124–142 (1954)
18. Gourley, S.A., Kuang, Y.: A stage structured predator-prey model and its dependence on maturation delay and death rate. J. Math. Biol. 49, 188–200 (2004)
19. Guckenheimer, J., Holmes, P.: Nonlinear Oscillations, Dynamical Systems and Bifurcations of Vector Field. Springer, New York (1983)
20. Gurney, W.S.C., Blythe, S.P., Nisbet, R.M.: Nicholson's blowflies revisited. Nature 287, 17–21 (1980)
21. Gurney, W.S.C., Nisbet, R.M.: Fluctuating periodicity, generation separation, and the expression of larval competition. Theo. Popu. Biol. 28, 150–180 (1985)
22. Hale, J.K.: Theory of Functional Differential Equations. Springer, New York (1997)
23. Hastings, A.: Age-dependent predation is not a simple process, I, continuous time models. Theo. Popu. Biol. 23, 347–362 (1983)
24. Hastings, A.: Delay in recruitment at different tropic levels: Effects on stability. J. Math. Biol. 21, 35–44 (1984)
25. Hofbauer, J., Sigmund, K.: Evolutionary Games and Population Dynamics and Dynamical Systems. Cambridge University Press, Cambridge (1998)
26. Huo, H., Li, W., Agarwal, R.P.: Optimal harvesting and stability for two species stage-structured system with cannibalism. Int. J. Appl. Math. 6, 59–79 (2001)
27. Jiao, J.J., Chen, L.S.: Global attractivity of a stage-structure variable coefficients predator-prey system with time delay and impulsive perturbations on predators. Int. J. Biomath. 1(2), 197–208 (2008)
28. Kar, T.K., Matsuda, H.: Controllability of a harvested prey-predator with time delay. J. Biol. Syst. 14(2), 243–254 (2006)
29. Kolesov, Y.S.: Properties of solutions of a class of equations with lag which describe the dynamics of change in the population of a species with the age structure taken into account. Math. USSR Sbo. Tom. 45, 91–100 (1983)

30. Kot, M.: Elements of Mathematical Biology. Cambridge University Press, Cambridge (2001)
31. Landahl, H.D., Hanson, B.D.: A three stage population model with cannibalism. Bull. Math. Biol. 37, 11–17 (1995)
32. Liu, B., Zhang, Y.J., Chen, L.S.: Dynamic complexities in a Lotka-Volterra predator-prey model concerning impulsive control strategy. Int. J. Bifur. Chaos. 15(2), 517–531 (2005)
33. Liu, C., Zhang, Q.L., Zhang, Y., Duan, X.D.: Bifurcation and control in a differential-algebraic harvested prey-predator model with stage structure for predator. Int. J. Bifur. Chaos. 18(10), 3159–3168 (2008)
34. Liu, S.Q., Chen, L.S.: Extinction and permanence in competitive stage-structured system with time delay. Non. Anal. 51, 1347–1361 (2002)
35. Liu, S.Q., Chen, L.S., Agarwal, R.: Recent progress on stage-structured population dynamics. Math. Compu. Model 36, 1319–1360 (2002)
36. Liu, S.Q., Chen, L.S., Liu, Z.: Extinction and permanence in nonautonomous competitive system with stage structure. J. Math. Anal. Appl. 274, 667–684 (2002)
37. Liu, S.Q., Chen, L.S., Luo, G.L., Jiang, Y.L.: Asymptotic behavior of competive Lotka-Volterra system with stage structure. J. Math. Anal. Appl. 271, 124–138 (2002)
38. Liu, S.Q., Kouche, M., Tatar, N.: Permanence and global asymptotic stability in a stage structured system with distributed delays. J. Math. Anal. Appl. 301, 187–207 (2005)
39. Magnusson, J.G.: Destabilizing effect of cannibalism on age structured predator-prey system. Math. Bios. 155, 195–200 (1999)
40. Ou, L., Luo, G.L., Jiang, Y.L., Li, Y.P.: The asymptotic behavior of a stage-structured autonomous predator-prey system with time delay. J. Math. Anal. Appl. 283, 534–548 (2003)
41. Song, X.Y., Chen, L.S.: Optimal harvesting and stability for a two species competitive system with stage structure. Math. Bios. 170, 173–186 (2001)
42. Song, X.Y., Chen, L.S.: A predator-prey system with stage structure and harvesting for predator. Anna. Diff. Equa. 18, 264–277 (2002)
43. Song, X.Y., Chen, L.S.: Modelling and analysis of a single species system with stage structure and harvesting. Math. Compu. Model 36, 67–82 (2002)
44. Song, X.Y., Chen, L.S., Neumann, A.U.: Ratio-dependent predator-prey system with stage structure for prey. Disc. Conti. Dyna. Syst. B 4, 747–758 (2004)
45. Song, X.Y., Cui, J.: The stage-structured predator-prey system with delay and harvesting. Appl. Anal. 81, 1127–1142 (2002)
46. Song, X.Y., Guo, H.J.: Global stability of a stage-structured predator-prey system. Int. J. Biomath. 1(3), 313–326 (2008)
47. Tang, S.Y., Chen, L.S.: Density-dependent birth rate, birth pulses and their population dynamic consequences. J. Math. Biol. 44, 185–199 (2002)
48. Tognetti, K.: The two stage stochastic model. Math. Bios. 25, 195–204 (1975)
49. Venkatasubramanian, V., Schaettler, H., Zaborszky, J.: Local bifurcations and feasibility regions in differential-algebraic systems. IEEE Trans. Auto. Contr. 40(12), 1992–2013 (1995)
50. Venkatasubramanian, V.: Singularity induced bifurcation and the van den Pol oscillator. IEEE Trans. Cir. Syst. I 41, 765–769 (1994)
51. Wang, W.D.: Global dynamics of a population model with stage structure for predator. Advanced Topics in Biomathmatics, pp. 253–257. World Scientific, Singapore (1998)
52. Wang, W.D., Chen, L.S.: A predator-prey system with stage-structure for predator. Compu. Math. Appl. 33, 83–91 (1997)
53. Wood, S.N., Blythe, S.P., Gurney, W.S.C., Nisbet, R.M.: Instability in mortality estimation schemes related to stage-structure population models. IMA J. Math. Appl. Medi. Biol. 6, 47–68 (1989)

54. Xu, D., Zhao, X.: Dynamics in a periodic competitive model with stage structure. J. Math. Anal. Appl. 311(2), 417–438 (2005)
55. Xu, R., Chaplin, M.A., Davidson, F.A.: Persistence and stability of a stage-structured predator-prey model with time delays. Appl. Math. Comp. 150, 259–277 (2004)
56. Yang, L.J., Tang, Y.: An improved version of the singularity induced bifurcation theorem. IEEE Tran. Auto. Contr. 49(6), 1483–1486 (2001)
57. Zhang, H., Georgescu, P., Chen, L.S.: An impulsive predator-prey system with Beddington-DeAngelis functional response and time delays. Int. J. Biomath. 1(1), 1–17 (2008)
58. Zhang, X., Chen, L., Neumann, U.A.: The stage-structured predator-prey model and optimal harvesting policy. Math. Bios. 168, 201–210 (2000)
59. Zhang, X., Zhang, Q.L., Zhang, Y.: Bifurcations of a class of singular biological economic models. Chaos Soli. Frac. 42(3), 1485–1494 (2009)
60. Zhang, Y., Zhang, Q.L.: Chaotic control based on descriptor bioeconomic systems. Contr. Deci. 22(4), 445–452 (2007)
61. Zhang, Y., Zhang, Q.L., Zhao, L.C.: Bifurcations and control in singular biological economical model with stage structure. J. Syst. Engin. 22(3), 232–238 (2007)

Chapter 5
Chaos and Control in Singular Biological Economic Systems

5.1 Preliminaries for Chaotic Behavior and Chaotic Control

5.1.1 Poincaré Surface-of-Section Technique

A traditional approach to gain preliminary insight into the properties of the dynamical system is to carry out a one-dimensional bifurcation analysis. One-dimensional bifurcation diagrams of Poincaré maps present information about the dependence of the dynamics on a certain parameter. The analysis is expected to reveal the type of attractor to which the dynamics will ultimately settle down after passing the initial transient phase and within which the trajectory will then remain forever. On a Poincaré surface of section, the dynamical behavior can be described by a discrete map whose phase-space dimension is less than that of the original continuous flow. Chaotic flows can be understood based on concepts that are convenient for maps such as unstable orbits. The limit sets of the Poincaré map correspond to long-term solutions of the underlying continuous dynamical system in the following way (see references [27, 22]).

- Case I: A finite number of points corresponds to a periodic solution; that is, one point corresponds to a solution of period equal to that of the forcing term, namely, $\frac{2\pi}{\omega}$ (period one); and n points correspond to a solution (sub-harmonic) of period $\frac{2n\pi}{\omega}$ (period n or a sub-harmonic of order $\frac{1}{n}$);
- Case II: A closed curve corresponds to a quasi-periodic solution, that is, a solution consisting of two incommensurate frequencies or, equivalently, having a trajectory that is dense on a torus;
- Case III: A collection of points that is fractal corresponds to chaos, namely, a stranger in phase space; a collection of points that form a cloud that is disorganized, partially organized, or fuzzy may (or may not) correspond to strange attractors (chaos).

Q. Zhang et al.: Complex., Analy. and Cntrl. of Singular Biological Systems, LNCIS 421, pp. 67–88.
springerlink.com

5.1.2 Lyapunov Exponents

In a given embedding dimension, the Lyapunov exponent is a measure of the speeds at which initially nearby trajectories of the system diverge. There is a Lyapunov exponent for each dimension of the process, which together constitutes the Lyapunov spectrum for the dynamical system. The Lyapunov exponent is related to the predictability of the system, and the largest Lyapunov exponent of a stable system does not exceed zero. However, a chaotic system has at least one positive Lyapunov exponent.

A bounded system with a positive Lyapunov exponent is one operational definition of chaotic behaviors, which presents a quantitative measure of the average rate of separation of nearby trajectories on the attractor. Over the years, a number of methods have been introduced for computation of Lyapunov exponents (see references [22, 11, 21, 30]). The sum of all Lyapunov exponents of a chaotic system will be negative, consistent with the idea that the chaotic attractor is globally stable. The more positive the largest Lyapunov exponent, the more unpredictable the system is. When the equations governing a system are known, the definitive test for chaos is one positive Lyapunov exponent with a negative sum of Lyapunov exponents.

Sensitive dependence on initial conditions is the most relevant property of chaos and its characterization in terms of Lyapunov exponents is the most satisfactory from a computable perspective. Lyapunov exponents measure average exponential divergence or convergence between trajectories that differ only in having an infinitesimally small difference in their initial conditions and remain well defined for a noisy system. A positive Lyapunov exponent implies that a chaotic process displays long term unpredictability, with the output being sensitively dependent on the initial conditions. Even slightly different initial values can lead to vastly different system outputs. Furthermore, another characteristic for chaos is the fractal dimension, which can be computed with the commutated Lyapunov exponents (see reference [16]).

5.1.3 Control Strategy for Chaotic Behavior

Considering the following discrete singular system,

$$\Xi X_{n+1} = F(X_n),$$

where $\Xi \in \begin{bmatrix} I & 0 \\ 0 & 0 \end{bmatrix} \in R^{N\times N}$, $I \in R^{N_1 \times N_1}$ is identical, then rank$E = N_1 < N$ (if $N_1 = N$, then above model system is normal system), and $X = (X_1^T, X_2^T)^T \in R^N (N \geq 2)$ represents state vector, $X_1 \in R^{N_1}$.

$$\begin{cases} X_{1(n+1)} = F_1(X_{1(n)}, X_{2(n)}), \\ 0 = F_2(X_{1(n)}, X_{2(n)}). \end{cases} \tag{5.1}$$

If model system (5.1) is regular and

$$\text{rank}(\partial F_2(X_1, X_2)/\partial X_2) = N - N_1,$$

F is a sufficiently continuous vector function. The model system with control input $u_n \in R^N$ takes the following form:

$$\Xi X_{n+1} = F(X_n) + u_n. \tag{5.2}$$

Without loss of generality, supposing there is an unstable fixed point X_0^* for model system (5.1), which satisfies $\Xi(X_0^*) = F(X_0^*)$ as the target orbit, then it is also the fixed point of controlled system (5.2) in the case of $u_n = 0$. The fixed point X_* of controlled system (5.2) satisfies the following system $\Xi X^* = F(X^*) + u_n$.

Theorem 5.1. *For controlled system (5.3), the chaotic orbit can be stabilized into the unstable fixed point* X_0^* *under control*

$$u_n = M^{-1}(lI - J)(X_n - X_0^*) \tag{5.3}$$

where $J = \partial F/\partial X|_{X=X_0^*}, M = \partial F/\partial u|_{u=0}$, *the eigenvalue of* J *is greater or less than* 1, l *is constant, and* $|l| < 1$.

Proof. Since the control is tiny, the fixed point of model system (5.1) under tiny control u_n is still around the fixed point of model system (5.1). It follows that

$$\Xi(X^* - X_0^*) = J(X^* - X_0^*) + Mu_n, \tag{5.4}$$

By substituting (5.3) into (5.4), it gives

$$\Xi(X^* - X_0^*) = J(X^* - X_0^*) + (lI - J)(X_n^* - x_0^*). \tag{5.5}$$

By linearizing (5.5), we get

$$\Xi(X_{n+1} - x^*) = J(X_n - X^*). \tag{5.6}$$

By adding (5.5) and (5.6), it can be obtained that

$$\Xi(X_{n+1} - x_0^*) = l(X_n - X_0^*). \tag{5.7}$$

It implies that the chaotic orbit is stabilized to target orbit X_0^* under the control law (5.3).

Remark 5.1. Compared with the OGY control strategy, the chaotic orbit can be stabilized to the target orbit without the stable manifold, and the duration when reaching the ideal orbit is short.

5.2 Chaotic Behavior and Control in Singular Logistic System

The logistic model is widely used to investigate the growth law of various biological ecosystems such as some kind of single-cell population (microzyme, paramecium, etc.) marine population (whale, sea bird, etc.). Furthermore, population variation of many birds and insects living on continent can be studied by virtue of the logistic model [5]. In this section, a discrete singular biological economic system is proposed. The dynamics of the model system, especially chaotic behavior, are discussed. Moreover, chaotic controlling strategy is designed to stabilize chaotic behavior in the discrete biological economic system.

5.2.1 Model Formulation

The growth law of various biological species is usually described by the classic logistic model. Compared with the continuous model, the dynamics of the discrete logistic model with one dimension are abundant [3, 14].

Consider the logistic discrete model

$$x_{n+1} = x_n + rx_n(1 - x_n/k), \tag{5.8}$$

where $x_n > 0$ represents population density, $r > 0$ represents the intrinsic growth rate, and $k > 0$ represents the environment capacity. There are two fixed points for model system (5.8): $x_1^* = 0, x_2^* = k$.

According to [13], the following model system is proposed to investigate the dynamics of the biological system and the economic interest of the harvest effort on the biological population:

$$\begin{cases} x_{n+1} = x_n + rx_n(1 - x_n/k) - E_n x_n, \\ 0 = E_n(px_n - c) - m, \end{cases} \tag{5.9}$$

where x_n, r, k share the same biological interpretations as in (5.8), E_n represents the harvest effort on population, p is the unit price of the harvested population, and c is the united cost of the harvest effort. r, k, p, and c are all constant.

For model system (5.9), the fixed point $P^*(x_0^*, E_0^*)$ satisfies the following equation:

$$\begin{cases} x_0^* + rx_0^*(1 - x_0^*/k) - E_0^* x_0^* = x_0^*, \\ E_0^*(px_0^* - c) - m = 0. \end{cases} \tag{5.10}$$

Fig. 5.1 shows the dynamics of model system (5.9) against the variation of parameter r. It should be noted that the values of parameters are set in appropriate units as follows: $k = 8.05 \times 10^7, p = 5 \times 10^{-3}, c = 8.75 \times 10^4$.

According to Fig. 5.1, it can be shown that model system (5.9) undergoes dynamical variation against the increase of intrinsic growth of population. Model system (5.9) is stable when $0 < r < 2$, which implies that there is a stable fixed point. When

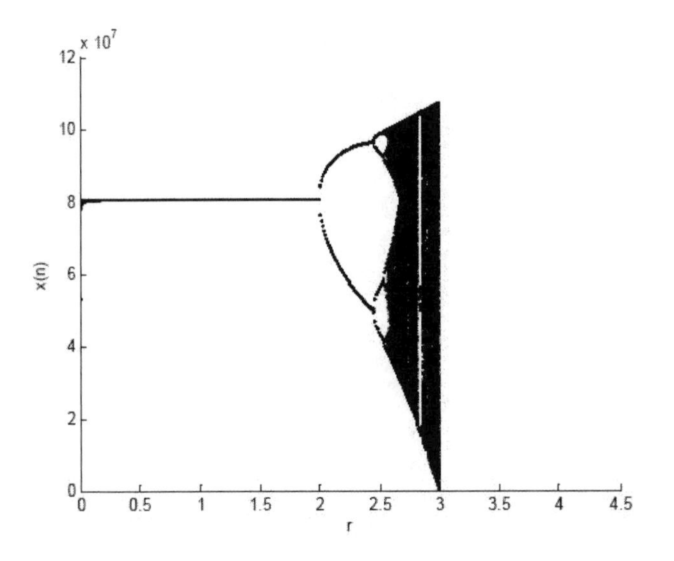

Fig. 5.1 Bifurcation diagrams of Poincaré for model system (5.9) against variation of parameter r.

$r \approx 2.01$, model system (5.9) bifurcates into periodical orbits and finally admits chaotic behavior when r increases through 2.58. Fig. 5.2 shows the chaotic attractor when $r = 2.8$.

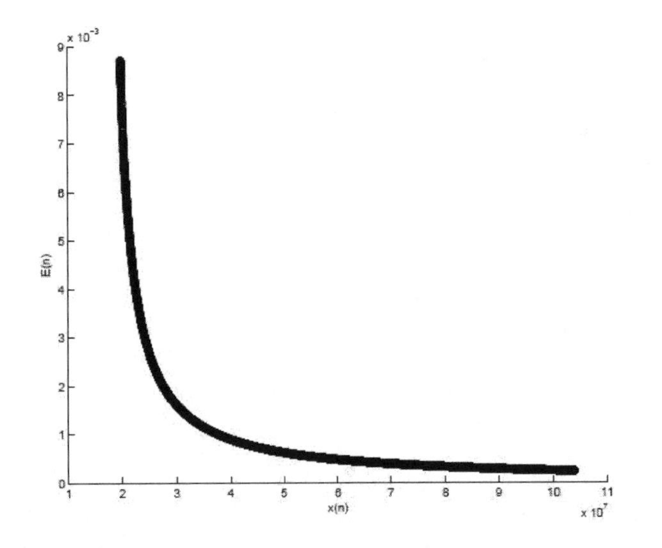

Fig. 5.2 Chaotic attractor for model system (5.9) in the case of $r = 2.8$.

By using the above control strategy, model system (5.9) can be expressed as follows:

$$\begin{cases} x_{n+1} = x_n + rx_n(1 - x_n/k) - E_n x_n + u_{1n}, \\ 0 = E_n(px_n - c) - m + u_{2n}, \end{cases} \tag{5.11}$$

where $u_n = (u_{1n}, u_{2n})^T$ is the input control. Model system (5.11) is equal to model system (5.9). The following chaotic orbit can be stabilized to the ideal unstable fixed point $P^*(x_0^*, E_0^*)$:

$$F(X_n) = \begin{bmatrix} x_n + rx_n(1 - x_n/k) - E_n x_n \\ E_n(px_n - c) - m \end{bmatrix},$$

$$\Xi = \begin{bmatrix} 1 & 0 \\ 0 & 0 \end{bmatrix}, X_1 = x, X_2 = E.$$

It follows that

$$J = \begin{bmatrix} 1 - \frac{r}{k}x_0^* & -x_0^* \\ pE_0^* & px_0^* - c \end{bmatrix}, M = \begin{bmatrix} 1 & 0 \\ 0 & 0 \end{bmatrix},$$

$$\begin{aligned} u_n^* &= (u_{1n}^*, u_{2n}^*)^T \\ &= \begin{bmatrix} l - 1 + \frac{r}{k}x_0^* & x_0^* \\ -pE_0^* & l - px_0^* + c \end{bmatrix} \begin{bmatrix} x_n - x_0^* \\ E_n - E_n^* \end{bmatrix}, \end{aligned} \tag{5.12}$$

where $|l| < 1$.

Remark 5.2. The biological interpretations of the controller are also obvious. The chaotic behavior of the model system can be stabilized into an unstable periodic fixed point with the introduction of input control. It is recommended that government formulate appropriate policies to regulate harvest effort, such as taxation and other regulatory measures.

5.2.2 Numerical Simulation

By virtue of related population data of pleuronectids in Pacific [26], a numerical simulation is carried out to show the effectiveness of the designed controller; the values of parameters are as follows: $r = 2.8$, $k = 8.05 \times 10^7, p = 5 \times 10^{-3}$, $c = 8.75 \times 10^4$, and $m = 1 \times 10^2$.

It follows from Fig. 5.2 that model system (5.11) admits chaotic behavior when $(u_{1n}, u_{2n})^T = (0, 0)^T$, and there are three fixed points $P_0^*(0, -1.143 \times 10^{-3})$, $P_1^*(8.049 \times 10^7, 3.175 \times 10{-}4)$, and $P_2^*(1.751 \times 10^7, 2.191)$.

Suppose the ideal target orbit is stable fixed point $P_1^*(8.049 \times 10^7, 3.175 \times 10^{-4})$. It should be noted that the referred fixed point P_1^* is contained in the chaotic attractor, which can be shown in Fig. 5.3. Hence, the chaotic orbit can be stabilized to P_1^*.

On the other hand, suppose the ideal target orbit is unstable fixed point $P_2^*(1.751 \times 10^7, 2.191)$. The chaotic orbit can be stabilized to P_2^*, which can be shown in

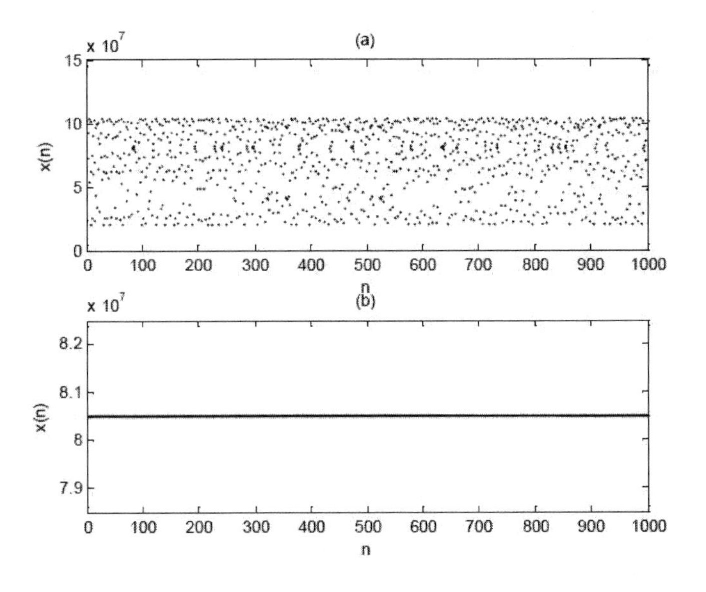

Fig. 5.3 The response of x_n in system (5.11) for the case of uncontrolled and controlled (the controlled object embodied in the chaotic attractor).

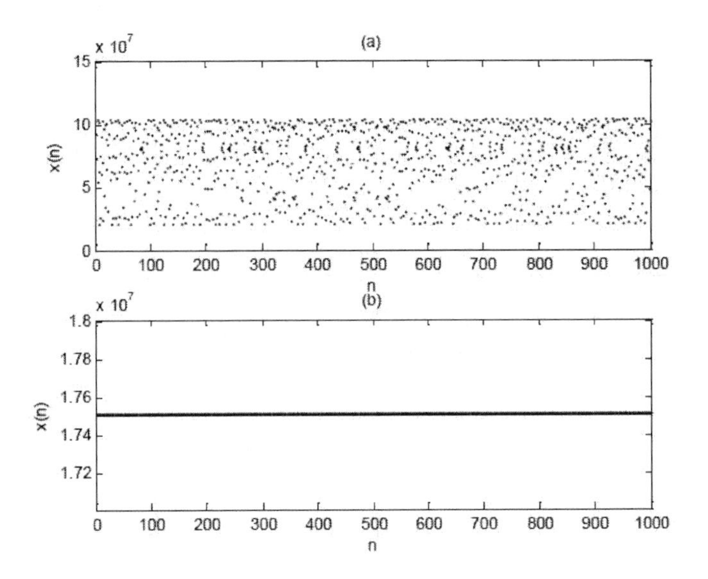

Fig. 5.4 The response of x_n in system (5.11) for the case of uncontrolled and controlled (the controlled object is not embodied in the chaotic attractor).

Fig. 5.4. It implies that the designed control strategy is also effective when the referred fixed point is not contained in the chaotic attractor.

5.3 Chaotic Behavior and Control in Singular Holling-Tanner Biological System

Considering the following Holling-Tanner model [5]

$$\begin{cases} \dfrac{dX(T)}{dT} = r_0 X(T)[1 - \dfrac{X(T)}{K}] - \dfrac{BX(T)}{A+X(T)}Y(T), \\ \dfrac{dY(T)}{dT} = sY(T)[1 - \dfrac{hY(T)}{X(T)}], \end{cases} \tag{5.13}$$

where $X(T), Y(T)$ represents the population density of prey and predator, respectively. The initial values are as follows: $X(0) > 0, Y(0) > 0$. K and r_0 are the environment capacity for prey and intrinsic growth rate of prey population, respectively. A is half saturation constant; B denotes consumption rate; and s is the intrinsic growth rate of predator; the environment capacity is proportional to prey population, and h denotes the consumption quantity per predator.

It should be noted that model system (5.13) can be utilized to investigate the interaction and coexistence mechanism of the prey-predator ecosystem. It should be noted that countless organisms live in a seasonal environment, and many parameters of such a system may oscillate simultaneously and not necessarily in plane. Recently, the model systems of an ecosystem with periodic external force have gathered new attention because of their complex dynamical behavior, especially chaotic behavior. There is already considerable literature on forced nonlinear oscillators on the ecosystem (see references [24, 7, 4, 2, 19, 23, 20, 28, 32, 31, 6, 12] and the references therein). It is pertinent to note that the intrinsic growth rate of the population in an ecosystem is usually periodically adjusted, which may be caused by periodic variation of fecundity of harvested biological resources and some related changes of policies from regulatory agencies. Based on these discussions, in the second section of this chapter, a periodical intrinsic growth rate on the prey population, which takes the form of a periodically external perturbation, is incorporated into the model proposed in model system (5.13), and a prey-predator model system with periodical harvesting for predator is established. We would like to investigate the complex dynamical behavior, especially chaotic behavior and control of the proposed system, which may be of constructive guidance to maintain continuous prosperity of biological resources within a periodical growth environment.

5.3.1 Model Formulation

Considering the seasonal periodical growth of the prey population, the intrinsic growth of the prey population can be expressed as follows:

$$r = r_0 + r_1(1 - \cos\omega T), \tag{5.14}$$

where $r > 0$, $\omega > 0$, r_1, and ω represents the amplitude and frequency of the seasonal factor, respectively. Substituting (5.14) into (5.13), then model (5.13) can be rearranged by non-scaling,

$$t = r_0 T, x = \frac{X}{K}, y = \frac{BY}{r_0 K}, \delta = \frac{s}{r_0}, \beta = \frac{sh}{B}, \gamma = \frac{A}{K}, \lambda = \frac{r_1}{r_0}, \theta = \frac{\omega}{r_0}.$$

Model system (5.13) takes the following form:

$$\begin{cases} \frac{\mathrm{d}x(t)}{\mathrm{d}t} = x(t)(1-x(t)) - \frac{x(t)y(t)}{\gamma + x(t)} + \lambda(1-\cos\theta t)x(t), \\ \frac{\mathrm{d}y(t)}{\mathrm{d}t} = y(t)[\delta - \beta\frac{y(t)}{x(t)}]. \end{cases} \tag{5.15}$$

According to the theory proposed by Gordon [13], a differential-algebraic model system is established to investigate the harvested biological system with seasonal intrinsic growth. Based on (4.3) and (5.14), a differential-algebraic model that consists of three differential equations and an algebraic equation can be established as follows:

$$\begin{cases} \frac{\mathrm{d}x(t)}{\mathrm{d}t} = x(t)(1-x(t)) - \frac{x(t)y(t)}{\gamma + x(t)} + \lambda(1-\cos\theta t)x(t), \\ \frac{\mathrm{d}y(t)}{\mathrm{d}t} = y(t)[\delta - \beta\frac{y(t)}{x(t)}], \\ 0 = E(t)(wy(t) - c) - v, \end{cases} \tag{5.16}$$

where $x(t), y(t), r_0, K, A, B, s$ and h share the same biological interpretations mentioned in (5.15). w represents the unit price of harvested predator, c represents the unit cost of harvest effort on predator and, v represents the economic interest of harvesting.

5.3.2 Analysis and Tracking Control for Chaotic Behavior

For the phase portrait of model system (5.16), the values of parameters are given with the appropriate unit, which are as follows: $\gamma = 0.15, \delta = 0.1, \beta = 0.03, \lambda = 2, \theta = 2.88, w = 1, c = 200$, and $v = 100$.

According to the preliminaries in Section 3.1.2, the Lyapunov exponents can be calculated as follows, $L_1 = 0.058883$, $L_2 = 0$, and $L_3 = -0.25052$, which follows that model system (5.15) admits chaotic behavior when $\gamma = 0.15$ and $v = 100$.

The controlled model system is as follows:

$$\begin{cases} \frac{\mathrm{d}x(t)}{\mathrm{d}t} = x(t)(1-x(t)) - \frac{x(t)y(t)}{\gamma+x(t)} + \lambda(1-\cos\theta t)x(t) + u(t), \\ \frac{\mathrm{d}y(t)}{\mathrm{d}t} = y(t)[\delta - \beta\frac{y(t)}{x(t)}], \\ 0 = E(t)(wy(t)-c) - v. \end{cases} \tag{5.17}$$

By substituting $z(t) = \theta_t$, the non-autonomous model system (5.16) is rewritten as an autonomous model system,

$$\begin{cases} \frac{\mathrm{d}x(t)}{\mathrm{d}t} = x(t)(1-x(t)) - \frac{x(t)y(t)}{\gamma+x(t)} + \lambda(1-\cos zt)x(t) + u(t), \\ \frac{\mathrm{d}y(t)}{\mathrm{d}t} = y(t)[\delta - \beta\frac{y(t)}{x(t)}], \\ \frac{\mathrm{d}z(t)}{\mathrm{d}t} = \theta, \\ 0 = E(t)(wy(t)-c) - v. \end{cases} \tag{5.18}$$

Definition 5.1. (M derivative [29]) Considering some kind of nonlinear differential-algebraic system,

$$\begin{cases} \frac{\mathrm{d}X(t)}{\mathrm{d}t} = x(t)(1-x(t)) - \frac{x(t)y(t)}{\gamma+x(t)} + \lambda(1-\cos zt)x(t) + u(t), \\ \frac{\mathrm{d}y(t)}{\mathrm{d}t} = y(t)[\delta - \beta\frac{y(t)}{x(t)}], \\ \frac{\mathrm{d}z(t)}{\mathrm{d}t} = \theta, \\ 0 = E(t)(wy(t)-c) - v, \end{cases} \tag{5.19}$$

where $X(t) = (x_1(t), x_2(t), \cdots, x_n(t))^T$, $Y(t) = (y_1(t), y_2(t), , y_m(t))^T$, $u(t) \in R$ represents n dimensional state vector, m dimensional constraint vector, and control input, respectively. f, g and σ are field of smooth vector, and $\mathrm{rank}(\frac{\partial\sigma}{\partial Y(t)}) = m$. Let $M_f(h(X(t),Y(t)))$ and $M_g(h(X(t),Y(t)))$ represent the derivative of vector field f, g with respect to $h(X(t),Y(t))$, and the definition is as follows:

$$\begin{cases} M_f(h(X(t),Y(t)) = \phi(h(X(t),Y(t)))f, \\ M_g(h(X(t),Y(t)) = \phi(h(X(t),Y(t))g, \\ \phi(h(X(t),Y(t)) = \frac{\partial h}{\partial X(t)} - \frac{\partial h}{\partial Y(t)} \times [\frac{\partial\sigma}{\partial Y(t)}]^{-1}\frac{\partial\sigma}{\partial X(t)}. \end{cases}$$

Definition 5.2. (Relevance degree [29]) Let $h(X(t),Y(t))$ represent the output function of model system (5.19), if the following constraints are satisfied in $(X(t),Y(t)) \in \Omega$(open connected set Ω):

$$M_g M_f^k h(X(t),Y(t)) = 0, 0 \leq k \leq r-2, M_g M_f^{r-1} h(X(t),Y(t)) \neq 0,$$

then the relevance degree of model system (5.19) in $(X(t),Y(t)) \in \Omega$ is r.

Model system (5.18) can be expressed as follows:

$$f(X(t),Y(t)) = \begin{bmatrix} x(t)(1-x(t)) - \dfrac{x(t)y(t)}{\gamma+x(t)} + \lambda(1-\cos z(t))x(t) \\ y(t)[\delta - \beta\dfrac{y(t)}{x(t)}] - y(t)E(t) \\ \theta \end{bmatrix},$$

$$g(X(t),Y(t)) = \begin{bmatrix} 1 \\ 0 \\ 0 \end{bmatrix},$$

$$\sigma(X(t),Y(t)) = E(t)(wy(t)-c) - v,$$

where $X(t) = (x(t),y(t),z(t))^T$ and $Y(t) = E(t)$, the output of model system (5.18) is $h(X(t),Y(t)) = y(t)$.

$$M_g M_f^0 h(X(t),Y(t)) = M_g M_f^0 y(t) = 0, M_g M_f h(X(t),Y(t)) \neq 0.$$

The relevancy degree of the model system is $2 < r < 3$ at every point satisfying $\dfrac{\beta y^2(t)}{x^2(t)} \neq 0$. It is obvious that g is not linear relevant to $M_f g$.

By using the coordinate transformation, $\begin{cases} \xi_1(t) = y(t), \\ \xi_2(t) = y(t)[\delta - \beta\dfrac{y(t)}{x(t)}] - y(t)E(t), \\ \eta(t) = \theta, \end{cases}$

model system (5.18) can be rewritten as follows:

$$\begin{cases} \dot{\xi}_1(t) = \xi_2(t), \\ \dot{\xi}_2(t) = \tilde{v}(t), \\ \dot{\eta}(t) = \theta, \\ 0 = E(t)(wy(t)-c) - v, \end{cases} \tag{5.20}$$

where $\tilde{v}(t) = \dfrac{\beta y^2(t)}{x(t)}[1 - x(t) - \dfrac{y(t)}{\gamma+x(t)} + \lambda(1-\cos z(t))] + [\delta - \frac{2\beta y(t)}{x(t)} + \dfrac{cE(t)}{wy(t)-c}] \times$ $[y(t)(\delta - \beta\dfrac{y(t)}{x(t)}) - y(t)E(t)] + \dfrac{\beta y^2(t)}{x^2(t)} u(t).$

Obviously, the differential part of model system (5.18) can be decomposed as a two-dimensional differential subsystem (the first two differential equations of model system (5.20)) and one-dimensional differential subsystem (the third differential equation of model system (5.20)). It should be noted that the output is not affected by the subsystem. Hence, only the first two differential equations and linear constraint are considered when investigating the output tracking. The ideal output track of model system (5.18) is $y_d(t)$.

Theorem 5.2. *If the controller $u(t)$ takes the following form, then the output of controlled model system (5.18) satisfies $y(t) \to y_d(t)$ as $t \to \infty$.*

$$u(t) = \frac{x^2(t)}{\beta y^2(t)}(-k_0 e_1(t) - k_1 e_2(t)) - x(t) \times [1 - x(t) - \frac{y(t)}{\gamma + x(t)} + \lambda(1 - \cos z(t))]$$
$$-\frac{x^2(t)}{\beta y^2(t)}[\delta - \tfrac{2\beta y(t)}{x(t)} + \frac{cE(t)}{wy(t) - c}] \times [y(t)(\delta - \beta\frac{y(t)}{x(t)}) - y(t)E(t)] + \frac{x^2(t)}{\beta y^2(t)\ddot{y}_d(t)} \tag{5.21}$$

the ideal target orbit of model system (5.18) $y_d(t)$ satisfying $y_d(t)$ satisfying $\dot{y}_d(t)$ and $\ddot{y}_d(t)$ is bounded when $t \geq 0$, $\ddot{y}_d(t)$ is a piecewise continuous function. k_0 and k_1 are constants that satisfy that all the roots of $q^2 + k_1 q + k_0 = 0$ are in the left half plane of q plane.

$$e_1(t) = y(t) - y_d(t), e_2(t) = y(t)[\delta - \beta\frac{y(t)}{x(t)}] - y(t)E(t) - \dot{y}_d(t).$$

Proof. Let $e(t) = \xi(t) - \xi_d(t)$, where $e(t) = (e_1(t), e_2(t))^T$, $\xi(t) = (\xi_1(t), \xi_2(t))^T$, $\xi_d(t) = (\xi_{1d}(t), \xi_{2d}(t))^T = (y_d(t), \dot{y}_d(t))^T$. The error system is as follows:

$$\begin{cases} \dot{e}(t) = \xi_1(t) - \dot{y}_d(t) = \xi_2 t - \dot{y}_d(t) = e_2(t), \\ \dot{e}_2(t) = \xi_2(t) - \ddot{y}_d(t) = \bar{v}(t) - \ddot{y}_d(t), \\ \dot{\eta}(t) = \theta, \\ 0 = E(t)(wy(t) - c) - v. \end{cases} \tag{5.22}$$

Substituting (5.21) into (5.22), then we have

$$\begin{cases} \dot{e}_1(t) = e_2(t), \\ \dot{e}_2(t) = v(t), \end{cases} \tag{5.23}$$

where $v(t) = -k_0 e_1(t) - k_1 e_2(t)$. According to [15], if all the roots of $q^2 + k_1 q + k_0 = 0$ are in the left half plane of q plane, then the controlled subsystem (5.21) is asymptotically stable $e(t) \to 0$ as $t \to \infty$, and then $y(t) \to y_d(t)$ as $t \to \infty$, which implies that the ideal target orbit of model system (5.18) is $y_d(t)$.

5.3.3 Numerical Simulation

By virtue of hypothetical values of parameters, numerical simulations are carried out to illustrate the theoretical results obtained in this section: $\gamma = 0.15, \lambda = 2, \theta = 2.88, \delta = 0.1, \beta = 0.03, w = 1, c = 200, v = 100, k_0 = 0.015$, and $k_1 = 0.1$. Select the ideal output as constant orbit, s.t., $y_d(t) = 2$. By using Theorem 5.2, the chaotic orbit can be stabilized to the given constant orbit.

Select the ideal output as periodical orbit, s.t., $y_d(t) = 0.1566 + 0.06\sin t$. By using Theorem 5.2, the chaotic orbit can be stabilized to the given periodical orbit.

Fig. 5.5(a) and Fig. 5.5(b) show the effectiveness of the controller, respectively.

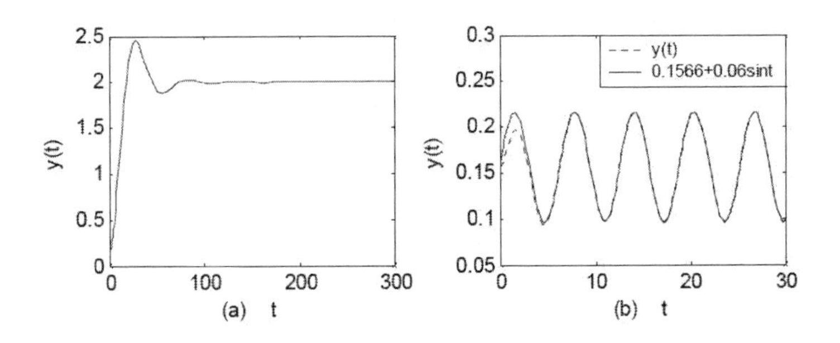

Fig. 5.5 Control of chaos to track; (a) dynamical response of $y(t)$ is stable under control, (b) $y(t)$ is stabilized to periodical orbit $y_d(t) = 0.1566 + 0.06\sin t$.

5.4 Chaotic Behavior in Singular Food Web with Harvesting

In recent years, there is a growing interest in the research field of the food web with multi-species. Rich dynamical behavior has been found in such a system [9, 17, 25]. Azar et al. [1] and Kumar et al. [18] have investigated the harvesting of predator species hunting two prey. The constant harvesting rate is treated as a control parameter and the system changes its stability to a limit cycle when harvesting exceeds a certain limit. Gakkhar and Naji et al. [10] have obtained the existence of chaotic dynamics in the food when comprising two preys and a predator without harvesting. Gakkhar et al. [8] found that the density dependent harvesting of predator is shown to control the chaos in the food web system. The model system investigated in [8] is as follows:

$$\begin{cases} \dfrac{dy_1}{dt} = y_1(1 - y_1 - \dfrac{w_2 y_3}{1 + w_3 y_1 + w_4 y_2}) = y_1 f_1(y_1, y_2, y_3), \\ \dfrac{dy_2}{dt} = y_2((1 - y_2)w_5 - \dfrac{w_7 y_3}{1 + w_4 y_2 + w_3 y_1}) = y_2 f_2(y_1, y_2, y_3), \\ \dfrac{dy_3}{dt} = y_3(\dfrac{w_8 y_1 + w_9 y_2}{1 + w_3 y_1 + w_4 y_2} - w_{10}) - w_{11} y_3 = y_3 f_3(y_1, y_2), \end{cases} \tag{5.24}$$

where $y_i(t)$, $i = 1, 2, 3$ represents the density of prey 1, prey 2, and predator, respectively. The constants w_j, $j = 1, 2, \cdots, 10$ are model parameters assuming only positive values. $w_1 1$ denotes the catch-ability-coefficient of the predator species. $E(t)$ represents the harvest effort on predator.

It is well known that harvesting has a strong impact on the dynamic evolution of a population. Nowadays, the biological resources in the prey-predator ecosystem are commercially harvested and sold with the aim of achieving economic interest. By considering the economic interest of the harvest effort on predator, a differential-algebraic model is proposed. It is utilized to investigate the dynamical behavior of the prey-predator ecosystem due to the variation of economic interest of harvesting, which extends the previous work done in [8] from an economic perspective. We aim to obtain some results that are theoretically beneficial to maintaining the sustainable development of the ecosystem as well as keeping the persistent prosperity of commercial harvesting.

5.4.1 Model Formulation

In 1954, Gordon proposed the economic theory of a common-property resource [13], which studied the effect of the harvest effort on the ecosystem from an economic perspective. In reference [13], an equation is proposed to investigate the economic interest of the yield of the harvest effort, which takes form as follows:

$$\text{Net Economic Revenue (NER)} = \text{Total Revenue (TR)} - \text{Total Cost (TC)}. \quad (5.25)$$

Let $E(t)$ and $Y(t)$ represent the harvest effort and the density of the harvested population, respectively. $\mathrm{TR} = p(t)E(t)Y(t)$ and $\mathrm{TC} = c(t)E(t)Y(t)$, where $p(t)$ represents the unit price of the harvested population, and $c(t)$ represents the unit cost of the harvest effort.

It should be noted that it is more realistic that the unit price of the harvested population is influenced by the fluctuation of supply and demand in the market, and the unit cost of the harvest effort is influenced by fecundity of the harvested population. Consequently, the unit price and unit cost are not always constant but time-varying. As introduced in the above section, the harvested predator is usually sold as a commodity in the market in order to achieve the economic interest of harvesting. Assuming that there is always constant demand for the harvested predator, then the unit price of the harvested predator and the cost of harvest effort can be expressed as $p(t) = \dfrac{a}{b+s(t)}$, $c(t) = \dfrac{d}{y_3(t)}$, respectively, where a, b, and d are all positive constants, $s(t)$ represents the supply amount of harvested predator, and $y_3(t)$ represents the population density of harvested predator. It is obvious that

$$\lim_{s(t)\to 0} p(t) = \lim_{s(t)\to 0} \frac{a}{b+s(t)} = \frac{a}{b} \quad \text{and} \quad \lim_{s(t)\to +\infty} s(t) = \lim_{s(t)\to \infty} \frac{a}{b+s(t)} = 0,$$

which imply that the unit price will decrease if the supply of the harvested predator is larger than the demand for the harvested predator, and the minimum unit price is zero; on the other hand, the unit price will increase if the supply of the harvested predator can not meet the demand for the harvested predator, and the maximum unit price is $\dfrac{a}{b}$.

It is also obvious that

$$\lim_{y_3(t)\to 0} c(t) = \lim_{y_3(t)\to 0} \frac{d}{y_3(t)} = \infty \quad \text{and} \quad \lim_{y_3(t)\to +\infty} c(t) = \lim_{y_3(t)\to \infty} \frac{d}{y_3(t)} = 0,$$

which imply that the unit cost of the harvest effort is inversely proportional to the population density of the harvested predator. The richer the population density of harvested predator becomes, the lower the unit cost of the harvest effort is, and the minimum price is zero; on the other hand, the unit cost of the harvest effort will increase if the population density of the harvested predator becomes relatively small.

Associated with model system (5.24), the supply amount can be expressed as $s(t) = E(t)y_3(t)$. Hence, for Eq. (5.25), TR and TC can be expressed as TR $= \frac{a}{b+E(t)y_3(t)}E(t)y_3(t)$ and TC $= \frac{d}{y_3(t)}E(t)y_3(t)$, respectively. Consequently, Eq. (5.25) can be rewritten as follows:

$$E(t)y_3(t)(\frac{a}{b+E(t)y_3(t)} - \frac{d}{y_3(t)}) = v, \tag{5.26}$$

where a, b, and d are all positive constants, and v represents the economic interest of harvesting.

In this section, we will establish a differential-algebraic model of two species (prey-predator) to study a harvested prey-predator system with harvest effort on predator. Based on (5.24) and (5.26), a differential-algebraic model that consists of three differential equations and an algebraic equation can be established as follows:

$$\begin{cases} \frac{dy_1}{dt} = y_1(1 - y_1 - \frac{w_2 y_3}{1 + w_3 y_1 + w_4 y_2}) = y_1 f_1(y_1, y_2, y_3), \\ \frac{dy_2}{dt} = y_2((1 - y_2)w_5 - \frac{w_7 y_3}{1 + w_4 y_2 + w_3 y_1}) = y_2 f_2(y_1, y_2, y_3), \\ \frac{dy_3}{dt} = y_3(\frac{w_8 y_1 + w_9 y_2}{1 + w_3 y_1 + w_4 y_2} - w_{10}) - w_{11} y_3 = y_3 f_3(y_1, y_2) \\ 0 = E(t)y_3(t)(\frac{a}{b+E(t)y_3(t)} - \frac{d}{y_3(t)}) - v, \end{cases} \tag{5.27}$$

where parameters have the same interpretations as mentioned in (5.24) and (5.26).

5.4.2 Numerical Experiments for Complex Dynamical Behavior

A traditional approach to gain preliminary insight into the properties of the dynamical system is to carry out a one-dimensional bifurcation analysis. One-dimensional bifurcation diagrams of Poincaré maps present information about the dependence of the dynamics on a certain parameter. The analysis is expected to reveal the type of attractor to which the dynamics will ultimately settle down after passing the initial transient phase and within which the trajectory will then remain forever.

According to the economic theory of common-property resource proposed in [13], there is a bio-economic equilibrium when the economic interest of harvesting is zero. In the following part, simulation work with a hypothetical set of parameters is performed to discuss the complex dynamical behavior of model system (5.27) in the case of bio-economic equilibrium. Hypothetical values of parameters are set in appropriate units as follows: $w_2 = 3, w_3 = 1.5, w_4 = 2, w_5 = 1.125, w_7 = 3.5, w_8 = 1.35, w_9 = 1.925$, and $w_{11} = 0$. Transmission coefficient w_{10} is a varied parameter in the numerical experiments. The bifurcation diagram of the Poincaré section for the logarithm of the prey $y_1(t)$ is plotted under the initial values $(0.1, 0.1, 0.15, 0.01)$.

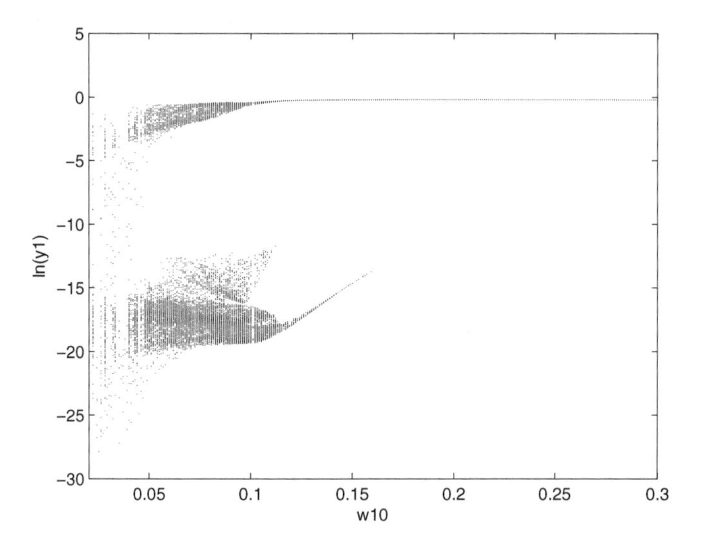

Fig. 5.6 Bifurcation diagram of Poincaré section for the logarithm of the sound prey population $y_1(t)$ and w_{10} is varied in [0.02,0.3].

Fig. 5.6 is the bifurcation diagram of the Poincaré section for the sound prey population in model system (5.27), w_{10} is a varied parameter, and the logarithm of prey density is plotted on the ordinate.

As w_{10} increases from 0, the phase portrait of model system (5.27) with $w_{10} = 0.02, 0.04, 0.06, 0.08$ is plotted in Fig. 5.7. The corresponding Poincaré points 5000-15000 are respectively plotted in Fig. 5.8, which is an indication of case III of the behavior characteristic of the Poincaré section for the chaotic attractor. When w_{10} further increases, the phase portrait of model system (5.27) with $w_{10} = 0.09, 0.1, 0.11, 0.12$ is plotted in Fig. 5.9, whose corresponding Poincaré points 5000-15000 plotted in Fig. 5.10 coincide with case II of behavior characteristic of the Poincaré section for the quasi-periodic attractor. Consequently, it implies that there exists a range of quasi-periodicy features. It can be observed from the Fig. 5.6 that the quasi-periodicy region ($c \in (0.082, 0.127)$) appears as a collapse of the invariant circle. When w_{10} increases beyond 0.127, the phase portrait of model system (5.27) with $w_{10} = 0.13, 0.14, 0.16, 0.2$ is plotted in Fig. 5.11. Since these

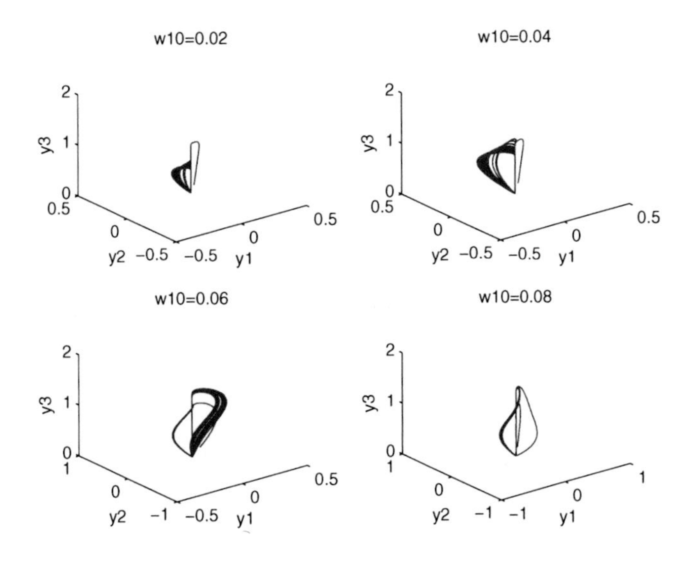

Fig. 5.7 The trajectories over the time interval from $t = 20000$ to $t = 40000$ of model system (5.27) with $w_{10} = 0.02, 0.04, 0.06, 0.08$.

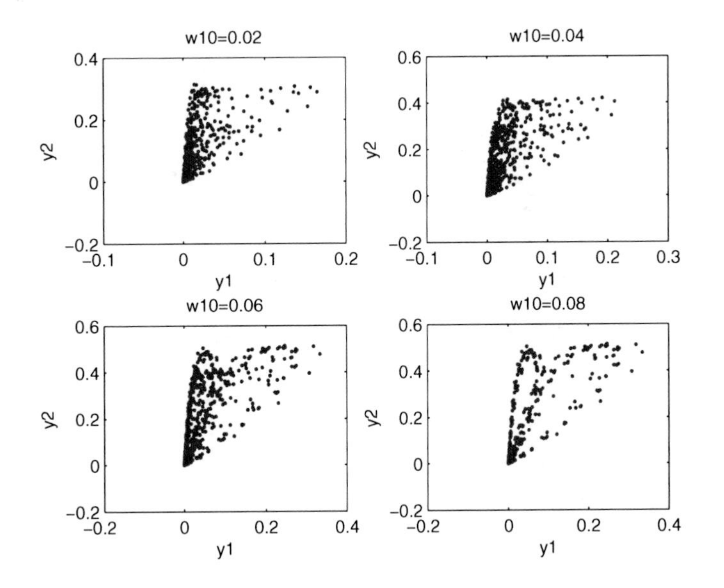

Fig. 5.8 Poincaré points 5000-15000 of model system (5.27) with $w_{10} = 0.02, 0.04, 0.06, 0.08$.

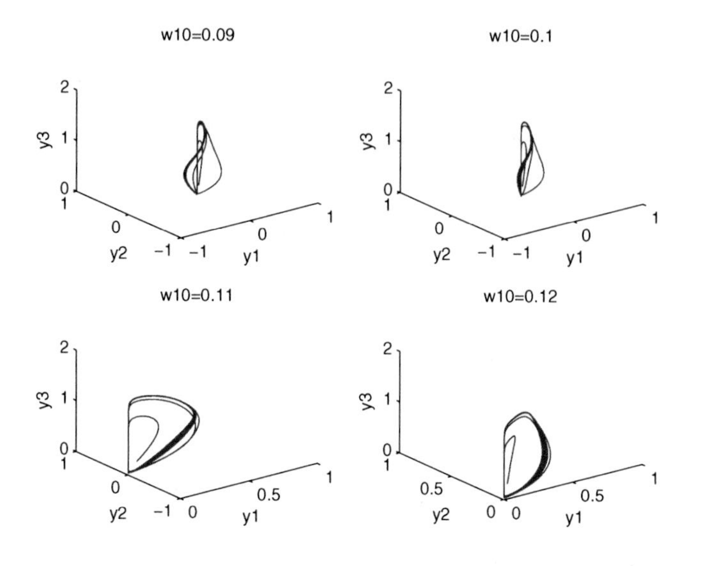

Fig. 5.9 The trajectories over the time interval from $t = 20000$ to $t = 40000$ of model system (5.27) with $w_{10} = 0.09, 0.1, 0.11, 0.12$.

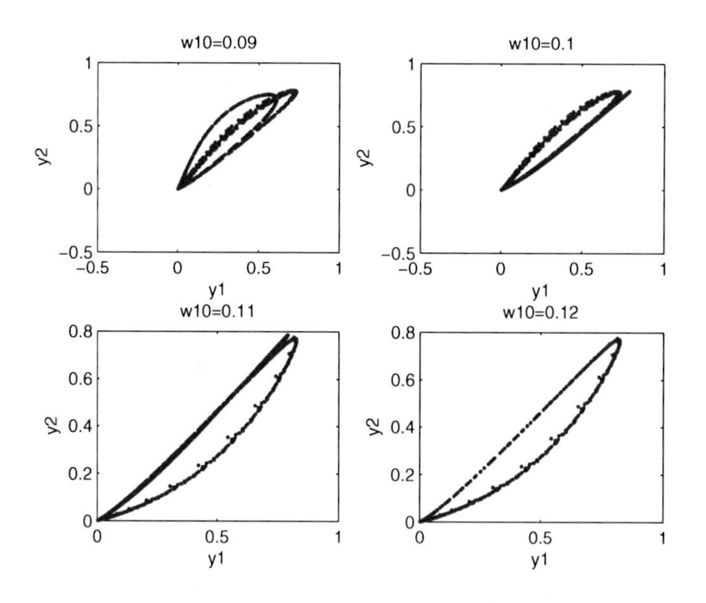

Fig. 5.10 Poincaré points 5000-15000 of model system (5.27) with $w_{10} = 0.09, 0.1, 0.11, 0.12$.

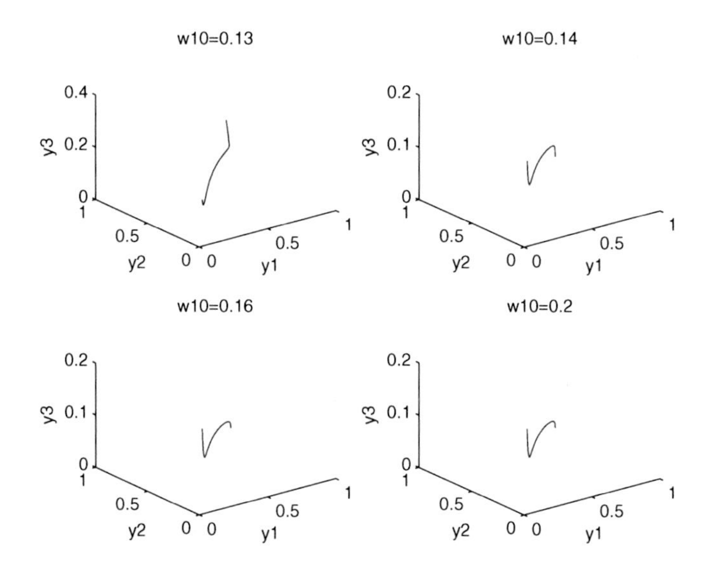

Fig. 5.11 The trajectories over the time interval from $t = 20000$ to $t = 40000$ of model system (5.27) with $w_{10} = 0.13, 0.14, 0.16, 0.2$.

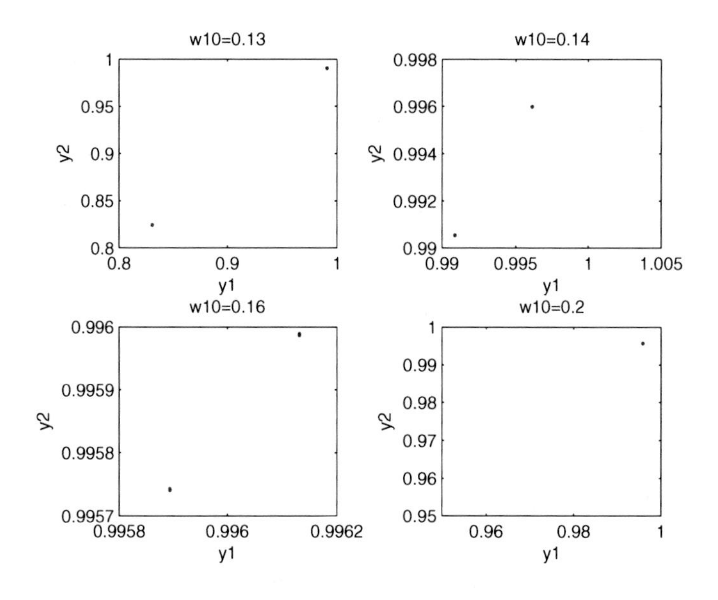

Fig. 5.12 Poincaré points 5000-15000 of model system (5.27) with $w_{10} = 0.13, 0.14, 0.16, 0.2$.

corresponding Poincaré points 5000-15000 plotted in Fig. 5.12 satisfy case I of the behavior characteristic of the Poincaré section for periodic attractor, it can be concluded that quasi-periodic attractor disappears and eventually returns in the form of a periodic attractor. Furthermore, Lyapunov exponents are utilized to show the existence of a strange attractor, and the changes of Lyapunov exponents with time of model system (5.27) are computed. $w_{10} = 0.03, 0.05, 0.07$ are arbitrarily selected from the interval $(0.02, 0.082)$, where it denotes the range of chaotic behavior.

Table 5.1 Lyapunov exponents and corresponding Lyapunov dimensions of model system (5.27) with $w_{10} = 0.03, 0.05, 0.07$

w_{10}	0.03	0.05	0.07
L_1	0.06789	0.08369	0.056743
L_2	0	0	0
L_3	-0.0359	-0.0153	-0.04151
D	2.0131	2.0112	2.6754

It follows from Table 5.1 that there is a positive Lyapunov exponent for model system (5.27) with $w_{10} = 0.03, 0.05, 0.07$, respectively. According to the characterization of strange attractors (chaos) introduced in Section 9.4 of reference [16], a bounded system with a positive Lyapunov exponent is one operational definition of chaotic behavior. Furthermore, it follows from Table 5.1 that the Lyapunov dimensions for model system (5.27) with $w_{10} = 0.03, 0.05, 0.07$ are all fractals.

It should be noted that pitchfork bifurcations and tangent bifurcations are abundantly in Fig. 5.3. Many different kinds of period attractors and strange attractors (chaos) have been observed, and the size of a strange attractor (chaos) changes as w_{10} smoothly varies.

References

1. Azar, C., Holmberg, J., Lindgren, K.: Stability analysis of harvesting in predator-prey model. J. Theo. Biol. 174, 13–19 (1995)
2. Allen, J.C.: Chaos and phase-locking in predator-prey models in relation to functional response. Flor. Ento. 13, 100–110 (1990)
3. Alligood, K., Sauer, T., Yorke, J.: An Introduction to Dynamical Systems. Springer, New York (1997)
4. Bardi, M.: Predator-prey model in periodically fluctuating environment. J. Math. Biol. 12, 127–140 (1981)
5. Clark, C.W.: Mathematical Bioeconomics: The Optimal Management of Renewable Resource, 2nd edn. John Wiley and Sons, New York (1990)
6. Croisier, H., Dauby, P.C.: Continuation and bifurcation analysis of a periodically forced excitable system. J. Theo. Biol. 246(3), 430–448 (2007)

7. Cushing, J.M.: Two species competition in a periodic environment. J. Math. Biol. 10, 364–380 (1980)
8. Gakkhar, S., Singh, B.: The dynamics of a food web consisting of two preys and a harvesting predator. Chaos Soli. Frac. 34, 1346–1356 (2007)
9. Gakkhar, S., Naji, R.K.: On a food web consisting of a specialist and a generalist predator. J. Biol. Syst. 11(4), 365–376 (2003)
10. Gakkhar, S., Naji, R.K.: Existence of chaos in two-prey, one-predator system. Chaos Soli. Frac. 17(4), 639–649 (2003)
11. Gencay, R., Dechert, W.D.: An algorithm for the n-Lyapunov exponents of an n-dimensional unknown dynamical system. Phys. D 59, 142–157 (1992)
12. Gomes, A.A., Manica, E., Varriale, M.C.: Applications of chaos control techniques to a three-species food chain. Chaos Soli. Frac. 35(3), 432–441 (2008)
13. Gordon, H.S.: The economic theory of a common property resource: The fishery. J. Polit. Econ. 62(2), 124–142 (1954)
14. Guckenheimer, J., Holmes, P.: Nonlinear Oscillations, Dynamical Systems and Bifurcations of Vector Field. Springer, New York (1983)
15. Isidori, A.: Nonlinear Control Systems, 3rd edn. Springer, New York (1995)
16. Kaplan, J., Yorke, J.: Chaotic behavior of multidimensional difference equations. In: Functional Differential Equations and Approximation of Fixed Points. Lecture Notes in Math. Springer, Berlin (1979)
17. Klebanoff, A., Hasting, A.: Chaos in one predator two prey model: General results from bifurcatin theory. Math. Bios. 112, 221–223 (1994)
18. Kumar, S., Srivastava, S.K., Chingakham, P.: Hopf bifurcation and stability analysis in a harvested one-predator-two-prey model. Appl. Math. Comput. 129, 107–118 (2002)
19. Kot, M., Schultz, T.W.: Complex dynamics in a model microbial system. Bull. Math. Biol. 54, 619–648 (1992)
20. Liu, X.N., Chen, L.S.: Complex dynamics of Holling type II Lotka-Volterra predator-prey system with impulsive perturbations on the predator. Chaos Soli. Frac. 16, 311–320 (2003)
21. Nychka, D.W., Ellner, S., Gallant, R.A., McCaffrey, D.: Finding chaos in noisy systems. The Roy. Stat. Soci. Seri. B 54, 399–426 (1992)
22. Parker, T.S., Chua, L.O.: Practical Numerical Algorithms for Chaotic Systems. Springer, New York (1989)
23. Sabin, C.W.: Chaos in a periodically forced predator-prey ecosystem model. Math. Bio. 113, 91–113 (1993)
24. Sunita, G., Raid, K.N.: Chaos in seasonally perturbed ratio-dependent prey-predator system. Chaos Soli. Frac. 15(1), 107–118 (2003)
25. Takeuchi, Y.: Global dynamical properties of Lotka-Volterra systems. World Scientific Publishing Co. Pte. Ltd. (1996)
26. Tindell, K., Burns, A., Wellings, A.J.: Calculating controller area network message response time. Cont. Engi. Prac. 3(8), 1163–1169 (1995)
27. Takens, F.: Detecting strange attractor in turbulence. Dynamical Systems and Turbulence. Lect. Math. Springer, New York (1981)
28. Venkatesan, A., Parthasarathy, S., Lkshmannan, M.: Occurrence of multiple period-doubling route to chaos in periodically pulsed chaotic dynamical systems. Chaos Soli. Frac. 18(4), 891–898 (2003)
29. Wang, J., Chen, C.: Nonlinear control of differential-algebraic model in power systems. Proc. CSEE 21(8), 15–18 (2001)

30. Wolf, A., Swift, J.B., Swinney, H.L., Vastano, J.A.: Determining Lyapunov exponents from a time series. Phys. D 16, 285–317 (1985)
31. Zhang, S., Tan, D., Chen, L.: Chaos in periodically forced Holling type IV predator-prey system with impulsive perturbations. Chaos Soli. Frac. 27(4), 980–990 (2006)
32. Zhang, S., Tan, D., Chen, L.: Chaos in periodically forced Holling type II predator-prey system with impulsive perturbations. Chaos Soli. Frac. 28(2), 367–376 (2006)

Chapter 6
T-S Fuzzy Control for Singular Biological Systems

6.1 Fuzzy State Feedback Control for Singular Biological Systems

6.1.1 Model Formulation and Local Stability Analysis

A model describing the dynamics of a single species with stage structure is proposed by Zhang et al. in [13] can be expressed as follows:

$$\begin{cases} \dot{x}_1(t) = \alpha x_2(t) - r_1 x_1(t) - \beta x_1(t) - \eta x_1^2(t), \\ \dot{x}_2(t) = \beta x_1(t) - r_2 x_2(t), \end{cases} \tag{6.1}$$

where $x_1(t)$ and $x_2(t)$ represents population density of immature species and mature species, respectively. α, r_1, β denotes the intrinsic growth rate, death rate, and transition rate, respectively. r_2 is the death rate of the mature population, and the growth of the immature population is restricted by population density that is reflected by $-\eta x_1^2(t)$.

According to [1], considering the economic interest of harvest effort on the immature population, a bio-economic singular model system can be established as follows:

$$\begin{cases} \dot{x}_1(t) = \alpha x_2(t) - r_1 x_1(t) - \beta x_1(t) - \eta x_1^2(t) - E(t) x_1(t), \\ \dot{x}_2(t) = \beta x_1(t) - r_2 x_2(t), \\ 0 = E(t)(p x_1(t) - c) - m, \end{cases} \tag{6.2}$$

where $E(t)$ is the harvest effort on the immature population, p is a price constant per the individual population, and m is a cost constant, m is the economic interest of harvesting.

Q. Zhang et al.: Complex., Analy. and Cntrl. of Singular Biological Systems, LNCIS 421, pp. 89–101.
springerlink.com

According to the economic theory of a common-property resource [1], there is a phenomenon of bio-economic equilibrium when the economic interest of harvesting is zero, i.e., $m = 0$. For model system (6.2), an interior equilibrium $P^*(\frac{c}{p}, \frac{c\beta}{pr_2}, \frac{\alpha\beta p - pr_1r_2 - \beta pr_2 - \eta cr_2}{pr_2})$ can be obtained in the case of the phenomenon of bio-economic equilibrium. According to the biological interpretation of the interior equilibrium, it follows that $\alpha\beta p - pr_1r_2 - \beta pr_2 - \eta cr_2 > 0$.

Theorem 6.1. *Model system (6.2) has a singularity induced bifurcation around the interior equilibrium P^*, and a stability switch occurs as m increases through 0.*

Proof. Let m be bifurcation parameter, $X(t) = (x_1(t), x_2(t))^T$,

$$f(x_1(t), x_2(t), E(t), m) = \begin{pmatrix} \alpha x_1(t) - r_1x_1(t) - \beta x_1(t) - \eta x_1^2(t) - E(t)x_1(t) \\ \beta x_1(t) - r_2x_2(t) \end{pmatrix},$$

$$g(x_1(t), x_2(t), E(t), m) = E(t)(px_1(t) - c) - m,$$

$$\Xi = \begin{pmatrix} 1\,0\,0 \\ 0\,1\,0 \\ 0\,0\,0 \end{pmatrix}, J = \begin{pmatrix} D_Xf\; D_Ef \\ D_Xg\; D_Eg \end{pmatrix}|_{P^*} = \begin{pmatrix} -\frac{\alpha\beta}{r_2} - \frac{c\eta}{p} & \alpha & -\frac{c}{p} \\ \beta & -r_2 & 0 \\ -\frac{(\alpha\beta p - pr_1r_2 - \beta pr_2 - \eta cr_2)}{r_2} & 0 & 0 \end{pmatrix},$$

$$\operatorname{trace}[D_Ef\operatorname{adj}(D_Eg)(D_Xg)]|_{P^*} = -\frac{c(\alpha\beta p - pr_1r_2 - \beta pr_2 - \eta cr_2)}{pr_2} \neq 0, \quad (6.3)$$

$$\det\begin{pmatrix} D_Xf\; D_Ef \\ D_Xg\; D_Eg \end{pmatrix}|_{P^*} = -\frac{c(\alpha\beta p - pr_1r_2 - \beta pr_2 - \eta cr_2)}{p}. \quad (6.4)$$

Let $\Delta = \det(D_Eg) = px_1(t) - c$,

$$\det\begin{pmatrix} D_Xf\; D_Ef\; D_mf \\ D_Xg\; D_Eg\; D_mg \\ D_X\Delta\; D_E\Delta\; D_m\Delta \end{pmatrix}|_{P^*} = -cr_2 \neq 0. \quad (6.5)$$

Based on the above analysis, five items (i-iv) can be obtained as follows:

(i) According to (6.4), $\operatorname{trace}(D_Eh\operatorname{adj}(D_Eg)(D_Hg, D_Eg))|_{P^*} \neq 0$.

(ii) It follows from (6.5) that $\begin{pmatrix} D_Xf\; D_Ef \\ D_Xg\; D_Eg \end{pmatrix}$ is nonsingular at P^*.

(iii) By virtue of (6.6), it can be shown that $\begin{pmatrix} D_Xf\; D_Ef\; D_mf \\ D_Xg\; D_Eg\; D_mg \\ D_X\Delta\; D_E\Delta\; D_m\Delta \end{pmatrix}$ is nonsingular at P^*, hence rank $\begin{pmatrix} D_Xf\; D_Ef\; D_mf \\ D_Xg\; D_Eg\; D_mg \\ D_X\Delta\; D_E\Delta\; D_m\Delta \end{pmatrix} = 4$.

(iv) It is easy to show that $\operatorname{rank}(h(H(t), E(t), v)) = 3$ and $\operatorname{rank}(g(H(t), E(t), v)) = 1$. Hence,

$$\operatorname{rank}\begin{pmatrix} D_X f & D_E f & D_m f \\ D_X g & D_E g & D_m g \\ D_X \Delta & D_E \Delta & D_m \Delta \end{pmatrix}$$
$$= \operatorname{rank}(f(H(t),E(t),v)) + \operatorname{rank}(g(H(t),E(t),v)) + 1.$$

Singularity induced bifurcation, which is introduced in Section III (A) of reference [9], consists of three conditions, i.e., SI1, SI2, and SI3. According to the above items (i-iv), SI1, SI2 and SI3 are all satisfied, hence, model system (6.2) has a singularity induced bifurcation at the interior equilibrium P^* and the bifurcation value is $m = 0$.

Along with the line of the above proof, for the differential-algebraic model system (6.2), by simple computing,

$$B = -\operatorname{trace}[D_E f \operatorname{adj}(D_E g) D_X g]|_{P^*} = \frac{c(\alpha\beta p - pr_1r_2 - \beta pr_2 - \eta cr_2)}{pr_2},$$

$$C = -\begin{pmatrix} p & 0 & 0 \end{pmatrix} \begin{pmatrix} -\frac{\alpha\beta}{r_2} - \frac{\eta c}{p} & \alpha & -\frac{c}{p} \\ \beta & -r_2 & 0 \\ \frac{\alpha\beta p - pr_1r_2 - \beta pr_2 - \eta cr_2}{r_2} & 0 & 0 \end{pmatrix}^{-1} \begin{pmatrix} 0 \\ 0 \\ -1 \end{pmatrix},$$

the inequality $B/C > 0$ satisfies Theorem 3 of reference [9]. According to Theorem 3 of [9], when v increases through 0, one eigenvalue (denoted by λ_1) of differential-algebraic model system (6.2) without discrete time delay moves from C^- to C^+ along the real axis by diverging through infinity; the movement behavior of this eigenvalue influences the stability of model system (6.2). Hence, there is a stability switch as m increases through 0.

6.1.2 Fuzzy Control for Model System

In this section, model system (6.2) is approached by using some local linear singular systems, and a series of local controllers is designed based on linear systematic analysis and control strategies. Finally, the obtained linear controllers can be integrated as a global controller by virtue of membership function. Let $z_1(t) = x_1(t) - \frac{c}{p}, z_2(t) = x_2(t) - \frac{c\beta}{pr_2}, z_3(t) = E(t) - \frac{\alpha\beta p - pr_1r_2 - \beta pr_2 - \eta cr_2}{pr_2}$.

By using the above transformation, model system (6.2) can be rewritten as follows:

$$\begin{cases} \dot{z}_1(t) = (-\frac{\alpha\beta}{r_2} - \frac{\eta c}{p})z_1(t) + \alpha z_2(t) - \frac{c}{p}z_3(t) - \eta z_1^2(t) - z_1(t)z_3(t), \\ \dot{z}_2(t) = \beta z_1(t) - r_2 z_2(t), \\ 0 = p(\frac{\alpha\beta}{r_2} - r_1 - \beta - \frac{\eta c}{p})z_1(t) + pz_1(t)z_3(t) - m. \end{cases} \tag{6.6}$$

The controlled system is as follows:

$$\begin{cases} \dot{z}_1(t) = (-\dfrac{\alpha\beta}{r_2} - \dfrac{\eta c}{p})z_1(t) + \alpha z_2(t) - \dfrac{c}{p}z_3(t) - \eta z_1^2(t) - z_1(t)z_3(t), \\ \dot{z}_2(t) = \beta z_1(t) - r_2 z_2(t), \\ 0 = p(\dfrac{\alpha\beta}{r_2} - r_1 - \beta - \dfrac{\eta c}{p})z_1(t) + p z_1(t) z_3(t) + u(t), \end{cases} \tag{6.7}$$

where $u(t)$ is the control input.

As analyzed above, the singularity induced bifurcation occurs in a certain neighborhood of interior equilibrium. Hence, it is feasible to suppose $z_1(t) \in [-d, d], d > 0$, and then the fuzzy state model can be written as follows, which is suitable for describing model system (6.7) as $z_1(t) \in [-d, d]$.

Rule 1: If $z_1(t)$ is M_1, then $\Xi \dot{Z}(t) = A_1 Z(t) + Bu(t)$.

Rule 2: If $z_2(t)$ is M_1, then $\Xi \dot{Z}(t) = A_2 Z(t) + Bu(t)$,

where $Z(t) = \begin{bmatrix} z_1(t) \ z_2(t) \ z_3(t) \end{bmatrix}^T$, $\Xi = \begin{pmatrix} 1 & 0 & 0 \\ 0 & 1 & 0 \\ 0 & 0 & 0 \end{pmatrix}$,

$$A_1 = \begin{pmatrix} -\frac{\alpha\beta}{r_2} - \frac{\eta c}{p} + \eta d & \alpha & -\frac{c}{p} + d \\ \beta & -r_2 & 0 \\ p(\frac{\alpha\beta}{r_2} - r_1 - \beta - \frac{\eta c}{p}) & 0 & -pd \end{pmatrix},$$

$$A_2 = \begin{pmatrix} -\frac{\alpha\beta}{r_2} & \alpha & -\frac{c}{p} - d \\ \beta & -r_2 & 0 \\ p(\frac{\alpha\beta}{r_2} - r_1 - \beta - \frac{\eta c}{p}) & 0 & pd \end{pmatrix}, B = \begin{pmatrix} 0 \\ 0 \\ 1 \end{pmatrix}.$$

$h_i(z_1(t)), (i = 1, 2)$ represent the membership function of $z_1(t)$ subject to fuzzy set $M_i, (i = 1, 2)$, and $h_1(z_1(t)) = \dfrac{1}{2}(1 - \dfrac{z_1(t)}{d})$, $h_2(z_1(t)) = \dfrac{1}{2}(1 + \dfrac{z_2(t)}{d})$.

The model system can be expressed as follows:

$$\Xi \dot{Z}(t) = \sum_{i=1}^{2} h_i(z_1(t))(A_i Z(t) + Bu(t)). \tag{6.8}$$

Definition 6.1. [8] For the following T-S singular system

$$\Xi \dot{x}(t) = \sum_{i=1}^{r} \lambda_i(\xi(t)) A_i x(t), \tag{6.9}$$

if the following conditions are satisfied, then model system (6.9) is quadrable stable.

(i) $\det[s\Xi - \sum_{i=1}^{r} \lambda_i(\xi(t)) A_i x(t)] \neq 0$, open-loop and impulse free; and

(ii) there exists $\mu > 0$ and an inverse matrix Q satisfying $\Xi^T Q = Q^T \Xi \geq 0$, $\dot{V}(x(t)) \leq -\mu x^T(t) x(t)$ where $V(x(t)) = x^T(t) \Xi^T Q x(t)$.

In the following part, controllers are designed to eliminate the singularity induced bifurcation and stability switch.

Theorem 6.2. *For model system (6.8), if there exist matrices M_i, Q, and Y_{ij} where $det(Q) \neq 0$, $Y_{ji} = Y_{ij}^T, i \neq j, i, j = 1,2$, and the following matrix inequalities are satisfied,*

$$
\begin{aligned}
&(a) \quad \Xi^T Q = Q^T \Xi \geq 0,\\
&(b) \quad Q^T A_i^T + M_i^T B^T + A_i Q + BM_i < Y_{ii},\\
&(c) \quad A_i Q + A_j Q + Q^T A_i^T + Q^T A_j^T + BM_i + M_j^T B^T + M_i^T B^T \geq Y_{ij} + Y_{ij}^T,\\
&(d) \quad [Y_{ij}]_{2\times 2},
\end{aligned}
$$

then following closed loop system is quadrable stable with $u(t) = \Sigma_{j=1}^2 h_j(z_1(t))K_j Z(t)$

$$
\Xi \dot{Z}(t) = \sum_{i=1}^{2}\sum_{j=1}^{2} h_i(z_1(t))h_j(z_1(t))((A_i + BK_j)Z(t)), \tag{6.10}
$$

where $K_j = M_j Q^{-1}, (j = 1,2)$.

Proof. Let $P = Q^{-1}$, then $K_i = M_i P$. By pre-multiplying P^T and post-multiplying P into inequality (b) and (c), respectively, and by pre-multiplying diag(P^T, P^T) and post-multiplying diag(P, P), it can be obtained that

$$
\begin{aligned}
&\text{(a)'} \quad (A_i^T + K_i^T B^T)P + P^T(A_i + BK_i) < P^T Y_{ii} P,\\
&\text{(b)'} \quad P^T(A_i + BK_j + A_j + BK_i) + (A_i^T + K_j^T B^T + A_j^T + K_i^T B^T)P \leq P^T Y_{ij} P + P^T Y_{ij}^T P,\\
&\text{(c)'} \quad [P^T Y_{ij} P]_{2\times 2} < -\mu I,
\end{aligned}
$$

where $\mu > 0$.

Constructing singular Lyapunov function $V(t) = Z^T(t)\Xi^T Z(t)$, then

$$
\begin{aligned}
\dot{V}(t) &= \dot{Z}^T \Xi^T P Z(t) + Z^T(t)\Xi^T P \dot{Z}(t)\\
&= \sum_{i=1}^{2} h_i^2(z_1(t))Z^T(t)((A_i^T + K_i^T B^T)P + P^T(A_i + BK_i))Z(t)\\
&+ \sum_{i=1}^{2}\sum_{i<j}^{2} h_i(z_1(t))h_j(z_1(t))Z^T(t)((A_i^T + K_j^T B^T + A_j^T + K_i^T B^T))P\\
&+ P^T(A_i + BK_j + A_j + BK_i)Z(t)\\
&\leq \sum_{i=1}^{2} h_i^2(z_1(t))^T Z^T(t)P^T Y_{ii} P Z(t)\\
&+ \sum_{i=1}^{2}\sum_{i<j}^{2} h_i(z_1(t))Z^T(t)(P^T Y_{ij} P + P^T Y_{ij}^T P)Z(t)\\
&\leq -\mu \sum_{i=1}^{2} h_i^2(z_1(t))Z^T(t)Z(t) \leq -\mu Z^T(t)Z(t).
\end{aligned}
$$

Based on the above analysis and Definition 6.1, model system (6.10) is quadrable stable.

6.1.3 Numerical Simulation

By using hypothetical values of parameters, numerical simulation is carried out to show consistency with theoretical analysis obtained in the above sections of this chapter. The parameters of values are given as follows: $\alpha = 0.15, r_1 = 0.2, \beta = 0.5, \eta = 0.001, r_2 = 0.1, p = 1, c = 40$, and $d = 10$.

According to Theorem 6.1, matrices Q, M_1, and M_2 can be obtained by virtue of LMI Toolbox in MATLAB, which are as follows:

$$Q = \begin{pmatrix} 77.6 & -31.9 & 0 \\ -31.9 & 4193.2 & 0 \\ 67.9 & 7.0 & 181.9 \end{pmatrix},$$

$$M_1 = \begin{pmatrix} 3601.1 & -171.5 & -525.3 \end{pmatrix}, M_2 = \begin{pmatrix} 9817.3 & -481.5 & -3264.8 \end{pmatrix},$$

and corresponding control law can be obtained as follows,

$$K_1 = \begin{pmatrix} 49.0862 & 0.3375 & -2.8873 \end{pmatrix}, K_2 = \begin{pmatrix} 142.6625 & 1.0008 & -17.9443 \end{pmatrix}.$$

The effectiveness of controller can be shown in Fig. 6.1, s.t., $z_1 \to 0, z_2 \to 0, z_3 \to 0$ as $t \to \infty$.

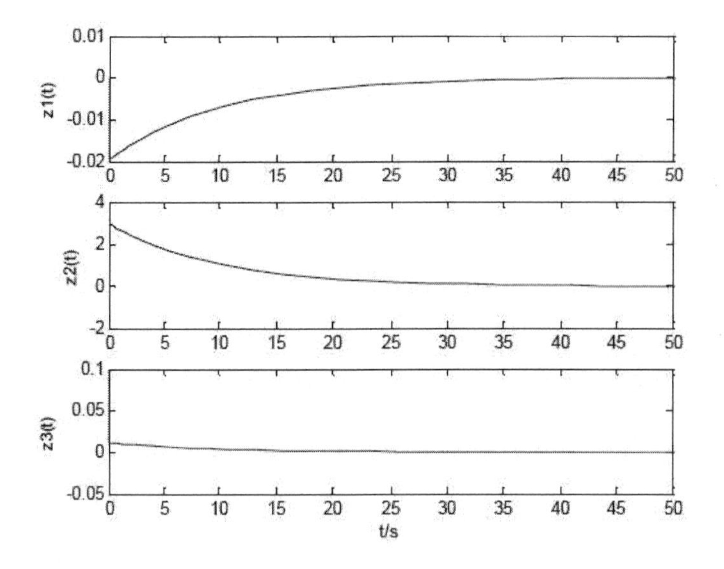

Fig. 6.1 Dynamical behavior of model system (6.7).

6.2 H_∞ Control for Singular Biological Systems

It is well known that any system can't operate without the dependence on the surrounding environment, and a case in point is a biological system. In recent years, scholars have concentrated on how to guarantee the model system to operate safely and minimize the influence of perturbation to the state and output of the system.

Generally, it is feasible to solve the above mentioned problems by utilizing H_∞ control theory, and there are some research results on utilization of H_∞ control into biological systems in [12, 2, 3, 6, 5, 10, 11]. The H_∞ control problem for a three-dimensional web system is investigated in [12], and the analysis of the macro-economic model of Liaoning province and its H_∞ control strategies is investigated in [2]. However, as far as H_∞ control for the biological economic system is concerned, related research results are few. In the following section, H_∞ control for singular biological systems with perturbation is studied. Furthermore, H_∞ state controller for a T-S fuzzy singular system is designed with less input energy, which may guarantee the model system to operate safely.

6.2.1 Model Formulation

Model system (6.2) is extended by incorporating perturbation. Considering the following singular biological economic system with perturbation,

$$\begin{cases} \dot{x}_1(t) = \alpha x_2(t) - r_1x_1(t) - \beta x_1(t) - \eta x_1^2(t) + b_{11}w(t) - E(t) - x_1(t), \\ \dot{x}_2(t) = \beta x_1(t) - r_2x_2(t) + b_{12}w(t), \\ 0 = E(t)(px_1(t) - c) - m + b_{13}w(t), \end{cases} \tag{6.11}$$

where $w(t)$ represents perturbation input, let $w(t) \in L^2(0,\infty;R)$, b_{11}, b_{12}, b_{13} denotes the perturbation coefficient. Obviously, model system (6.11) is equivalent to model system (6.2) when $b_{11} = b_{12} = b_{13} = 0$. Let $\xi_1(t) = x_1(t) - \frac{c}{p}, \xi_2(t) = x_2(t) - \frac{c\beta}{pr_2}, \xi_3(t) = E(t) - \frac{\alpha\beta p - r_1r_2 - \beta pr_2 - \eta cr_2}{pr_2}$.

In order to facilitate investigation, model system (6.11) can be written by the above transformation,

$$\begin{cases} \dot{\xi}_1(t) = (-\frac{\alpha\beta}{r_2} - \frac{\eta c}{p})\xi_1(t) + \alpha\xi_2(t) - \frac{c}{p}\xi_3(t) - \eta\xi_1^2(t) - \xi_1(t)\xi_3(t) + b_{11}w(t), \\ \dot{\xi}_2(t) = \beta\xi_1(t) - r_2\xi_2(t) + b_{12}w(t), \\ 0 = p(\frac{\alpha\beta}{r_2} - r_1 - \beta - \frac{\eta c}{p})\xi_1(t) + p\xi_1(t)\xi_3(t) - m + b_{13}w(t). \end{cases} \tag{6.12}$$

According to Theorem 6.2 of this chapter, when $b_{11} = b_{12} = b_{13} = 0$, model system (6.12) has a singularity induced bifurcation around the interior equilibrium, and there is a stability switch as m increases through 0.

In order to eliminate the singularity induced bifurcation and stability switch, the controlled model system is as follows:

$$\begin{cases} \dot{\xi}_1(t) = (-\frac{\alpha\beta}{r_2} - \frac{\eta c}{p})\xi_1(t) + \alpha\xi_2(t) - \frac{c}{p}\xi_3(t) - \eta\xi_1^2(t) - \xi_1(t)\xi_3(t) + b_{11}w(t), \\ \dot{\xi}_2(t) = \beta\xi_1(t) - r_2\xi_2(t) + b_{12}w(t), \\ 0 = p(\frac{\alpha\beta}{r_2} - r_1 - \beta - \frac{\eta c}{p})\xi_1(t) + p\xi_1(t)\xi_3(t) - m + b_{13}w(t) + u(t), \end{cases} \tag{6.13}$$

where $u(t)$ is the control input representing regulation control for a biological resource. In this section, we concentrate only on the case $b_{11} \neq 0, b_{12} = b_{13} = 0$, which implies that only growth of the immature population is under perturbation.

As analyzed above, the singularity induced bifurcation occurs in a certain neighborhood of interior equilibrium. Hence, it is feasible to suppose $\xi_1(t) \in [-d,d], d > 0$, and then the fuzzy state model can be written as follows, which is suitable for describing model system (6.13) as $\xi_1(t) \in [-d,d]$.

Rule 1: If $\xi_1(t)$ is M_1, then $\Xi\dot{\xi}(t) = A_1\xi(t) + B_{11}w(t) + B_{21}u(t)$.

Rule 2: If $\xi_1(t)$ is M_2, then $\Xi\dot{\xi}(t) = A_2\xi(t) + B_{12}w(t) + B_{22}u(t)$,

where $\xi(t) = \begin{pmatrix} \xi_1(t) & \xi_2(t) & \xi_3(t) \end{pmatrix}^T$

$$\Xi = \begin{pmatrix} 1 & 0 & 0 \\ 0 & 1 & 0 \\ 0 & 0 & 0 \end{pmatrix}, A_1 = \begin{pmatrix} -\frac{\alpha\beta}{r_2} - \frac{\eta c}{p} + \eta d & \alpha & -\frac{c}{p} + d \\ \beta & -r_2 & 0 \\ p(\frac{\alpha\beta}{r_2} - r_1 - \beta - \frac{\eta c}{p}) & 0 & -pd \end{pmatrix},$$

$$A_2 = \begin{pmatrix} -\frac{\alpha\beta}{r_2} - \frac{\eta c}{p} - \eta d & \alpha & -\frac{c}{p} - d \\ \beta & -r_2 & 0 \\ p(\frac{\alpha\beta}{r_2} - r_1 - \beta - \frac{\eta c}{p}) & 0 & pd \end{pmatrix},$$

$$B_{11} = B_{12} = \begin{pmatrix} b_{11} \\ 0 \\ 0 \end{pmatrix}, B_{21} = B_{22} = \begin{pmatrix} 0 \\ 0 \\ 1 \end{pmatrix}.$$

$\lambda_i(\xi_1(t)), (i = 1,2)$ represent the membership function of $\xi_1(t)$ subject to fuzzy set $M_i, (i = 1,2)$, which follows that

$$\lambda_1(\xi_1(t)) = \frac{1}{2}(1 - \frac{\xi_1(t)}{d}), \lambda_2(\xi_1(t)) = \frac{1}{2}(1 + \frac{\xi_2(t)}{d}), \Sigma_{i=1}^2 \lambda_i(\xi_1(t)) = 1.$$

The model system can be expressed as follows:

$$\Xi\dot{\xi}(t) = \sum_{i=1}^{2} \lambda_i(\xi_1(t))(A_i\xi(t) + B_1w(t) + B_2u(t)), \tag{6.14}$$

where $B_1 = B_{11} = B_{12}, B_2 = B_{21} = B_{22}$.

6.2.2 Design of H_∞ Controller

Our objective is to minimize the following optimization problem by adopting appropriate strategies,

$$\min \Omega = \left(\int_0^\infty u^2(t)\mathrm{d}t\right)^{\frac{1}{2}}. \tag{6.15}$$

It should be noted that it is difficult to obtain the minimum value of (6.15) due to the existence of perturbation $w(t)$. Hence, the following inequality w.r.t. given an optimization object should be satisfied:

$$\min \Omega = \left(\int_0^\infty u^2(t)\mathrm{d}t\right)^{\frac{1}{2}} \geq \gamma\left(\int_0^\infty w^2(t)\mathrm{d}t\right)^{\frac{1}{2}} = \gamma\|w(t)\|_2,$$

where $\gamma > 0$ is constant.

If the output of the system is $z(t) = \Sigma_{i=1}^2 \lambda_i(\xi_1(t))D_i u(t)$, then the system can be rewritten as

$$\begin{cases} \Xi\dot{\xi}(t) = \Sigma_{i=1}^2 \lambda_i(\xi_1), \\ z(t) = \Sigma_{i=1}^2 \lambda_i(\xi_1(t))D_i u(t). \end{cases} \tag{6.16}$$

When $D_i = 1, i = 1,2$, the above problem is transformed as $\|z(t)\|_2^2 \leq \gamma^2\|w(t)\|_2^2$, which is a H_∞ control problem.

In the following, H_∞ controller for the T-S fuzzy singular system is designed. It is utilized to eliminate the singularity induced bifurcation and stability switch, which can guarantee the model system to operate safely.

Theorem 6.3. *For given positive constant γ, if there exist matrices M_i, Q, and Z_{ij}, where $det(Q) \neq 0$, $Z_{ji} = Z_{ij}^T, i \neq j, i,j = 1,2$, and the following matrix inequalities are satisfied,*

$$\begin{aligned}
&(a)\quad \Xi^T Q = Q^T \Xi, \\
&(b)\quad Q^T A_i^T + M_i^T B_2^T + A_i Q + B_2 M_i + \tfrac{1}{\gamma^2} B_1 B_1^T < Z_{ii}, \\
&(c)\quad S_{ij} + S_{ji}^T \leq Z_{ij} + Z_{ij}^T, i \neq j, \\
&(d)\quad H_k = \begin{pmatrix} Z_{11} & Z_{12} & V_{1k}^T \\ Z_{21} & Z_{22} & V_{2k}^T \\ V_{1k} & V_{2k} & -1 \end{pmatrix} < 0,
\end{aligned}$$

where $S_{ij} = A_i Q + B_2 M_j + A_j Q + B_2 M_i + \frac{1}{\gamma^2} B_1 B_1^T, V_{ik} = D_i M_k$, then the H_∞ norm is less than γ and the following closed loop system is quadrable stable with $u(t) = \Sigma_{j=1}^2 \lambda_j(\xi_1(t))K_j\xi(t)$ in the case of $w(t) = 0$.

$$\begin{cases} \Xi\dot{\xi}(t) = \Sigma_{i=1}^2 \Sigma_{j=1}^2 \lambda_i(\xi_1(t))\lambda_j(\xi_1(t))((A_i + B_2 K_j)Z(t)) + \Sigma_{i=1}^2 \lambda_i(\xi_1(t))B_1 w(t), \\ z(t) = \Sigma_{i=1}^2 \Sigma_{j=1}^2 \lambda_i(\xi_1(t))\lambda_j(\xi_1(t))D_i K_j \xi(t), \end{cases} \tag{6.17}$$

where $K_j = M_j Q^{-1}, (j = 1,2)$.

Proof. Let $P = Q^{-1}$, then $K_i = M_i P$. By left multiplying P^T and post-multiplying P into inequality (b) and (c), respectively, and prey-multiplying $\mathrm{diag}(P^T, P^T)$ and right multiplying $\mathrm{diag}(P,P)$, it can be obtained that

(a)' $(A_i^T + K_i^T B_2^T)P + P^T(A_i + B_2 K_i) + \frac{1}{\gamma^2} P^T B_1 B_1^T P < P^T Z_{ii} P$,

(b)' $P^T(A_i + B_2 K_j + A_j + B_2 K_i) + (A_i^T + K_j^T B_2^T + A_j^T + K_i^T B_2^T + \frac{2}{\gamma^2} P^T B_1 B_1^T)P \leq P^T Z_{ij} P + P^T Z_{ij}^T P$.

According to (d) and $\sum_{i=1}^2 \lambda_i = 1$,

$$H_k = \begin{pmatrix} Z_{11} & Z_{12} & \sum_{k=1}^2 \lambda_k V_{1k}^T \\ Z_{21} & Z_{22} & \sum_{k=1}^2 \lambda_k V_{2k}^T \\ \sum_{k=1}^2 \lambda_k V_{1k} & \sum_{k=1}^2 \lambda_k V_{2k} & -1 \end{pmatrix} < 0,$$

by using Schur compliment theory, it follows that

$$\begin{pmatrix} Z_{11} & Z_{12} \\ Z_{21} & Z_{22} \end{pmatrix} < -\begin{pmatrix} \sum_{k=1}^2 \lambda_k V_{1k}^T \\ \sum_{k=1}^2 \lambda_k V_{2k}^T \end{pmatrix} \begin{pmatrix} \sum_{k=1}^2 \lambda_k V_{1k}^T \\ \sum_{k=1}^2 \lambda_k V_{2k}^T \end{pmatrix}^T \leq 0,$$

by pre-multiplying $\mathrm{diag}(P^T, P^T)$ and post-multiplying $\mathrm{diag}(P,P)$, it gives that

$$[P^T Z_{ij} P]_{2\times 2} < -\begin{pmatrix} \sum_{k=1}^2 \lambda_k V_{1k}^T \\ \sum_{k=1}^2 \lambda_k V_{2k}^T \end{pmatrix} \begin{pmatrix} \sum_{k=1}^2 \lambda_k V_{1k}^T \\ \sum_{k=1}^2 \lambda_k V_{2k}^T \end{pmatrix}^T \leq 0,$$

and

$$[P^T Z_{ij} P]_{2\times 2} < -\mu I, \mu > 0.$$

For any $\xi(t) \neq 0, t > 0$, it gives that

$$\begin{aligned} \sum_{i=1}^2 \sum_{j=1}^2 \lambda_i \lambda_j \xi^T(t) P^T Z_{ij} P \xi(t) &< -\sum_{i=1}^2 \sum_{j=1}^2 \sum_{u=1}^2 \sum_{v=1}^2 \lambda_i \lambda_j \lambda_u \lambda_v \xi^T(t) P^T V_{iu}^T V_{jv} P \xi(t) \\ &= -Z^T(t) z(t), \end{aligned} \tag{6.18}$$

$$\sum_{i=1}^2 \sum_{j=1}^2 \lambda_i \lambda_j \xi^T(t) P^T Z_{ij} P \xi(t) < -\mu \xi^T(t) \xi(t). \tag{6.19}$$

Constructing singular Lyapunov function $V(t) = \xi^T(t) \Xi^T \xi(t)$, then

$$
\begin{aligned}
\dot{V}(t) &= \dot{\xi}^T \Xi^T P\xi(t) + \xi^T(t)P^T \Xi\dot{\xi}(t) \\
&= \sum_{i=1}^{2} \lambda_i^2 \xi^T(t)((A_i + B_2^T K_i)^T P + P^T(A_i + B_2 K_i) + \frac{1}{\gamma^2} P^T B_1 B_1^T P)\xi(t) \\
&+ \sum_{i=1}^{2}\sum_{i<j}^{2} \lambda_i\lambda_j \xi^T(t)(P^T(A_i + B_2K_i) + (A_i^T + K_i^T B_2^T + K_j^T B_2^T + A_j^T + K_i^T B_2^T \\
&+ \frac{2}{\gamma^2} P^T B_1 B_1^T)P)\xi(t) - \sum_{i=1}^{2}\sum_{j=1}^{2} \lambda_i\lambda_j \frac{1}{\gamma^2}\xi^T(t)P^T B_1 B_1^T P\xi(t) \\
&+ \sum_{i=1}^{2} \lambda_i(w^T(t)(t)B_1^T P\xi(t) + \xi^T(t)P^T B_1 w(t)) \\
&< \sum_{i=1}^{2}\sum_{i=1}^{2} \lambda_i^2 \xi^T(t)P^T Z_{ii} P\xi(t) + \sum_{i=1}^{2}\sum_{i<j}^{2} \lambda_i\lambda_j \xi^T(t)(P^T Z_{ij} P + P^T Z_{ij}^T P)\xi(t) \\
&+ \gamma^2 w^T(t)w(t) - (\gamma w(t) - \frac{1}{\gamma}\sum_{i=1}^{2} \lambda_i B_1^T P\xi(t))^T(\gamma w(t) - \frac{1}{\gamma}\sum_{i=1}^{2} \lambda_i B_1^T P\xi(t)).
\end{aligned}
$$

When $w(t) = 0$, it follows that

$$
\dot{V}(t) \le \sum_{i=1}^{2}\sum_{j=1}^{2} \lambda_i\lambda_j \xi^T(t)P^T Z_{ij} P\xi(t) < -\mu\xi^T(t)\xi(t).
$$

Based on the above analysis and Definition 6.1, model system (6.17) is quadrable stable.

When $w(t) \neq 0$, based on (6.18), it follows that

$$
\dot{V}(t) < -z^T(t)z(t) + \gamma^2 w^T(t)w(t) - (\gamma w(t) - \frac{\sum_{i=1}^{2} \lambda_i B_1^T P\xi(t)}{\gamma})^T(\gamma w(t) - \frac{\sum_{i=1}^{2} \lambda_i B_1^T P\xi(t)}{\gamma}).
$$

Without loss of generality, supposing $V(\xi(0)) = 0$ as $\xi(0) = 0$, as integrating both sides of above inequality from 0 to ∞ against t, it gives that,

$$
0 \le V(T) < -\|z\|_2^2 + \gamma^2\|W\|_2^2 - \gamma^2\|w - \frac{1}{\gamma^2}\sum_{i=1}^{2} \lambda_i B_1^T P\xi(t)\|_2^2,
$$

hence, it can be obtained that $\|z\|_2^2 \ge \gamma^2\|w\|_2^2$.

Remark 6.1. Based on the above analysis, for the model system with perturbation input and impulse behavior, it can be stabilized around the interior equilibrium by adopting appropriate control strategies. In the meanwhile, singularity induced bifurcation can be eliminated and the energy of control can be maintained within a bounded range. The biological interpretations of the above theoretical analysis are as follows. In order to control the growth rate and transmission rate of fish, the government may take measures to regulate harvest effort, which reduces the effect of perturbation and guarantees the ecological balance of the ecosystem.

6.2.3 Numerical Simulation

By using hypothetical values of parameters, numerical simulations are carried out to show the effectiveness of H_∞ controller, which eliminates the stability switch phenomenon as well as guarantees that the norm of H_∞ is less than $\gamma = 0.05$. The values of parameters are as follows: $\alpha = 0.15, r_1 = 0.2, \beta = 0.5, \eta = 0.001, r_2 = 0.1, p = 1, c = 40, b_{11} = 0.1$, and $d = 10$.

By using LMI Toolbox in MATLAB, some matrices that follow Theorem 6.3 can be obtained as follows,

$$Q = \begin{pmatrix} 4.1041 & -2.6063 & 0 \\ -2.6063 & 12.2893 & 0 \\ 0.0801 & 0.1069 & 0.0226 \end{pmatrix},$$

$$M_1 = \begin{pmatrix} 0.0307 & 0.1336 & -0.8257 \end{pmatrix},$$

$$M_2 = \begin{pmatrix} 0.0003 & -0.2694 & -0.9389 \end{pmatrix},$$

and the corresponding control parameters are as follows,

$$K_1 = \begin{pmatrix} 1.1503 & 0.5728 & -36.5611 \end{pmatrix},$$

$$K_2 = \begin{pmatrix} 1.1866 & 0.5913 & -41.5721 \end{pmatrix}.$$

References

1. Gordon, H.S.: The economic theory of a common property resource: The fishery. J. Polit. Econ. 62(2), 124–142 (1954)
2. Huang, X., Zhong, M.: The analysis of macro-economic model of Liaoning province and its H_∞ control strategies. Contr. Theo. Appl. 17(5), 781–783 (2000)
3. Huang, Y.J., Chen, F.D., Zhong, L.: Stability analysis of a prey-predator model with Holling type III response function incorporating a prey refuge. Appl. Math. Compu. 182, 672–683 (2006)
4. Jun, Y., Ichikawa, A.: H_∞ control for Takagi-Sugeno fuzzy descriptor systems. In: Proc. IEEE Int. Conf. Syst. Man and Cyber., Tokyo, Japan, pp. 28–33 (1999)
5. Liu, Z.J., Tan, R.H.: Impulsive harvesting and stocking in a Monod-Haldane functional response predator-prey system. Chaos Soli. Frac. 34, 454–464 (2007)
6. Peng, R., Wang, M.X.: Global stability of the equilibrium of a diffusive Holling-Tanner prey-predator model. Appl. Math. Lett. 20, 664–670 (2007)
7. Taniguchi, T., Tanaka, K., Yamafuji, K., Wang, H.O.: Fuzzy descriptor systems stability analysis and design via LMIs. In: Proc. Amer. Cont. Conf., San Diego, California, pp. 1827–1931 (1999)
8. Taniguchi, T., Tanaka, K., Yamafuji, K., Wang, H.O.: Fuzzy descriptor systems and nonlinear model following control. IEEE Trans. Fuz. Syst. 8, 442–452 (2000)
9. Venkatasubramanian, V., Schaettler, H., Zaborszky, J.: Local bifurcations and feasibility regions in differential-algebraic systems. IEEE Trans. Auto. Contr. 40(12), 1992–2013 (1995)

10. Wang, F.Y., Zhang, S.W., Chen, L.S., Sun, L.H.: Bifurcation and complexity of Monod type predator-prey system in a pulsed chemostat. Chaos Soli. Frac. 27, 447–458 (2006)
11. Wang, F.Y., Hao, C.P., Chen, L.S.: Bifurcation and chaos in a Monod-Haldane type food chain chemostat with pulsed input and washout. Chaos Soli. Frac. 32, 181–194 (2007)
12. Zhao, L., Zhang, Q., Yang, Q.: Dissipation control of three-species food chain systems. J. Biomath. 18(1), 82–92 (2003)
13. Zhang, X.A., Chen, L.S., Avidan, U.N.: The stage-structured predator-prey model and optimal harvesting policy. Math. Bios. 168, 201–210 (2000)

Chapter 7
Dynamic Analysis and Optimal Control of a Singular Biological Economic Model

7.1 Model Formulation

The logistic equation (sometimes called the Verhulst model or logistic growth curve) is a model of population growth first published by Pierre Verhulst [12, 2, 5, 8, 10, 14, 13, 16, 9]. The model is continuous in time, but a modification of the continuous equation to a discrete quadratic recurrence equation known as the logistic map is also widely used.

$$\dot{x}(t) = rx(t)(1 - \frac{x(t)}{k}), \tag{7.1}$$

where $x(t)$ represents population density at time t, r represents the intrinsic rate of population, and k is environment capacity.

Consider the economic interest of harvest effort on the population, the following singular biological model can be established and utilized to investigate the dynamical mechanism of the harvested ecosystem due to variation of economic interest of harvesting:

$$\begin{cases} \dot{x}(t) = rx(t)(1 - \frac{x(t)}{k}) - E(t)x(t), \\ 0 = E(t)(px(t) - c) - m, \end{cases} \tag{7.2}$$

where p represents price of harvested population, c represents cost of harvesting, and m represents economic interest of harvesting.

The biological interpretations of interior equilibrium imply that population and harvest effort on population all exist, which are relevant to our study.

In this chapter, local stability of differential-algebraic model system (7.2) at the interior equilibrium will be investigated. A stability switch caused by the variation of economic interest of harvest effort on the predator is also studied.

Q. Zhang et al.: Complex., Analy. and Cntrl. of Singular Biological Systems, LNCIS 421, pp. 103–111.
springerlink.com

Furthermore, optimal controllers are designed to stabilize the model system at the interior equilibrium.

7.2 Local Stability Analysis

Theorem 7.1. *Singular model system (7.2) has a singularity induced bifurcation at the interior equilibrium, and $m = 0$ is a bifurcation value. Furthermore, a stability switch occurs as m increases through* 0.

Proof. According to the economic theory of a common-property resource [6], there is a phenomenon of bio-economic equilibrium when the economic interest of harvesting is zero, i.e., $m = 0$. For model system (7.2), an interior equilibrium $P^*(x^*, E^*)$ can be obtained in the case of the phenomenon of bio-economic equilibrium, where $x^* = \frac{c}{p}$, $E^* = r(1 - \frac{c}{pk})$. According to the biological interpretation of the interior equilibrium, it follows that $x^* > 0$ and $E^* > 0$. In order to guarantee the existence of the interior equilibrium, some inequalities are satisfied,

$$pk - c > 0. \tag{7.3}$$

Let m be a bifurcation parameter, D be a differential operator (i.e., $D_E h$ represents the partial differential operator of $h(x(t), E(t), v)$ with respect to $x(t), E(t)$). $h(x(t), E(t), m) = rx(t)(1 - \frac{x(t)}{k}) - E(t)x(t)$, and $g(H(t), E(t), v) = E(t)(px(t) - c) - m$.

By simple computing, $g(x(t), E(t), m)|_{P^*} = \begin{cases} 0, & \text{v=0;} \\ \text{nonzero}, & \text{m} \neq 0. \end{cases}$,which implies that

$$\dim\ker(D_E g(x(t), E(t), m)|_{P^*}) = 1. \tag{7.4}$$

It can be calculated that

$$\mathrm{trace}(D_E f \mathrm{adj}(D_E g)(D_x g, D_E g))|_{P^*} = -cr(1 - \frac{c}{pk}),$$

and it follows from (7.3) that

$$\mathrm{trace}(D_E h \mathrm{adj}(D_E g)(D_x g, D_E g))|_{P^*} \neq 0. \tag{7.5}$$

Furthermore, it can also be calculated that

$$\begin{vmatrix} D_x h & D_E h \\ D_x g & D_E g \end{vmatrix}_{P^*} = \frac{cr(pk - c)}{pk},$$

and by virtue of (7.3), it is easy to show that

$$\begin{vmatrix} D_x h & D_E h \\ D_x g & D_E g \end{vmatrix}_{P^*} \neq 0. \tag{7.6}$$

According to Part A of Section IV in the reference [11], $\Delta(x(t),E(t),m)$ can be defined as follows:

$$\Delta(H(t),E(t),v) = \det(D_E g) = wx(t) - c.$$

When $m \neq 0$, it follows that $\det[D_E g(x(m),E(m),m)] \neq 0$, and $(x(m),E(m))$ denotes an equilibrium curve whose initial value is $P^*(x(0),E(0))$.

On the other hand, model system (7.2) can be rewritten as the Kronecker norm [4], and it is easy to show that the Kronecker index of matrix pencil $\{\Xi J\}$ of model system (7.2) is 1 as $m \neq 0$, and the Kronecker index of matrix pencil $\{\Xi J\}$ of model system (7.2) is 2 as $m = 0$, where

$$\Xi = \begin{pmatrix} 1 & 0 \\ 0 & 0 \end{pmatrix}, J = \begin{pmatrix} D_x h & D_E h \\ D_x g & D_E g \end{pmatrix} = \begin{pmatrix} r - \frac{2r}{k}x - E & -x \\ pE & px - c \end{pmatrix}.$$

Based on Theorem 2.1 in [15], model system (7.2) has a singularity induced bifurcation around the interior equilibrium, and the bifurcation value is $m = 0$. Furthermore, it can be calculated that

$$B = -\text{trace}[D_x f \text{adj}(D_E g) D_x g]|_{P^*} = cr(1 - \frac{c}{pk}) \neq 0,$$

$$C = D_m\delta - \begin{pmatrix} D_x\delta & D_E\delta \end{pmatrix} \begin{pmatrix} D_x f & D_E f \\ D_x g & D_E g \end{pmatrix}^{-1} \begin{pmatrix} D_m f \\ D_m g \end{pmatrix} |_{P^*} = \frac{kp}{r(pk-c)} \neq 0.$$

It is obvious that $\frac{B}{C} = \frac{cr^2(pk-c)^2}{p^2k^2}$, which follows from Theorem 2.1 in [15] that model system (7.2) is unstable around the interior equilibrium P^*.

Remark 7.1. Some preliminaries about the Kronecker canonical form and Kronecker index are introduced as follows.

Definition 7.1. [4] Let $(\hat{A},\hat{B}) \in L(R^N)$ be a square matrix pencil, and $s \longmapsto s\hat{A} - \hat{B}$ induces an affine map: $R \to L(R^N)$, R^N is an n dimensional real vector space, then $(\hat{A},\hat{B})$ associated with the induced map is called a matrix pencil. If there exists a $s_0 \in C$ satisfying $\det(s_0\hat{A} - \hat{B} \neq 0)$, then the spectrum of this matrix is $\sigma\hat{A} - \hat{B} = \{s \in C | \det(s\hat{A} - \hat{B}) = 0\}$, where $L(*)$ represents the linear matrix space in vector space, $\sigma(*)$ represents the spectrum of the matrix pencil, and $\longmapsto$ represents the map.

Lemma 7.1. *[1](Kronecker canonical form and Kronecker index) If $(\hat{A},\hat{B})$ is a regular matrix pencil in R^N and $R^N = U \oplus V$, and there exists map: $P \in G_L(U \oplus V, R^N), Q \in L(U)$, which satisfies $P\hat{A}Q = \begin{pmatrix} I_{n-q} & 0 \\ 0 & N \end{pmatrix}$, $P\hat{B}Q = \begin{pmatrix} \hat{C} & 0 \\ 0 & N \end{pmatrix}$, and $N^q =$*

$0, N^{q-1} = 0 (q \geq 1), I_q$ is an identity matrix, $I_{n-q} \in L(U), N, I_q \in L(V)$, and $\sigma(\hat{A},\hat{B}) = \sigma(\hat{C}), B_d\sigma(\hat{A},\hat{B}) = dimU$, then the Kronecker index of the matrix pencil $(\hat{A},\hat{B})$ is q, where $\oplus$ represents the direct sum of set, $G_L(*)$ represents all invertible linear map spaces, and dim represents the dimension of matrix or space.

7.3 Optimal Control for Model System

Based on the above dynamic analysis of model system (7.2), it can be concluded that model system (7.2) has a singularity induced bifurcation around the interior equilibrium, which causes model system (7.2) to be unstable around the interior equilibrium when the economic interest of harvesting $m > 0$. In Chapter 4 of this book, a state feedback controller is designed to eliminate the singularity induced bifurcation and stabilize the model system around the interior equilibrium. Compared with the controllers designed in Chapter 4 of this book, optimal controllers are designed, which not only guarantee the elimination of singularity induced bifurcation, but also stabilize the model system around the interior equilibrium with less control energy.

It is an optimal control problem. Considering the following model system

$$\Xi \dot{X}(t) = f(X(t)) + g(X(t))u(t), \tag{7.7}$$

where

$$\Xi = \begin{pmatrix} 1 & 0 \\ 0 & 0 \end{pmatrix}, X(t) = \begin{pmatrix} x(t) & E(t) \end{pmatrix}^T,$$

$$f(X(t)) = \begin{pmatrix} rx(t)(1 - \frac{x(t)}{k}) - E(t)x(t) \\ E(t)(px(t) - c) \end{pmatrix}, g(X(t)) = \begin{pmatrix} 1 & 0 \\ 0 & 1 \end{pmatrix},$$

$u(t) \in R^2$ is a control input.

Definition 7.2. [7] If model system (7.7) is regular, and rank$\Xi < n$, then the optimal control problem of model system (7.7) is as follows: Find a control law $u^*(t)$ subject to the minimum value of the following function index,

$$\Omega = \int_0^{\infty} u^2(t)\mathrm{d}t, \tag{7.8}$$

and guarantee that model system (7.7) is stable and impulse free.

By using the following transformation,

$$\tilde{x}(t) = x(t) - x^*, \tilde{E}(t) = E(t) - E^*,$$

where $x^* = \frac{c}{p}, E^* = r(1-\frac{c}{pk})$, model system (7.2) can be rewritten as follows:

$$\begin{cases} \dot{\tilde{x}}(t) = -\frac{rc}{pk}\tilde{x}(t) - \frac{c}{p}\tilde{E}(t) - \frac{r}{k}\tilde{x}^2(t) - \tilde{x}(t)\tilde{E}(t), \\ 0 = rp(1-\frac{c}{pk})\tilde{x}(t) + p\tilde{x}(t)\tilde{E}(t). \end{cases} \tag{7.9}$$

By virtue of the above transformations, the interior equilibrium of model system (7.2) $P^*(\frac{c}{p}, r(1-\frac{c}{pk}))$ is transformed as the interior equilibrium of model system (7.9) $\tilde{P}^*(0,0)$, and the optimal control problem of (7.7) can be transformed as follows:

$$\Xi\dot{\tilde{x}}(t) = \tilde{f}(\tilde{X}(t)) + \tilde{g}(\tilde{X}(t))\tilde{u}(t), \tag{7.10}$$

where

$$\Xi = \begin{pmatrix} 1 & 0 \\ 0 & 0 \end{pmatrix}, \tilde{x}(t) = \begin{pmatrix} \tilde{x}(t) & \tilde{E}(t) \end{pmatrix}^T,$$

$$\tilde{f}(\tilde{X}(t)) = \begin{pmatrix} -\frac{rc}{pk}\tilde{x}(t) - \frac{c}{p}\tilde{E}(t) - \frac{r}{k}\tilde{x}^2(t) - \tilde{x}(t)\tilde{E}(t) \\ rp(1-\frac{c}{p})\tilde{x}(t) + p\tilde{x}\tilde{E}(t) \end{pmatrix}, g(X(t)) = \begin{pmatrix} 1 & 0 \\ 0 & 1 \end{pmatrix},$$

$\tilde{u}(t) \in R^2$ is the control input, which guarantees the minimum value of the following system performance specification

$$\tilde{\Omega} = \int_0^{\infty} \tilde{u}^2(t)\mathrm{d}t. \tag{7.11}$$

Select the following state feedback controller,

$$\tilde{u}(t) = \tilde{H}\tilde{X}(t) + \tilde{v}(t), \tilde{H} = \begin{pmatrix} 0 & 0 \\ 0 & h \end{pmatrix}, \tilde{v}(t) = \begin{pmatrix} \tilde{v}_1(t) & 0 \end{pmatrix}^T.$$

Since the singularity induced bifurcation and impulse phenomenon occur in the neighborhood of interior equilibrium $\tilde{P}_2^*(0,0)$, the linearized system of model system (7.6) around interior equilibrium $\tilde{P}_2^*$ is as follows:

$$\Xi\dot{\tilde{x}}(t) = A_0\tilde{X}(t) + B_0\tilde{u}(t), \tag{7.12}$$

where

$$\Xi = \begin{pmatrix} 1 & 0 \\ 0 & 0 \end{pmatrix}, A_0 = \begin{pmatrix} -\frac{rc}{pk} & -\frac{c}{p} \\ \frac{r(pk-c)}{k} & 0 \end{pmatrix}, B_0 = \begin{pmatrix} 1 & 0 \\ 0 & 1 \end{pmatrix}.$$

It is obvious that model system (7.12) is impulse-controllable [4].

By substituting $\tilde{u}(t) = \tilde{H}\tilde{X}(t) + \tilde{v}(t)$ into model system (7.12), it can be obtained that

$$\Xi\dot{\tilde{x}}(t) = \tilde{A}_0\tilde{X}(t) + \tilde{B}_0\tilde{v}_1(t), \tag{7.13}$$

where

$$\Xi = \begin{pmatrix} 1 & 0 \\ 0 & 0 \end{pmatrix}, \tilde{A}_0 = \begin{pmatrix} -\dfrac{rc}{pk} & -\dfrac{c}{p} \\ \dfrac{r(pk-c)}{k} & h \end{pmatrix}, \tilde{B}_0 = \begin{pmatrix} 1 \\ 0 \end{pmatrix}.$$

It can be shown that the Kronecker index of model system (7.13) is 1, and then the optimal control problem of model system (7.12) is transformed to discuss the control law $\tilde{v}(t)$, which minimizes the following system performance specification:

$$\tilde{\Omega} = \int_0^{\infty} \tilde{u}^2(t)\mathrm{d}t = \int_0^{\infty} [\tilde{v}_1^2(t) + h^2\tilde{E}^2(t)]\mathrm{d}t. \tag{7.14}$$

By utilizing quadratic form theory of singular system, the minimum control law subject to model system (7.12) can be obtained as follows:

$$\tilde{v}_1^*(t) = -\tilde{B}_0^T M\tilde{X}(t),$$

where M represents an admissible solution of the following singular Ricatti equation,

$$\tilde{A}_0^T M + M^T\tilde{A}_0 - M^T\tilde{B}_0\tilde{B}_0^T M + \begin{pmatrix} 0 \\ h \end{pmatrix} \begin{pmatrix} 0 & h \end{pmatrix} = 0,$$

$$\Xi^T M = M^T\Xi \geq 0.$$

By solving the above equation, it can be obtained that

$$M = \begin{pmatrix} \bar{M} & 0 \\ \dfrac{c}{pk}\bar{M} + \dfrac{rp}{h^2}(1-\dfrac{c}{pk}) - \dfrac{rp}{2h}(1-\dfrac{c}{pk}) & -\dfrac{h}{2} \end{pmatrix},$$

where

$$\bar{M} = \frac{rc}{h}(1-\frac{c}{pk}) - \frac{rc}{pk} + \sqrt{[\frac{rc}{h}(1-\frac{c}{pk}) - \frac{rc}{pk}]^2 + r^2p^2(1-\frac{c}{pk})^2}.$$

It follows that

$$\tilde{v}_1^*(t) = -\bar{M}\tilde{x}(t). \tag{7.15}$$

By substituting (7.15) into $\tilde{u}(t) = \tilde{H}\tilde{X}(t) + \tilde{v}(t)$, the optimal control law of model system (7.12) can be obtained,

$$\tilde{u}^*(t) = \begin{pmatrix} \bar{M} & 0 \\ 0 & h \end{pmatrix} \begin{pmatrix} \tilde{x}(t) \\ \tilde{E}(t) \end{pmatrix} \tag{7.16}$$

and then the optimal trajectory satisfies the following simultaneous equations:

$$\begin{aligned} \dot{\tilde{x}}_{opt}^*(t) &= -(\tfrac{rc}{pk} + \bar{M})\tilde{x}_{opt}^*(t) - \tfrac{c}{p}\tilde{E}_{opt}^*(t), \\ 0 &= rp(1-\tfrac{c}{pk})\tilde{x}_{opt}^*(t) + h\tilde{E}_{opt}^*(t). \end{aligned} \tag{7.17}$$

Based on the above analysis, the local optimal control law subject to (7.3) can be derived as follows,

$$u^*(t) = \begin{pmatrix} -\bar{M} & 0 \\ 0 & h \end{pmatrix} \begin{pmatrix} x(t) - x^* \\ E(t) - E^* \end{pmatrix}, \tag{7.18}$$

and then the optimal trajectory satisfies the following simultaneous equations:

$$\begin{aligned} \dot{x}^*_{opt}(t) &= -(\tfrac{rc}{pk} + \bar{M})x^*_{opt}(t) - \tfrac{c}{p}(E^*_{opt}(t) - E^*), \\ 0 &= rp(1 - \tfrac{c}{pk})\tilde{x}^*_{opt}(t) + h(\tilde{E}^*_{opt}(t) - E^*). \end{aligned} \tag{7.19}$$

The optimal system performance specification takes the following form:

$$\Omega^* = \bar{M}[x^*_{opt}(0) - x^*]^2. \tag{7.20}$$

Based on the above analysis, it can be concluded that local optimal control $u^*(t)$ can be designed based on (7.18), which minimizes system performance specification (7.8) for model system (7.7) around interior equilibrium $P^*(\dfrac{c}{p}, r(1 - \dfrac{c}{pk}))$.

Remark 7.2. Based on the above analysis, the local optimal control $u^*(t)$ can be designed based on (7.18), which minimizes the function performance (7.8) of model system (7.2) around the interior equilibrium $P^*(\dfrac{c}{p}, r(1 - \dfrac{c}{pk}))$, and then optimal track X^*_{opt} can be derived according to (7.19), where $h > 0$.

Remark 7.3. It follows from (7.20) that the optimal performance specification increases with the decrease of h. However, the mediating transient time for achieving a stable state increases with the increase of h, which can eliminate the impulse phenomenon as well as stabilize the model system.

7.4 Numerical Simulation

With the help of practical data from Pacific plaice, numerical simulation is carried out to show the feasibility of optimal control designed in the previous section of this chapter. Parameter values used in this section are from a research report about Pacific plaice by H.S. Mohring in 1973 [3], $r = 0.71, K = 80.5 \times 10^6$, and $c : p = 17.5 \times 10^6$. Let $p = 0.005$, and then $c = 8.75 \times 10^4$. By using these parameter values, it can be obtained that the interior equilibrium of model system (7.2) is $P^*(17.5 \times 10^6, 0.5557)$, which derives the optimal law

$$u^*(t) = \begin{pmatrix} -[(\frac{48620}{h} - \frac{71}{460}) + \sqrt{(\frac{48620}{h} - \frac{71}{460})^2 + \frac{7^2}{2500^2}}] & 0 \\ 0 & h \end{pmatrix} \begin{pmatrix} x(t) - 17.5 \times 10^6 \\ E(t) - 0.5557 \end{pmatrix}.$$

By utilizing the above optimal controllers, model system (7.7) can be stabilized around P^* and there is no impulse phenomenon in the neighborhood of interior equilibrium P^*, which can be shown in Fig. 7.1 and Fig. 7.2. For model system (7.13) with controller, the mediating transient time for achieving a stable state increases with the increase of h. Compared with Fig. 5.3 where $h = 20$, the model

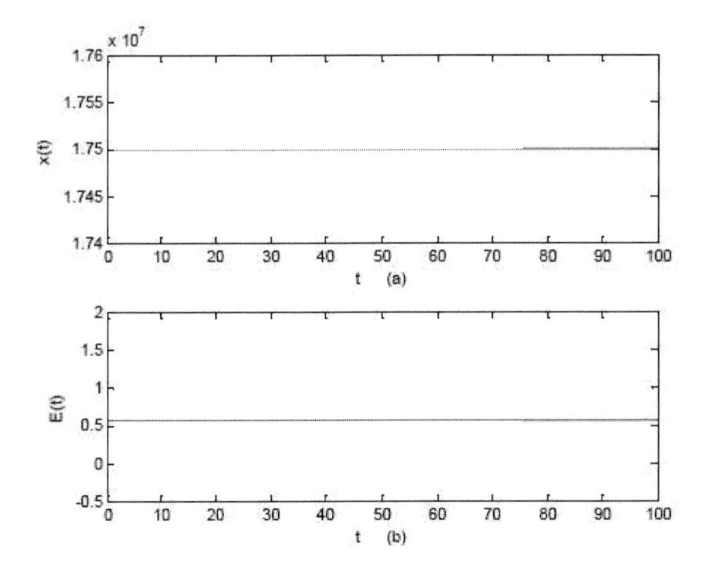

Fig. 7.1 Dynamical response of $x(t)$ and $E(t)$ of model system (7.13) as $h = 3$.

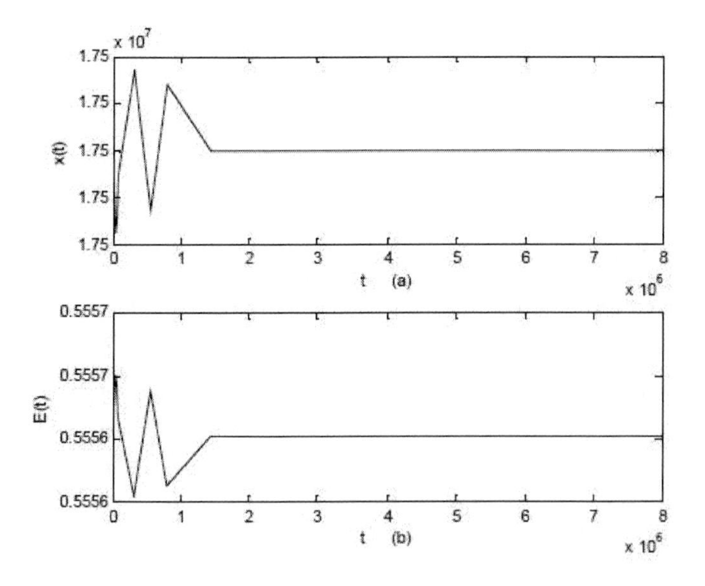

Fig. 7.2 Dynamical response of $x(t)$ and $E(t)$ of model system (7.13) as $h = 20$.

system can be stabilized into an ideal state quickly as $h = 3$ and the mediating transient time is short, which can be shown in Fig. 7.1.

References

1. Beardmore, R.E.: The singularity-induced bifurcation and its Kronecker normal form. SIAM J. Matr. Ana. Appl. 23(1), 126–137 (2001)
2. Bohner, M., Fan, M., Zhang, J.M.: Existence of periodic solutions in predator-prey and competition dynamic systems. Nonl. Anal. 7, 1193–1204 (2006)
3. Clark, C.W.: Mathematical Bioeconomics: The Optimal Management of Renewable Resource, 2nd edn. John Wiley and Sons, New York (1990)
4. Dai, L.: Singular Control System. Springer, New York (1989)
5. Freedman, H., Rao, V.S.H.: The trade-off between mutual interference and time lags in predator-prey systems. Bull. Math. Biol. 45, 991–1004 (1983)
6. Gordon, H.S.: The economic theory of a common property resource: The fishery. J. Polit. Econ. 62(2), 124–142 (1954)
7. Hale, J.K.: Theory of Functional Differential Equations. Springer, New York (1997)
8. Kooij, R.E., Zegeling, A.: A predator-prey model with Ivlev's functional response. J. Math. Anal. Appl. 198, 473–489 (1996)
9. Kot, M.: Elements of Mathematical Biology. Cambridge University Press, Cambridge (2001)
10. Sugie, J.: Two-parameter Bifurcation in a predator-prey system of Ivlev type. J. Math. Anal. Appli. 217, 349–371 (1998)
11. Venkatasubramanian, V., Schaettler, H., Zaborszky, J.: Local bifurcations and feasibility regions in differential-algebraic systems. IEEE Trans. Auto. Contr. 40(12), 1992–2013 (1995)
12. Verhulst, P.F.: Notice sur la loi que la population poursuit dans son accroissement. Corr. Math. Mati. Phy. 10, 113–121 (1838)
13. Xu, C.J., Wang, M.J.: Permanence for a delayed discrete three-level food-chain model with Beddington-DeAngelis functional response. Appl. Math. Comp. 187, 1109–1119 (2007)
14. Xu, R., Chaplain, M.A.J., Davidson, F.A.: Periodic solutions for a predator-prey model with Holling-type functional response and time delays. Appl. Math. Comp. 161, 637–654 (2005)
15. Yang, L.J., Tang, Y.: An improved version of the singularity induced-bifurcation theorem. IEEE Tran. Auto. Contr. 49(6), 1483–1486 (2001)
16. Zhang, J.M., Wang, J.: Periodic solutions for discrete predator-prey systems with the Beddington-DeAngelis functional response. Appl. Math. Lett. 19, 1361–1366 (2006)

Chapter 8
Bifurcations of a Singular Predator-Prey Model with Holling-II Functional Response

8.1 Introduction

At present, humankind is facing the problems of a shortage of resources and a worsening environment. So there has been rapidly growing interest in the analysis and modelling of biological systems. From the view of human needs, the exploitation of biological resources and harvest of populations are commonly practiced in the fields of fishery, wildlife, and forestry management. T.K. Kar and U.K. Pahari [6], Dongmei Xiao et al. [12], and S. Kumar [7] have studied the dynamics in predator-prey models with harvesting and obtained complex dynamic behavior, such as stability of equilibrium points, Bogdanov-Takens bifurcation, Hopf bifurcation, limit cycle, heteroclinic bifurcation, and so on. References [3, 10, 5, 13] have discussed the predator-prey model with time delay and analyzed the effects of time delay on model dynamics, such as the time delay may change the stability of equilibrium points and even cause a switching of stabilities. In addition, there is also a considerable literature on discrete dynamical systems, e.g., see [9, 4, 8, 14] and references therein, modeling some species whose generations are non-overlapping by applying a Euler scheme to differential equations. Most of these discussions on biological models are based on normal systems governed by differential equations or difference equations.

In daily life, economic profit is a very important factor for governments, merchants, and even every citizen, so it is necessary to research biological economic systems, which are always described by differential-algebraic equations. However, to our best knowledge, the reports of such systems are few in biology. Considering the economic theory of fishery resource proposed by Gordon in 1954 [1], this chapter studies a class of predator-prey singular biological economic models with a Holling type II functional form as follows:

$$\begin{cases} \dfrac{\mathrm{d}\tilde{x}}{\mathrm{d}\tilde{t}} = \tilde{x}(r_1 - a_{11}\tilde{x} - \dfrac{\tilde{y}}{a+\tilde{x}}), \\ \dfrac{\mathrm{d}\tilde{y}}{\mathrm{d}\tilde{t}} = \tilde{y}(-r_2 + \dfrac{b\tilde{x}}{a+\tilde{x}} - a_{22}\tilde{y}) - \tilde{E}\tilde{y}, \\ 0 = \tilde{E}(\tilde{p}\tilde{y} - \tilde{c}) - m, \end{cases} \tag{8.1}$$

Q. Zhang et al.: Complex., Analy. and Cntrl. of Singular Biological Systems, LNCIS 421, pp. 113–125.
springerlink.com

where $\tilde{x}$ and $\tilde{y}$ represent the prey density and predator density at time $\tilde{t}$, respectively; the Holling type II function response given by $\frac{b\tilde{x}}{a+\tilde{x}}$ represents the attack rate per predator; $r_1 > 0$ and $-r_2 < 0$ are the intrinsic growth rate and intrinsic mortality rate of respective species; $a_{ii} > 0, i = 1,2$ represents the strength of respective intraspecific competition; $\tilde{E}$ represents capture capability; $\tilde{E}\tilde{y}$ indicates that the harvest of predator is proportional to its density at time $\tilde{t}$; and $\tilde{p} > 0$, $\tilde{c} > 0$ and $m > 0$ are the harvest reward, the cost, and the economic profit, respectively.

In this chapter, we assume reasonable biological conditions as follows:

$$b > r_2, \quad r_1 > aa_{11}. \tag{8.2}$$

In order to model the predator's gestation period, we will replace $\tilde{x}$ in the predator's equation by

$$\tilde{z}(\tilde{t}) = \int_{-\infty}^{\tilde{t}} \tilde{x}(\tau)\frac{1}{\tilde{T}}\exp(-\frac{1}{\tilde{T}}(\tilde{t}-\tau))\mathrm{d}\tau,$$

where the entire past history of $\tilde{x}$ is taken into account and the importance of individual events is measured by $\frac{1}{\tilde{T}}\exp(\frac{1}{\tilde{T}}(\tilde{t}-\tau))$, where $\tilde{T}$ is the measure of the influence of the past, $\tau < \tilde{t}$ is some particular time in the past, and $\tilde{t}$ is the current time. Therefore, we will study the following singular system:

$$\begin{cases} \frac{\mathrm{d}\tilde{x}}{\mathrm{d}\tilde{t}} = \tilde{x}(r_1 - a_{11}\tilde{x} - \frac{\tilde{y}}{a+\tilde{x}}), \\ \frac{\mathrm{d}\tilde{y}}{\mathrm{d}\tilde{t}} = \tilde{y}(-r_2 + \frac{b\tilde{z}}{a+\tilde{z}} - a_{22}\tilde{y}) - \tilde{E}\tilde{y}, \\ \frac{\mathrm{d}\tilde{z}}{\mathrm{d}\tilde{t}} = \frac{1}{\tilde{T}}(\tilde{x}-\tilde{z}), \\ 0 = \tilde{E}(\tilde{p}\tilde{y} - \tilde{c}) - m. \end{cases} \tag{8.3}$$

We nondimensionalize system (8.3) with the following scaling

$$x = \frac{\tilde{x}}{a}, \quad y = \frac{\tilde{y}}{ar_1}, \quad z = \frac{\tilde{z}}{a}, \quad E = \frac{\tilde{E}}{r_1}, \quad t = r_1\tilde{t},$$

and then obtain the form

$$\begin{cases} \frac{\mathrm{d}x}{\mathrm{d}t} = x(1 - a_1 x - \frac{y}{1+x}), \\ \frac{\mathrm{d}y}{\mathrm{d}t} = y(-r + \frac{b_1 z}{1+z} - a_2 y - E), \\ \frac{\mathrm{d}z}{\mathrm{d}t} = \frac{1}{T}(x-z), \\ 0 = E(py - c) - m, \end{cases} \tag{8.4}$$

where the non-dimensional parameters are defined as

$$a_1 = \frac{a_{11}a}{r_1}, \quad a_2 = aa_{22}, \quad b_1 = \frac{b}{r_1},$$
$$c = \tilde{c}r_1, \quad r = \frac{r_2}{r_1}, \quad p = \tilde{p}ar_1^2, \quad T = r_1\tilde{T}.$$

Now, the reasonable conditions (8.2) become

$$b_1 > r, \quad a_1 < 1. \tag{8.5}$$

For simplicity, let

$$f(X,E,\mu) = \begin{pmatrix} f_1(X,E,\mu) \\ f_2(X,E,\mu) \\ f_3(X,E,\mu) \end{pmatrix} = \begin{pmatrix} x(1 - a_1x - \frac{y}{1+x}) \\ y(-r + \frac{b_1z}{1+z} - a_2y - E) \\ \frac{1}{T}(x-z) \end{pmatrix},$$

$$g(X,E,\mu) = E(py - c) - m,$$

where $X = (x,y,z)^T$, μ is a bifurcation parameter, which will be defined in what follows.

In this chapter, we mainly discuss the effects of time delay and economic profit on the model dynamics in the region $R_+^3 = \{(x,y,z,E) | x \geq 0, y \geq 0, z \geq 0, E \geq 0\}$. In Sect. 8.2, under the condition of zero economic profit, we study the existence and stability of equilibrium points and give sufficient conditions of existence for transcritical bifurcation and singular induced bifurcation. Considering positive economic profit, Section 8.3 discusses Hopf Bifurcation and proves that the increase of time delay destabilizes the system and bifurcates into small amplitude periodic solution. In Sect. 8.4, numerical simulations verify the effectiveness of mathematical conclusions and existence of saddle-node bifurcation.

8.2 The Model with Zero Economic Profit

When the economic profit is zero, system (8.4) can be written by

$$\begin{cases} \frac{dx}{dt} = x(1 - a_1x - \frac{y}{1+x}), \\ \frac{dy}{dt} = y(-r + \frac{b_1z}{1+z} - a_2y - E), \\ \frac{dz}{dt} = \frac{1}{T}(x-z), \\ 0 = E(py - c). \end{cases} \tag{8.6}$$

By the analysis of roots for system (8.6), we obtain the following theorem:

Theorem 8.1. *(1) System (8.6) has two equilibria* $P_0(0,0,0,0)$ *and* $P_1(\frac{1}{a_1},0,\frac{1}{a_1},0)$ *for any positive parameters.*
(2) If $b_1 > r(1+a_1)$*, there exists another equilibrium point* $P_2(\bar{x},\bar{y},\bar{z},0)$*, where* $\bar{x}$ *is the intersection of the following null-clines:*

$$y = (1+x)(1-a_1x),$$

$$y = \frac{1}{a_2}(-r+\frac{b_1x}{1+x}),$$

and $\bar{y} = (1+\bar{x})(1-a_1\bar{x})$, $\bar{z} = \bar{x}$.
(3) (i) If $\Delta_1 > 0$ *and* $\frac{b_1x_3^{(1)}}{1+x_3^{(1)}} - r - \frac{a_2c}{p} > 0$*, there exist two positive equilibrium points* $P_3^{(1)}(x_3^{(1)},y^*,z_3^{(1)},E_3^{(1)})$ *and* $P_3^{(2)}(x_3^{(2)},y^*,z_3^{(2)},E_3^{(2)})$*;*
(ii) If $\Delta_1 > 0$, $\frac{b_1x_3^{(1)}}{1+x_3^{(1)}} - r - \frac{a_2c}{p} < 0$*, and* $\frac{b_1x_3^{(2)}}{1+x_3^{(2)}} - r - \frac{a_2c}{p} > 0$*, there exists a unique positive equilibrium point* $P_3(x_3^{(2)},y^*,z_3^{(2)},E_3^{(2)})$*. In both cases (i) and (ii),* $\Delta_1 = p^2(1-a_1)^2 - 4pa_1(c-p)$*, and* $x_3^{(i)}, i = 1,2$ *are roots of the following equation*

$$pa_1x^2 + p(a_1-1)x + c - p = 0$$

and satisfy $x_3^{(1)} < x_3^{(2)}$, $y^* = \frac{c}{p}$, $z_3^{(i)} = x_3^{(i)}$, $E_3^{(i)} = \frac{b_1x_3^{(i)}}{1+x_3^{(i)}} - r - \frac{a_2c}{p}$.
(iii) If $\Delta_1 = 0$*, and* $\frac{b_1(1-a_1)}{1+a_1} - r - \frac{a_2c}{p} > 0$*, the unique positive equilibrium point is* $P_3(\frac{1-a_1}{2a_1},\frac{c}{p},\frac{1-a_1}{2a_1},\frac{b_1(1-a_1)}{1+a_1} - r - \frac{a_2c}{p})$.

From system (8.6), we get

$$A = D_Xf - D_Ef(D_Eg)^{-1}D_Xg = \begin{pmatrix} A_{11} & -\frac{x}{1+x} & 0 \\ 0 & A_{22}+\frac{pEy}{py-c} & \frac{b_1y}{(1+z)^2} \\ \frac{1}{T} & 0 & -\frac{1}{T} \end{pmatrix},$$

where

$$A_{11} = 1 - 2a_1x - \frac{y}{(1+x)^2},$$

$$A_{22} = -r + \frac{b_1z}{1+z} - 2a_2y - E.$$

The characteristic polynomial of the matrix A is

$$|\lambda I - A| = \lambda^3 + p_1(X,E)\lambda^2 + p_2(X,E)\lambda + p_3(X,E) = 0,$$

where

$$p_1(X,E) = r - 1 + 2a_1x + \frac{y}{(1+x)^2} + \frac{1}{T} - \frac{b_1 z}{1+z} + 2a_2y + E - \frac{pEy}{py-c},$$

$$\begin{aligned} p_2(X,E) &= (r - \frac{b_1 z}{1+z} + 2a_2y + E - \frac{pEy}{py-c})(-1 + 2a_1x + \frac{y}{(1+x)^2} + \frac{1}{T}) \\ &\quad + \frac{1}{T}(-1 + 2a_1x + \frac{y}{(1+x)^2}) \end{aligned},$$

$$\begin{aligned} p_3(X,E) &= \frac{1}{T}(r - \frac{b_1 z}{1+z} + 2a_2y + E - \frac{pEy}{py-c}) \cdot (2a_1x + \frac{y}{(1+x)^2} - 1) \\ &\quad + \frac{b_1xy}{T(1+x)(1+z)^2}. \end{aligned}$$

So we obtain easily that the equilibrium point P_0 is a saddle for any positive parameters and if $b_1 < r(1+a_1)$, the equilibrium point P_1 is a stable focus or node; otherwise, P_1 is a saddle. For equilibrium points P_2 and P_3, if they satisfy $p_1 > 0$ and $p_1p_2 > p_3 > 0$, they are a stable focus or node.

From the analysis above, we can see that the stability of equilibrium point P_1 changes from stable to unstable when b_1 increases through $r(1+a_1)$. Therefore, if b_1 is regarded as the bifurcation parameter, i.e., $\mu = b_1$, the following result is obtained:

Theorem 8.2. *If $a_2(1+a_1)^2 + ra_1^2 \neq 0$, system (8.6) undergoes transcritical bifurcation at the equilibrium point P_1 when bifurcation parameter μ is increased through $r(1+a_1)$.*

Proof. When the bifurcation parameter $\mu = b_1 = r(1+a_1)$,

$$(D_X f_R)_{P_1} = A_{P_1} = \begin{pmatrix} -1 & -\frac{1}{1+a_1} & 0 \\ 0 & 0 & 0 \\ \frac{1}{T} & 0 & -\frac{1}{T} \end{pmatrix}$$

has a simple zero eigenvalue with right eigenvector $\omega_2 = (-1, 1+a_1, -1)^T$ and left eigenvector $\omega_1 = (0,1,0)$. And

$$\begin{aligned} &(\omega_1(D_X D_\mu f_R)\omega_2)_{P_1} \\ &= (0 \quad 1 \quad 0)\begin{pmatrix} 0 & 0 & 0 \\ 0 & \frac{z}{1+z} & \frac{y}{(1+z)^2} \\ 0 & 0 & 0 \end{pmatrix}\begin{pmatrix} -1 \\ 1+a_1 \\ -1 \end{pmatrix} = 1, \end{aligned}$$

$$
\begin{aligned}
\left(\omega_1 D_X^2 f_R(\omega_2,\omega_2)\right)_{P_1} &= \left(\omega_1 \Sigma_{i=1}^3 (e_i \omega_2^T D_X (D_X f_i)^T \omega_2)\right)_{P_1} \\
&= -2a_2(1+a_1)^2 - 2ra_1^2,
\end{aligned}
$$

where e_i is the unit vector.

According to the literature [2], system (8.6) undergoes transcritical bifurcation at the equilibrium point P_1. This completes the proof.

When the economic profit $m=0$, $\det D_E g = 0$ at the equilibrium point P_3. In succession, we discuss the bifurcation behavior regarding m as the bifurcation parameter, i.e., $\mu = m$.

Theorem 8.3. *Assume* $-1+\dfrac{2c}{p(1+a_1)} < 0$ *and* $-a_1+\dfrac{y^*}{(1+x^*)^2} \neq 0$. *When the bifurcation parameter* μ *increases through* 0, *system (8.6) undergoes singular induced bifurcation at the equilibrium point* P_3 *and the stability of the equilibrium point* P_3 *changes, i.e., from stable to unstable.*

Proof. Let $\Delta = D_E g = py - c$. It has a simple zero eigenvalue and

$$
\begin{aligned}
\begin{vmatrix} D_X f & D_E f \\ D_X g & D_E g \end{vmatrix}_{P_3} &= \begin{vmatrix} A_{11} & -\dfrac{x}{1+x} & 0 & 0 \\ 0 & A_{22} & \dfrac{b_1 y}{(1+z)^2} & -y \\ \dfrac{1}{T} & 0 & -\dfrac{1}{T} & 0 \\ 0 & pE & 0 & py-c \end{vmatrix}_{P_3} \\
&= -\frac{p}{T}E^* x^* y^* \left(-a_1 + \frac{y^*}{(1+x^*)^2}\right),
\end{aligned}
$$

$$
\begin{aligned}
\begin{vmatrix} D_X f & D_E f & D_\mu f \\ D_X g & D_E g & D_\mu g \\ D_X \Delta & D_E \Delta & D_\mu \Delta \end{vmatrix}_{P_3} &= \begin{vmatrix} A_{11} & -\dfrac{x}{1+x} & 0 & 0 & 0 \\ 0 & A_{22} & \dfrac{b_1 y}{(1+z)^2} & -y & 0 \\ \dfrac{1}{T} & 0 & -\dfrac{1}{T} & 0 & 0 \\ 0 & pE & 0 & py-c & -1 \\ 0 & p & 0 & 0 & 0 \end{vmatrix}_{P_3} \\
&= -\frac{p}{T}x^* y^* \left(-a_1 + \frac{y^*}{(1+x^*)^2}\right),
\end{aligned}
$$

$$
\begin{aligned}
\operatorname{trace}(D_E f \operatorname{adj}(D_E g) D_X g)_{P_3} &= \operatorname{trace}\left(\begin{pmatrix} 0 \\ -y \\ 0 \end{pmatrix} \begin{pmatrix} 0 & pE & 0 \end{pmatrix}\right)_{P_3} \\
&= -pE^* y^*.
\end{aligned}
$$

On the other hand, we get

$$c_1 = -\text{trace}(D_E f \text{adj}(D_E g) D_X g)_{P_3} = pE^* y^* > 0,$$

$$\begin{aligned} c_2 &= \left(D_\mu \Delta - (D_X \Delta \quad D_E \Delta) \begin{pmatrix} D_X f & D_E f \\ D_X g & D_E g \end{pmatrix}^{-1} \begin{pmatrix} D_\mu f \\ D_\mu g \end{pmatrix} \right)_{P_3} \\ &= \frac{1}{E^*} > 0. \end{aligned}$$

According to Theorem 3 in reference [11], system (8.6) undergoes singular induced bifurcation at the equilibrium point P_3 when the bifurcation parameter $\mu = m = 0$. When μ increases through 0, one eigenvalue of system (8.6) moves from C^- to C^+ along the real axis by diverging through ∞. It brings impulse, i.e., rapid expansion of the population in biological explanation. By a simple calculation, the others remain bounded and stay at C^- away from the origin. Therefore, the stability of system (8.6) changes from stable to unstable at the equilibrium point P_3 when the economic profit increases through 0. This completes the proof.

8.3 The Model with Positive Economic Profit

In this section, we investigate how time delay influences the stability of system (8.4). For simplicity, we discuss system (8.4) without intraspecific competition of the predator, i.e., $a_2 = 0$.

8.3.1 System (8.4) without Time Delay

We consider system (8.4) without time delay, which can be written as follows:

$$\begin{cases} \dfrac{dx}{dt} = x(1 - a_1 x - \dfrac{y}{1+x}), \\ \dfrac{dy}{dt} = y(-r + \dfrac{b_1 x}{1+x} - E), \\ 0 = E(py - c) - m. \end{cases} \tag{8.7}$$

If the equilibrium point of system (8.7) is $\tilde{P}_4(x,y,E)$, the coordinate of x is the root of the following equation:

$$f(x) = Ax^3 + Bx^2 + Cx + D = 0, \tag{8.8}$$

where
$A = pa_1(b_1 - r)$,

$B = -p(b_1 - r - a_1 b_1 + 2a_1 r)$,
$C = m - pb_1 + 2pr + cb_1 - cr - a_1 rp$,
$D = r(p-c) + m$.

By the analysis of the roots for Eq. (8.8), it follows if there exists $m = m^* > 0$ satisfying $B^2 > 3AC$ and $f(\frac{-B+\sqrt{B^2-3AC}}{3A}) = 0$, the number of equilibrium point changes from two to zero when m increases through m^*, i.e., there are two positive equilibrium points when $m < m^*$, no positive equilibrium points when $m > m^*$, and only one positive equilibrium point when $m = m^*$. It is difficult to compute analytically the value m^*, hence, we will study saddle-node bifurcation numerically in the next section.

For the equilibrium point $\tilde{P}_4$, we can get

$$A = \begin{pmatrix} -a_1 x + \dfrac{xy}{(1+x)^2} & -\dfrac{x}{1+x} \\ \dfrac{b_1 y}{(1+x)^2} & \dfrac{pEy}{py-c} \end{pmatrix}.$$

Its characteristic polynomial is

$$\left|\lambda I - A\right| = \lambda^2 - (r_1 + r_4)\lambda + r_1 r_4 - r_2 r_3 = 0,$$

where $r_1 = -a_1 x + \frac{xy}{(1+x)^2}$, $r_2 = -\frac{x}{1+x}$, $r_3 = \frac{b_1 y}{(1+x)^2}$, and $r_4 = \frac{pEy}{py-c}$.

Assume

$$r_1 + r_4 < 0 \quad \text{and} \quad r_1 r_4 - r_2 r_3 > 0, \tag{8.9}$$

then $\tilde{P}_4$ is an asymptotically stable equilibrium point of the system (8.7).

8.3.2 Dynamical Analysis of System (8.4)

Assume that the positive equilibrium point of system (8.4) is as follows, $P_4(x,y,z,E)$, where $z = x$ and other coordinates are the same as those of $\tilde{P}_4$; then the characteristic polynomial, at the equilibrium point P_4, is

$$\left|\lambda I - A\right| = \lambda^3 + m_1(x,y,z,E)\lambda^2 + m_2(x,y,z,E)\lambda + m_3(x,y,z,E) = 0,$$

where

$$m_1(x,y,z,E) = a_1 x - \frac{xy}{(1+x)^2} + \frac{1}{T} - \frac{pEy}{py-c}, \tag{8.10}$$

$$m_2(x,y,z,E) = -\frac{pEy}{T(py-c)} + x(a_1 - \frac{y}{(1+x)^2})(-\frac{pEy}{py-c} + \frac{1}{T}), \tag{8.11}$$

$$m_3(x,y,z,E) = -\frac{pExy}{T(py-c)}(a_1 - \frac{y}{(1+x)^2}) + \frac{b_1 xy}{T(1+x)^3}. \tag{8.12}$$

If condition (8.9), $r_1 r_4 < 0$ and $T < \dfrac{r_1 + r_4}{r_1 r_4}$ hold, we obtain $m_i > 0, i = 1,2,3$.

Furthermore,

$$m_1 m_2 - m_3 = \frac{1}{T^2}(n_1 T^2 + n_2 T + n_3),$$

where $n_1 = -r_1 r_4 (r_1 + r_4)$, $n_2 = (r_1 + r_4)^2 + r_2 r_3$, and $n_3 = -(r_1 + r_4)$.

Let

$$T^* = \frac{-n_2 - \sqrt{n_2^2 - 4n_1 n_3}}{2n_1}.$$

Then, according to the Routh-Hurwitz criterion, the equilibrium point P_4 is stable when $T < T^*$; when $T > T^*$, the equilibrium point P_4 is unstable.

Theorem 8.4. *As time delay T is increased through T^*, system (8.4) undergoes a Hopf bifurcation at equilibrium point P_4 if condition (8.9), $r_1 r_4 < 0$ and $T < \dfrac{r_1 + r_4}{r_1 r_4}$ hold, i.e., for $T > T^*$ and $\left| T - T^* \right| \ll 1$, an attracting invariant closed curve bifurcates from the equilibrium point P_4.*

Proof. For $T = T^*$, we have the following eigenvalues:

$$\lambda_1 = \sqrt{m_2} i, \bar{\lambda}_1 = -\sqrt{m_2} i, \lambda_2 = -m_1.$$

Assume

$$\lambda_1 = \alpha_1(T) + i\beta_1(T), \bar{\lambda}_1 = \alpha_1(T) - i\beta_1(T), \lambda_2 = \alpha_2(T),$$

then the transversality condition is

$$\left(\frac{\mathrm{d}\alpha_1(T)}{\mathrm{d}T} \right)_{T^*} = \left(\frac{m_3' - m_2' m_1 - m_2 m_1'}{2(m_1^2 + m_2)} \right)_{T^*}.$$

From Eqs. (8.10), (8.11), and (8.12), we get

$$\left(\frac{\mathrm{d}\alpha_1(T)}{\mathrm{d}T} \right)_{T^*} = \frac{(r_1 + r_4)(r_1 r_4 T^{*2} - 1)}{2T^*[(1 - (r_1 + r_4)T^*)^2 + (r_1 r_4 - r_2 r_3)T^*]} > 0.$$

Thus, system (8.4) undergoes a Hopf bifurcation at equilibrium point P_4. Moreover, for $T > T^*$ and $\left| T - T^* \right| \ll 1$, an attracting invariant closed curve bifurcates from the equilibrium point P_4. This completes the proof.

8.4 Numerical Simulation

To illustrate the results obtained, let us consider the following particular cases.

Case I. For the parameter values

$$a_1 = 0.8, \quad a_2 = 1, \quad b_1 = 6, \quad p = 7, \quad c = 1, \quad r = 2, \quad T = 0.2, \quad m = 0,$$

system (8.4) has a positive equilibrium point $P_1^*(1.1676, 0.1429, 1.1676, 1.0891)$. When economic profit $m = -0.001$, the eigenvalues are $-0.9, -5, -2721.8$, and then become $-0.899, -5, 3631.3$ when the parameter value $m = 0.001$. It is obvious that two eigenvalues remain almost constant and another moves from C^- to C^+ along the real axis by diverging through ∞.

Case II. Take $a_1 = 0.8, a_2 = 0, b_1 = 6, p = 7, c = 1, r = 2, m = 0.5$, for example; then the bifurcation value of the time delay is $T^* = 0.1494$.

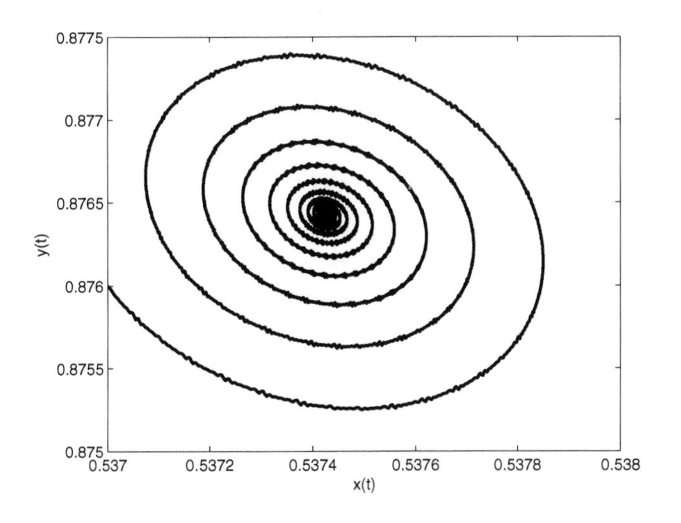

Fig. 8.1 When $T = 0.1 < T^*$, the equilibrium point P_2^* is stable.

Fig. 8.1 illustrates that $P_2^*(0.5374, 0.8764, 0.5374, 0.0974)$ of system (8.4) is a stable equilibrium point when the time delay $T < T^*$. When T passes through the critical value T^*, the equilibrium point P_2^* loses its stability and a Hopf bifurcation occurs. The bifurcating period solution from P_2^* at T^* is stable, which is depicted in Fig. 8.2 and Fig. 8.3

When $m = 0.2$ and other parameters remain invariable, both the prey-predator populations converge to their equilibrium values (see Fig. 8.4). This indicates that the economic profit has an effect of stabilizing the equilibrium point of the predator-prey system.

Case III. We discuss saddle-node bifurcation by numerical simulation. When $r = 2.723, T = 0.2, m = m^* = 0.5$ and the others are the same as those of Case II, there is a unique positive equilibrium point $P_3^*(0.9995, 0.4008, 0.9995, 0.2772)$. By numerical simulations, we can obtain that there are two positive equilibrium points as follows

$$P_3^{(1)}(0.9247, 0.5009, 0.9247, 0.1596)$$

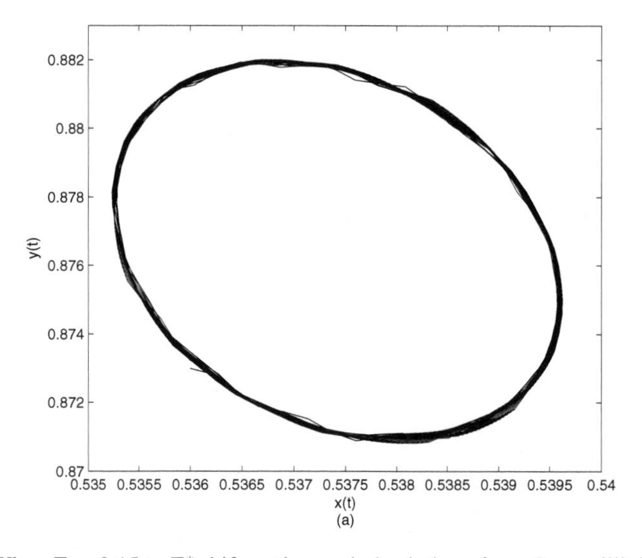

Fig. 8.2 When $T = 0.15 > T^*$, bifurcation period solutions from the equilibrium point P_2^* occur, and (a) is phase trajectories.

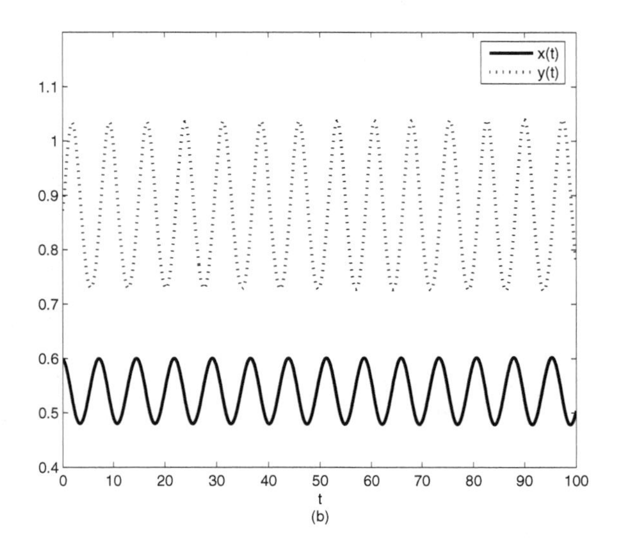

Fig. 8.3 When $T = 0.15 > T^*$, bifurcation period solutions from the equilibrium point P_2^* occur, and (b) is the time response.

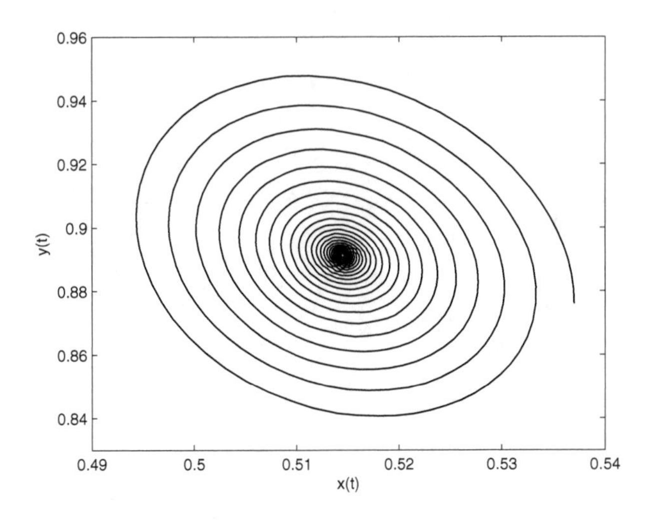

Fig. 8.4 Phase trajectories for $m = 0.2$; remaining other parameters are invariable.

$$P_3^{(2)}(1.0744, 0.2914, 1.0744, 0.3847)$$

when $m = 0.4$ and no positive equilibrium points when $m > m^*$. Furthermore,

$$\left(D_X f - D_E f (D_E g)^{-1} D_X g\right)_{P_3^*} = \begin{pmatrix} -0.6993 & -0.5001 & 0 \\ 0 & 0.4302 & 0.6015 \\ 5 & 0 & -5 \end{pmatrix},$$

which has a geometrically simple zero eigenvalue with the right eigenvector $v = (0.5028, -0.7032, 0.5028)^T$ and the left eigenvector $u = (1, 1.1625, 0.1399)$. And

$$\left(u D_\mu f_R\right)_{P_3^*} = \left(u(D_\mu f - D_E f (D_E g)^{-1} D_\mu g)\right)_{P_3^*} = -0.258,$$

$$\left(u D_X^2 f_R(v, v)\right)_{P_3^*} = \left(u \textstyle\sum_{i=1}^3 (e_i v^T D_X (D_X f_i)^T v)\right)_{P_3^*} = 1.4606,$$

where $e_i, i = 1, 2, 3$ is the unit vector.

According to the literature [11], system (8.6) undergoes saddle-node bifurcation at the equilibrium point P_3^*.

8.5 Conclusion

Consider that the predator growth rate depends on past quantities of prey in an exponentially decreasing way. This chapter discusses a class of predator-prey singular biological economic models with Holling type II functional form, which is a differential-algebraic system. The algebraic equation is inspired by the economic theory of fishery property resources proposed by Gordon [1]. It shows that this model exhibits two bifurcation phenomena when the economic profit is zero. One is transcritical bifurcation, which changes the stability of the system, and the other is singular induced bifurcation, which indicates that zero economic profit brings impulse, i.e., rapid expansion of the population in biological explanation. On the other hand, under the condition of positive economic profit, the combination of time delay and economic profit may change the stability of the system or cause the population and capture capability to oscillate periodically. At the same time, the system undergoes a saddle-node bifurcation if the value of economic profit goes through a critical value of positive economic profit.

References

1. Gordon, H.S.: The economic theory of a common property resource: The fishery. J. Polit. Econ. 62(2), 124–142 (1954)
2. Guckenheimer, J., Holmes, P.: Nonlinear Oscillations, Dynamical Systems and Bifurcations of Vector Field. Springer, New York (1983)
3. Huang, C.X., He, Y.G., Huang, L.H., Yuan, Z.H.: Hopf bifurcation analysis of two neurons with three delays. Nonlinear Anal.-Real 8(3), 903–921 (2007)
4. Jing, Z.J., Yang, J.P.: Bifurcation and chaos in discrete-time predator-prey system. Chaos Soliton. Fract. 27(1), 259–277 (2006)
5. Kar, T.K., Pahari, U.K.: Non-selective harvesting in prey-predator models with delay. Commun. Nonlinear Sci. 11(4), 499–509 (2006)
6. Kar, T.K., Pahari, U.K.: Modeling and analysis of a prey-predator system with stage-structure and harvesting. Nonlinear Anal.-Real 8(2), 601–609 (2007)
7. Kumar, S., Srivastava, S.K., Chingakham, P.: Hopf bifurcation and stability analysis in a harvested one-predator-two-prey model. Appl. Math. Comput. 129, 107–118 (2002)
8. Liao, X.Y., Ouyang, Z.G., Zhou, S.F.: Permanence and stability of equilibrium for a two-prey one-predator discrete model. Appl. Math. Comput. 186(1), 93–100 (2007)
9. Summers, D., Cranford, J.G., Healey, B.P.: Chaos in periodically forced discrete-time ecosystem models. Chaos Soliton. Fract. 11(14), 2331–2342 (2000)
10. Sun, C.J., Lin, Y.P., Han, M.A.: Stability and Hopf bifurcation for an epidemic disease model with delay. Chaos Soliton. Fract. 30(1), 204–216 (2006)
11. Venkatasubramanian, V., Schaettler, H., Zaborszky, J.: Local bifurcations and feasibility regions in differential-algebraic systems. IEEE T. Automat. Contr. 40(12), 1992–2013 (1995)
12. Xiao, D.M., Li, W.X., Han, M.A.: Dynamics in ratio-dependent predator-prey model with predator harvesting. J. Math. Anal. Appl. 324(1), 14–29 (2006)
13. Zhang, C.R., Zu, Y.G., Zheng, B.D.: Stability and bifurcation of a discrete red blood cell survival model. Chaos Soliton. Fract. 28(2), 386–394 (2006)
14. Zhang, Y., Zhang, Q.L., Zhao, L.C., Yang, C.Y.: Dynamical behaviors and chaos control in a discrete functional response model. Chaos Soliton. Fract. 34(4), 1318–1327 (2007)

Chapter 9
A Singular Bioeconomic Model with Diffusion and Time Delay

9.1 Introduction

At present, differential-algebraic equations have been studied widely in power systems, neural networks, and genetic networks [1, 10, 9, 3, 11]. A lot of analytic methods have been presented, such as local stability, optimal control, singularity induced bifurcation, feasibility regions, and so on. However, to our best knowledge, such reports of singular biological systems are few.

Recently, it is noted that some works [5, 13, 4, 2] have been focused on the effects of diffusion and obtained some sufficient conditions for permanence and stability of a dynamical system. In fact, the natural ecological environment has been separated into many patches due to the development of the economy and the construction of highways and railways. In order to survive, the biological population always diffuses from a patch with high density to another with low density. Hence, it is necessary to model the biological system with the dispersal process. This chapter proposes the following prey-predator singular bioeconomic model with diffusion and time delay in a habitat of two identical patches linked by diffusion:

$$\begin{cases} \dot{x}_1 = x_1(1 - x_1(t-\tau)) - \dfrac{a_0 x_1 y_1}{1+ax_1} + D_1(x_2 - x_1), \\ \dot{y}_1 = y_1(-d + \dfrac{b_0 x_1(t-\tau)}{1+ax_1(t-\tau)}) - E_1 y_1 + D_2(y_2 - y_1), \\ 0 = E_1(py_1 - c) - m, \\ \dot{x}_2 = x_2(1 - x_2(t-\tau)) - \dfrac{a_0 x_2 y_2}{1+ax_2} + D_1(x_1 - x_2), \\ \dot{y}_2 = y_2(-d + \dfrac{b_0 x_2(t-\tau)}{1+ax_2(t-\tau)}) - E_2 y_2 + D_2(y_1 - y_2), \\ 0 = E_2(py_2 - c) - m, \end{cases} \tag{9.1}$$

where x_i and y_i, $i = 1,2$ represent the prey density and predator density in the ith patch at time t, respectively. E_i, $i = 1,2$ is the harvesting effort for predator in the

Q. Zhang et al.: Complex., Analy. and Cntrl. of Singular Biological Systems, LNCIS 421, pp. 127–142.
springerlink.com

ith patch. a_0, a, d, b_0, τ, D_1, D_2, p, c, m are model parameters. These parameters are defined as follows: $d > 0$ is the intrinsic mortality rate of predator; $a_0 > 0$ is the capture rate of the predator; $b_0 > 0$ is the rate of conversion of nutrients into the reproduction rate of the predator; $a > 0$ is interpreted as a constant handling time for each prey captured; the predator takes time $\tau > 0$ to convert the food into its growth; $D_i > 0$, $i = 1,2$ represents the dispersal rate in the ith patch; and $p > 0$, $c > 0$, and $m \geq 0$ represent harvesting reward per unit harvesting effort for unit weight of the predator, harvesting cost per unit harvesting effort, and the economic profit per unit harvesting effort, respectively.

For the sake of simplicity, let

$$f_i(X_i, E_i, \mu) = \begin{pmatrix} x_i(1 - x_i(t-\tau)) - \dfrac{a_0 x_i y_i}{1 + a x_i} \\ y_i(-d + \dfrac{b_0 x_i(t-\tau)}{1 + a x_i(t-\tau)}) - E_i y_i \end{pmatrix},$$

$$g_i(X_i, E_i, \mu) = E_i(p y_i - c) - m,$$

where $X_i = (x_i, y_i)^T$, μ is a bifurcation parameter, which will be defined in what follows.

9.2 Model (9.1) without Diffusion

In the absence of diffusion, system (9.1) is separated into two independent and similar subsystems. Thus, we discuss the following subsystem:

$$\begin{cases} \dot{x}_1 = x_1(1 - x_1(t-\tau)) - \dfrac{a_0 x_1 y_1}{1 + a x_1}, \\ \dot{y}_1 = y_1(-d + \dfrac{b_0 x_1(t-\tau)}{1 + a x_1(t-\tau)}) - E_1 y_1, \\ 0 = E_1(p y_1 - c) - m. \end{cases} \tag{9.2}$$

9.2.1 Model (9.2) with Zero Economic Profit

By the analysis of roots for system (9.2), we obtain the following theorem:

Theorem 9.1. *When economic profit is zero, i.e.,* $m = 0$,
(1) *system (9.2) has three boundary equilibrium points, which are* $P_1(0,0,0)$, $P_2(1,0,0)$, *and* $P_3(\dfrac{d}{b_0 - ad}, \dfrac{b_0(b_0 - d - ad)}{a_0(b_0 - ad)^2}, 0)$ *for* $b_0 - d - ad > 0$.
(2) *the positive equilibrium point is* $P_4(x_1^{(4)}, y_1^{(4)}, E_1^{(4)})$, *where* $y_1^{(4)} = \dfrac{c}{p}$, $E_1^{(4)} =$

$\dfrac{b_0 x_1^{(4)}}{1+ax_1^{(4)}} - d$ *and* $x_1^{(4)}$ *is the root of the following equation:*

$$pax_1^2 + p(1-a)x_1 + a_0 c - p = 0.$$

In fact, Sunita Gakkhar et al. [6] have investigated the dynamic behavior of system (9.2) without harvesting. Thus, we obtain easily that the equilibrium point P_1 is a saddle for any positive parameter; the equilibrium point P_2 is a stable focus or node if $b_0 - d - ad < 0$ and $\tau < \dfrac{\pi}{2}$; Hopf bifurcation with respect to the equilibrium point P_3 occurs at the following critical values of time delay

$$\tau_{cr} = \frac{1}{\omega_{cr}}[\arccos(\frac{(\omega_{cr}^2 - A_0)B_0}{B_0^2 + \omega^2 B_1^2}) + 2k\pi], \quad k = 0, 1, 2, \cdots,$$

where $A_0 = -\dfrac{ad(b_0 - d - ad)}{b_0(b_0 - ad)}$, $B_1 = \dfrac{d}{b_0 - ad}$, $B_0 = \dfrac{d}{b_0}(b_0 - d - ad)$, and ω_{cr}^2 is the positive root of the following equation:

$$v^2 + (-2A_0 - B_1^2)v + (A_0^2 - B_0^2) = 0.$$

In what follows, we mainly discuss other bifurcation behavior of system (9.2) with zero economic profit.

The Jacobian matrix of the system (9.2) can be written as

$$A_1 = D_{X_1} f_1 - D_{E_1} f_1 (D_{E_1} g_1)^{-1} D_{X_1} g_1 = \begin{pmatrix} a_{11} - x_1 e^{-\lambda\tau} & a_{12} \\ a_{21} e^{-\lambda\tau} & a_{22} \end{pmatrix},$$

where $a_{11} = 1 - x_1 - \dfrac{a_0 y_1}{(1+ax_1)^2}$, $a_{12} = -\dfrac{a_0 x_1}{1+ax_1}$, $a_{21} = \dfrac{b_0 y_1}{(1+ax_1)^2}$, $a_{22} = -d + \dfrac{b_0 x_1}{1+ax_1} - E_1 + \dfrac{pE_1 y_1}{py_1 - c}$.

From the analysis above, we can see that the stability of equilibrium point P_2 becomes unstable when d decreases through $\frac{b_0}{1+a}$ and τ remains in the region $(0, \dfrac{\pi}{2})$. Therefore, if d is regarded as the bifurcation parameter, the following result is obtained:

Theorem 9.2. *If* $\tau < \dfrac{\pi}{2}$, *system (9.2) undergoes transcritical bifurcation at the equilibrium point* P_2 *when bifurcation parameter* d *decreases through* $\frac{b_0}{1+a}$.

Proof. When the bifurcation parameter $d = \dfrac{b_0}{1+a}$, the characteristic polynomial at the equilibrium point P_2 has a simple zero eigenvalue with left null vector $u = (0 \quad 1)$ and right null vector $v = (-a_0 \quad 1+a)^T$. And

$$u(D_X D_d f_R)v|_{P_2} = (0 \quad 1)\begin{pmatrix} 0 & 0 \\ 0 & -1 \end{pmatrix}\begin{pmatrix} -a_0 \\ 1+a \end{pmatrix} = -(1+a),$$

$$uD_X^2 f_R(v,v)|_{P_2} = -\frac{2a_0 b_0}{1+a}.$$

According to the transcritical bifurcation theorem [7], system (9.2) undergoes transcritical bifurcation at the equilibrium point P_2.

Remark 9.1. Transcritical bifurcation implies that the equilibrium point P_2 remains stable if the intrinsic mortality rate of predator d is larger than the critical point $\frac{b_0}{1+a}$, otherwise, the stability of P_2 is lost and the equilibrium point P_3 exists. On the other hand, stable equilibrium point P_2 means that the predator population is to be extinct, which results from larger the intrinsic mortality rate of predator d.

When the economic profit $m = 0$, $\det D_{E_1} g_1 = 0$ at the equilibrium point P_4. In succession, we discuss the bifurcation behavior regarding m as the bifurcation parameter.

Theorem 9.3. *Assume $aa_0 c \neq p(1+ax_1^{(4)})^2$. When bifurcation parameter m increases through zero, the system (9.2) undergoes singularity induced bifurcation at equilibrium point $P_4(x_1^{(4)}, y_1^{(4)}, E_1^{(4)})$ and the stability of equilibrium point P_4 changes, i.e., from stable to unstable.*

Proof. Let $\Delta = D_{E_1} g_1 = py_1 - c$. It has a simple zero eigenvalue at the equilibrium point P_4 and

$$\text{trace}(D_{E_1} f_1 \,\text{adj}(D_{E_1} g_1) D_{X_1} g_1)_{P_4} = -pE_1^{(4)} y_1^{(4)},$$

$$\begin{vmatrix} D_{X_1} f_1 & D_{E_1} f_1 \\ D_{X_1} g_1 & D_{E_1} g_1 \end{vmatrix}_{P_4} = \begin{vmatrix} -x_1^{(4)} + \dfrac{a_0 a x_1^{(4)} y_1^{(4)}}{(1+ax_1^{(4)})^2} & -\dfrac{a_0 x_1^{(4)}}{1+ax_1^{(4)}} & 0 \\ \dfrac{b_0 y_1^{(4)}}{(1+ax_1^{(4)})^2} & 0 & -y_1^{(4)} \\ 0 & pE_1^{(4)} & 0 \end{vmatrix}$$

$$= -pE_1^{(4)} x_1^{(4)} y_1^{(4)} \left(1 - \frac{aa_0 y_1^{(4)}}{(1+ax_1^{(4)})^2}\right),$$

$$\begin{vmatrix} D_{X_1}f_1 & D_{E_1}f_1 & D_mf_1 \\ D_{X_1}g_1 & D_{E_1}g_1 & D_mg_1 \\ D_{X_1}\Delta & D_{E_1}\Delta & D_m\Delta \end{vmatrix}_{P_4}$$

$$= \begin{vmatrix} -x_1^{(4)} + \dfrac{a_0ax_1^{(4)}y_1^{(4)}}{(1+ax_1^{(4)})^2} & -\dfrac{a_0x_1^{(4)}}{1+ax_1^{(4)}} & 0 & 0 \\ \dfrac{b_0y_1^{(4)}}{(1+ax_1^{(4)})^2} & 0 & -y_1^{(4)} & 0 \\ 0 & pE_1^{(4)} & 0 & -1 \\ 0 & p & 0 & 0 \end{vmatrix}$$

$$= -px_1^{(4)}y_1^{(4)}(1-\frac{aa_0y_1^{(4)}}{(1+ax_1^{(4)})^2}).$$

On the other hand, we get

$$\xi_1 = -\mathrm{trace}(D_{E_1}f_1\mathrm{adj}(D_{E_1}g_1)D_{X_1}g_1)_{P_4} = pE_1^{(4)}y_1^{(4)} > 0,$$

$$\xi_2 = \left| D_m\Delta - (D_{X_1}\Delta \quad D_{E_1}\Delta)\begin{pmatrix} D_{X_1}f_1 & D_{E_1}f_1 \\ D_{X_1}g_1 & D_{E_1}g_1 \end{pmatrix}^{-1}\begin{pmatrix} D_mf_1 \\ D_mg_1 \end{pmatrix}\right|_{P_4}$$

$$= \frac{1}{E_1^{(4)}} > 0.$$

According to the singularity induced bifurcation theorem [12], system (9.2) undergoes singularity induced bifurcation at equilibrium point P_4 when the bifurcation parameter $m = 0$. When m increases through zero, one eigenvalue of system (9.2) moves from C^- (the complex left half plane) to C^+ (the complex right half plane) along the real axis by diverging through ∞, which leads to an impulsive phenomenon of a singular system, i.e., rapid expansion of the population in biological explanation. By a simple calculation, the other eigenvalue remains bounded and stays at C^- away from the origin. Therefore, system (9.2) becomes unstable at equilibrium point P_4.

For system (9.2) without time delay, eigenvalues of Jacobian matrix A_1 at equilibrium point P_4 are roots of the following equation:

$$\lambda^2 - (a_{11} - x_1^{(4)} + a_{22})\lambda + (a_{11} - x_1^{(4)})a_{22} - a_{12}a_{21} = 0.$$

Since $a_{22} \to +\infty(-\infty)$ for $m \to 0^+(0^-)$, the matrix A_1 has two eigenvalues:

$$\begin{aligned} \lambda_1 &= \frac{1}{2}[(a_{11} - x_1 + a_{22}) + \sqrt{(a_{11} - x_1 - a_{22})^2 + 4a_{12}a_{21}})] \\ &\to +\infty(\approx a_{11} - x_1) \end{aligned}$$

and

$$\lambda_2 = \frac{1}{2}[(a_{11} - x_1 + a_{22}) - \sqrt{(a_{11} - x_1 - a_{22})^2 + 4a_{12}a_{21}})]$$
$$\approx a_{11} - x_1(\to -\infty).$$

If the time delay is considered, eigenvalues of A_1 at equilibrium point P_4 satisfy

$$\lambda^2 - (a_{11} + a_{22})\lambda + a_{11}a_{22} + e^{-\lambda\tau}(x_1^{(4)}\lambda - x_1^{(4)}a_{22} - a_{12}a_{21}) = 0.$$

We know that it is difficult to compute analytical solutions of the equation above. Hence, we need to obtain numerical solutions using some mathematical software.

Remark 9.2. A singular system can be modeled by a parameter dependent differential-algebraic equation (DAE) of the following form:

$$\begin{cases} \dot{x} = f(x,y,\alpha), & f : R^{n+m+q} \to R^n, \\ 0 = g(x,y,\alpha), & g : R^{n+m+q} \to R^m, \end{cases}$$

where $x \in R^n$ is the vector of dynamic state variables, $y \in R^m$ is the vector of algebraic variables, and $\alpha \in R^q$ is the vector of parameters. Local bifurcation analysis of the singular system often results in three different bifurcation types: saddle-node bifurcation, Hopf bifurcation, and singularity induced bifurcation (SIB). The SIB is a new type of bifurcation and does not occur in an ordinary differential equation system. It occurs when an equilibrium point crosses the following singular surface:

$$S := \{(x,y,\mu)|g(x,y,\mu) = 0, \det[D_y g(x,y,\mu)] = 0\}.$$

At the SIB, the equilibrium point undergoes stability exchanges and one eigenvalue of the system Jacobian matrix $(D_x f - D_y f(D_y g)^{-1} D_x g)$ becomes unbounded. One important implication of the occurrence of SIB is that it causes impulsive phenomena, which might yield catastrophic consequences, for example, hard fault [8] of electrical circuits and rapid expansion of the population for biological systems. More details of SIB can be found in [12].

9.2.2 Model (9.2) with Positive Economic Profit

When economic profit is positive , system (9.2) has positive equilibrium point $P_5(x_1^{(5)}, y_1^{(5)}, E_1^{(5)})$, where coordinate $E_1^{(5)} = \dfrac{m}{py_1^{(5)} - c}$, $y_1^{(5)} = \dfrac{1}{a_0}(1 - x_1^{(5)})(1 + ax_1^{(5)})$ and $x_1^{(5)}$ is the root of the following equation:

$$h(x) = Ax^3 + Bx^2 + Cx + D = 0,$$

where $A = ap(b_0 - ad)$, $B = -p[(a-1)(b_0 - ad) + ad]$, $C = aa_0m + dp(a-1) - (b_0 - ad)(p - a_0c)$, $D = a_0m + d(p - a_0c)$.

The characteristic polynomial of system (9.2) without time delay at equilibrium point P_5 has the following form:

$$Q_1(\lambda) = \lambda^2 - (r_1 + r_4)\lambda + r_1 r_4 - r_2 r_3 = 0,$$

where $r_1 = r_{11} + r_{12}$, $r_{11} = -x_1^{(5)}$, $r_{12} = \frac{a_0 a x_1^{(5)} y_1^{(5)}}{(1+ax_1^{(5)})^2}$, $r_2 = -\frac{a_0 x_1^{(5)}}{1+ax_1^{(5)}}$, $r_3 = \frac{b_0 y_1^{(5)}}{(1+ax_1^{(5)})^2}$, $r_4 = \frac{pE_1^{(5)} y_1^{(5)}}{py_1^{(5)} - c}$.

It is easy to see that P_5 is locally asymptotically stable for system (9.2) if

$$r_1 + r_4 < 0 \quad \text{and} \quad r_1 r_4 - r_2 r_3 > 0. \tag{9.3}$$

The characteristic polynomial of system (9.2) is given by

$$Q_2(\lambda) = \lambda^2 + p_1\lambda + p_2 + (p_3\lambda + p_4)e^{-\lambda\tau} = 0, \tag{9.4}$$

where $p_1 = -r_{12} - r_4$, $p_2 = r_{12} r_4$, $p_3 = -r_{11}$, $p_4 = r_{11} r_4 - r_2 r_3$.

Assume that a purely imaginary solution of the form $\lambda = i\omega$ exists in Eq. (9.4). Substituting it into (9.4) and equating its real and imaginary parts give

$$\begin{aligned} p_2 - \omega^2 + p_4 \cos\omega\tau + p_3\omega \sin\omega\tau = 0, \\ p_1\omega + p_3\omega\cos\omega\tau - p_4 \sin\omega\tau = 0, \end{aligned} \tag{9.5}$$

which leads to

$$\omega^4 + (p_1^2 - 2p_2 - p_3^2)\omega^2 + p_2^2 - p_4^2 = 0. \tag{9.6}$$

From condition (9.3), we have

$$p_1^2 - 2p_2 - p_3^2 < 0, \quad p_2 + p_4 > 0.$$

Therefore, Eq. (9.6) has a positive solution ω_0 if

$$p_2 - p_4 < 0. \tag{9.7}$$

The critical value of the delay corresponding to ω_0 is given by

$$\tau_k = \frac{1}{\omega_0} \arccos \frac{p_4\omega_0^2 - p_2 p_4 - p_1 p_3 \omega_0^2}{p_3^2\omega_0^2 + p_4^2} + \frac{2k\pi}{\omega_0}, \quad k = 0, 1, 2, \cdots.$$

Now, differentiating (9.4) with respect to τ, we obtain

$$\frac{\mathrm{d}\lambda}{\mathrm{d}\tau} = \frac{(p_3\lambda + p_4)\lambda e^{-\lambda\tau}}{2\lambda + p_1 + [p_3 - \tau(p_3\lambda + p_4)]e^{-\lambda\tau}}.$$

Substituting the eigenvalue $i\omega_0$ and noticing that the derivative of Eq. (9.6) at ω_0^2 is positive, it follows that

$$\operatorname{sign}\left(\frac{\mathrm{d}}{\mathrm{d}\tau}\mathrm{Re}(\lambda)\right)_{\tau_k} = \operatorname{sign}\left(\,\omega_0^2(2\omega_0^2+p_1^2-2p_2-p_3^2)\,\right) > 0.$$

Then the following result is obtained:

Theorem 9.4. *Under conditions (9.3) and (9.7), system (9.2) undergoes Hopf bifurcation at the equilibrium point P_5 when $\tau = \tau_k$, $k = 0,1,2,\cdots$. Furthermore, for $\tau > \tau_0$ and $\left|\,\tau - \tau_0\,\right| \ll 1$, an attracting invariant closed curve bifurcates from equilibrium point P_5.*

9.3 Dynamical Analysis of System (9.1)

In this section, we discuss the effects of time delay and diffusion on the stability of equilibrium point $P_5^D(x_1^{(5)},y_1^{(5)},E_1^{(5)},x_2^{(5)},y_2^{(5)},E_2^{(5)})$, where $x_1^{(5)} = x_2^{(5)}$, $y_1^{(5)} = y_2^{(5)}$, and $E_1^{(5)} = E_2^{(5)}$. The characteristic polynomial of system (9.1) at equilibrium point P_5^D is

$$\begin{aligned}
&\left|\lambda I - A_{P_5^D}\right| \\
&= \begin{vmatrix} \lambda + s_{11} & -r_2 & -D_1 & 0 \\ -r_3 e^{-\lambda\tau} & \lambda + s_{22} & 0 & -D_2 \\ -D_1 & 0 & \lambda + s_{33} & -r_2 \\ 0 & -D_2 & -r_3 e^{-\lambda\tau} & \lambda + s_{44} \end{vmatrix} \\
&= Q_2(\lambda)Q_3(\lambda) \\
&= 0,
\end{aligned} \tag{9.8}$$

where

$$Q_3(\lambda) = \lambda^2 + q_1\lambda + q_2 + (q_3\lambda + q_4)e^{-\lambda\tau},$$

with

$s_{11} = -r_{11}e^{-\lambda\tau} - r_{12} + D_1,$
$s_{22} = -r_4 + D_2,$
$s_{33} = -r_{11}e^{-\lambda\tau} - r_{12} + D_1,$
$s_{44} = -r_4 + D_2,$
$q_1 = 2D_1 + 2D_2 - r_{12} - r_4,$
$q_2 = (r_{12} - 2D_1)(r_4 - 2D_2),$
$q_3 = -r_{11},$
$q_4 = -r_2 r_3 + r_{11}(r_4 - 2D_2).$

For system (9.1) without time delay, characteristic polynomial (9.8) becomes

$$Q_1(\lambda)[\lambda^2 + (2D_1 + 2D_2 - r_1 - r_4)\lambda + (r_1 - 2D_1)(r_4 - 2D_2) - r_2 r_3] = 0. \tag{9.9}$$

For the sake of simplicity, let

$$Q_4(\lambda) = \lambda^2 + (2D_1 + 2D_2 - r_1 - r_4)\lambda + (r_1 - 2D_1)(r_4 - 2D_2) - r_2r_3,$$

where $(r_1 - 2D_1)(r_4 - 2D_2) - r_2r_3 = r_1r_4 - r_2r_3 - 2D_2r_1 - 2D_1(r_4 - 2D_2)$.

From (9.3), we have

$$2D_1 + 2D_2 - r_1 - r_4 > 0,$$

$$r_1r_4 - r_2r_3 - 2D_2r_1 > 0.$$

Then the following result is obtained:

Theorem 9.5. *Assume that (9.3) holds; then for system (9.1) without time delay,*

(1) *if the diffusion satisfies $2D_2 > r_4$, equilibrium point P_5^D is stable.*

(2) *if $2D_2 < r_4$, equilibrium point P_5^D becomes unstable for $D_1 > D_1^*$, where $D_1^* = \dfrac{r_1r_4 - r_2r_3 - 2D_2r_1}{2(r_4 - 2D_2)}$.*

In what follows, we discuss the effect of time delay on $Q_3(\lambda) = 0$. Assume that a purely imaginary solution of the form $\lambda = i\omega$ exists in $Q_3(\lambda) = 0$. Substitute it into $Q_3(\lambda) = 0$ and we have

$$\omega^4 + (q_1^2 - 2q_2 - q_3^2)\omega^2 + q_2^2 - q_4^2 = 0, \tag{9.10}$$

where
$q_1^2 - 2q_2 - q_3^2 = (r_{12} - 2D_1)^2 + (r_4 - 2D_2)^2 - r_{11}^2$,
$q_2^2 - q_4^2 = (r_{12} - 2D_1)^2(r_4 - 2D_2)^2 - [r_{11}(r_4 - 2D_2) - r_2r_3]^2$.

If $2D_2 > r_4$ and D_1 satisfies

$$(r_{12} - 2D_1)^2 > \max\{r_{11}^2 - (r_4 - 2D_2)^2,$$

$$r_{11}^2 + \frac{r_2r_3}{(r_4 - 2D_2)^2}[r_2r_3 - 2r_{11}(r_4 - 2D_2)]\},$$

there is no positive root for Eq. (9.10), i.e., the stability of equilibrium point P_5^D depends on $Q_2(\lambda)$. Hence, system (9.1) is stable at equilibrium point P_5^D for $\tau < \tau_0$.

If $2D_2 > r_4$ and $\left| r_{12} - 2D_1 \right| < \dfrac{r_{11}(r_4 - 2D_2) - r_2r_3}{2D_2 - r_4}$, there exists a positive solution ω_D for Eq. (9.6). The critical value of the delay corresponding to ω_D is given by

$$\tau_k^D = \frac{1}{\omega_D}\arccos\frac{q_4\omega_D^2 - q_2q_4 - q_1q_3\omega_D^2}{q_3^2\omega_D^2 + q_4^2} + \frac{2k\pi}{w_D}, \quad k = 0, 1, 2, \cdots.$$

thus,

$$\text{sign}\left(\frac{\mathrm{d}}{\mathrm{d}\tau}\text{Re}(\lambda)\right)_{\tau_k^D} = \text{sign}\left(\omega_D^2(2\omega_D^2 + q_1^2 - 2q_2 - q_3^2)\right) > 0.$$

Theorem 9.6. *Assume that condition (9.3) holds. For $2D_2 > r_4$,*
(1) *if D_1 satisfies*

$$(r_{12}-2D_1)^2 > max\{r_{11}^2-(r_4-2D_2)^2, r_{11}^2+\frac{r_2r_3}{(r_4-2D_2)^2}[r_2r_3-2r_{11}(r_4-2D_2)]\},$$

system (9.1) is stable at the equilibrium point P_5^D for $\tau < \tau_0$.
(2) *if* $|r_{12}-2D_1| < \frac{r_{11}(r_4-2D_2)-r_2r_3}{2D_2-r_4}$, *system (9.1) undergoes Hopf bifurcation at equilibrium point P_5^D when $\tau = \tau_k$, τ_k^D, $k = 0,1,2,\cdots$.*

9.4 Numerical Simulation

In this section, numerical simulations are carried out to investigate singularity induced bifurcation and Hopf bifurcation for the fixed parameters $a = 1.25, a_0 = 5, b_0 = 7.5, d = 2, p = 35$, and $c = 1$.
Case I: Singularity induced bifurcation.

When $m = 0$, i.e., zero economic profit, system (9.2) has a positive equilibrium point $Q_1(0.9341, 0.0286, 1.2317)$. In the neighbor of $m = 0$, two values of economic profit are chosen: For $m = -0.01$, the eigenvalues of system (9.2) at Q_1 are -0.9001 and -150.9777; for $m = 0.01$, the eigenvalues are -0.8971 and 152.6869. It is obvious that the first eigenvalue remains almost constant and the second moves from C^- to C^+ along the real axis by diverging through ∞.
Case II: Hopf bifurcation.

When $m = 0.5$, equilibrium point $Q_2(0.4299, 0.1753, 0.0974)$ is stable for system (9.2) without time delay. Since $p_2 - p_4 = -0.7045 < 0$, there is a positive root $\omega_0 = 0.8922$ for Eq. (9.6). The critical value of the delay corresponding to ω_0 is $\tau_0 = 0.1521$.

For system (9.1) without time delay, a stable equilibrium point can be obtained $Q_3(0.4299, 0.1753, 0.0974, 0.4299, 0.1753, 0.0974)$ if the diffusion satisfies $2D_2 > r_4(= 0.1164)$, which can be seen in Fig. 9.1 and Fig. 9.2 with $D_1 = 3$ and $D_2 = 1$. When $D_1 = 20 > D_1^*(= 10.5781)$ and $D_2 = 0.04$ (satisfying $2D_2 < r_4$), the instability of equilibrium point Q_3 can be seen in Fig. 9.3 and Fig. 9.4, which show that the prey population of two patches stays at a stable state, the predator population of the first patch increases, and the second patch decreases. When $m = 0.2$, the prey population, the predator population, and harvesting effort converge to their equilibrium values, which can be seen in Fig. 9.5 and Fig. 9.6. This indicates that the economic profit has an effect on stabilizing the positive equilibrium point of the prey-predator singular bioeconomic system.

If $D_2 = 1$ and $D_1 > 0.521$, system (9.1) is stable at equilibrium point Q_3 for $\tau < \tau_0(= 0.1521)$. Fig. 9.7 illustrates that a stable periodic solution occurs around equilibrium point Q_3 when $\tau = 0.155$. If $D_2 = 1$ and $D_1 < 0.521$, equilibrium point Q_3 of system (9.1) undergoes Hopf bifurcation at the positive critical values of time delay $\tau_0^D \in (1.4716, 283.9)$.

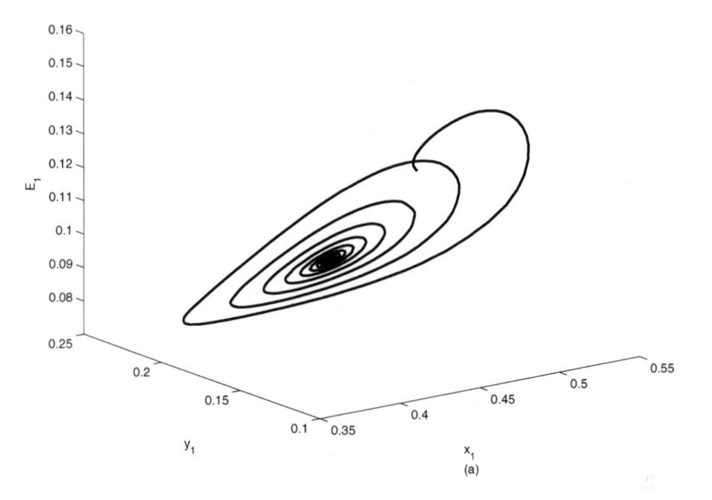

Fig. 9.1 Dynamical behavior of system (9.1) without time delay for $D_1 = 3$ and $D_2 = 1$ in $(x_1 - y_1 - E_1)$ plane.

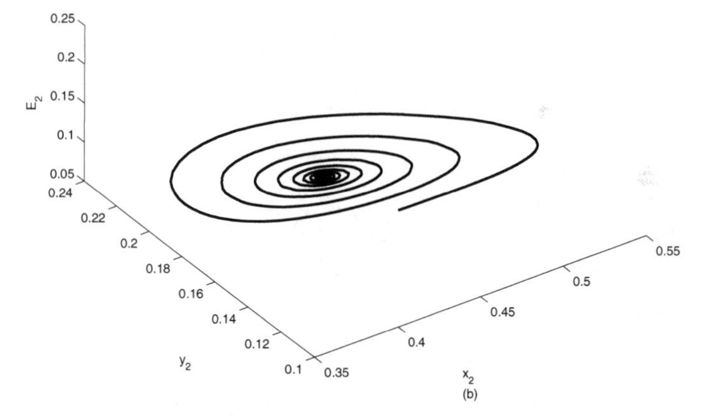

Fig. 9.2 Dynamical behavior of system (9.1) without time delay for $D_1 = 3$ and $D_2 = 1$ in $(x_2 - y_2 - E_2)$ plane.

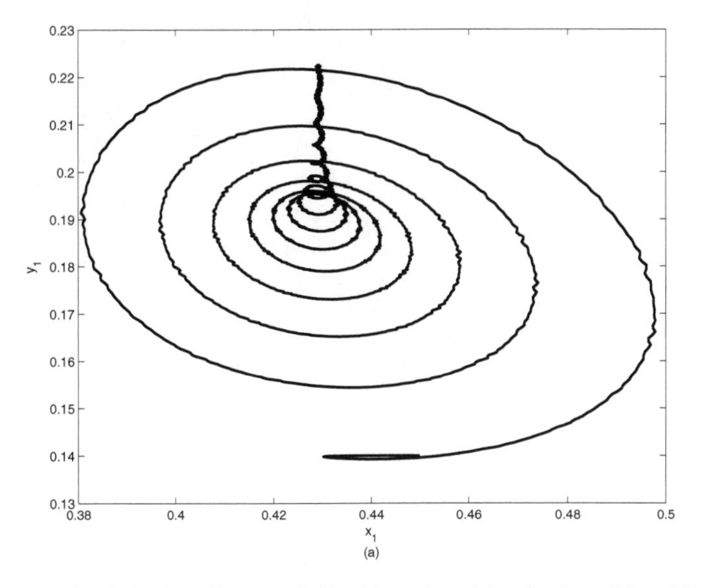

Fig. 9.3 Dynamical behavior of system (9.1) without time delay for $D_1 = 20$ and $D_2 = 0.04$ in $(x_1 - y_1)$ plane.

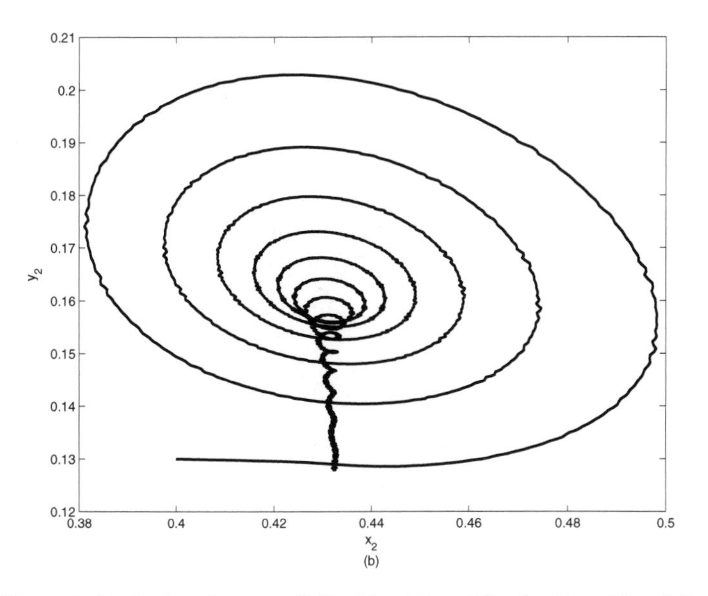

Fig. 9.4 Dynamical behavior of system (9.1) without time delay for $D_1 = 20$ and $D_2 = 0.04$. in $(x_2 - y_2)$ plane.

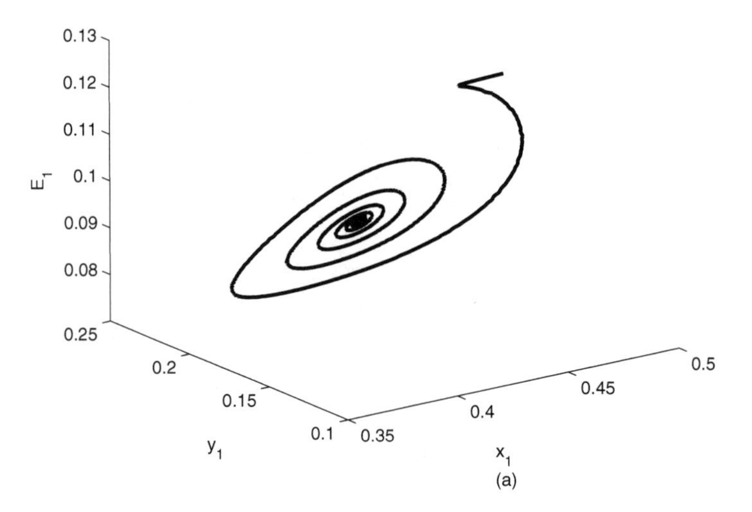

Fig. 9.5 Behavior of the populations for $m = 0.2$; remaining other parameters are the same as those in Fig. 9.3.

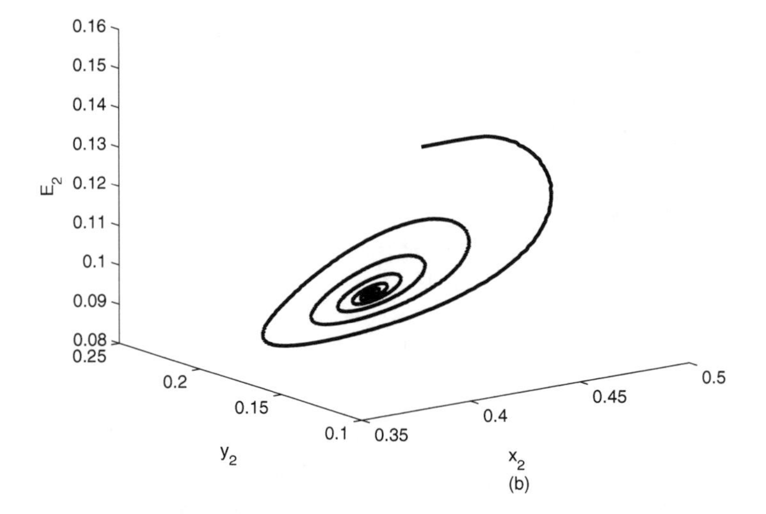

Fig. 9.6 Behavior of the populations for $m = 0.2$; remaining other parameters are the same as those in Fig. 9.4.

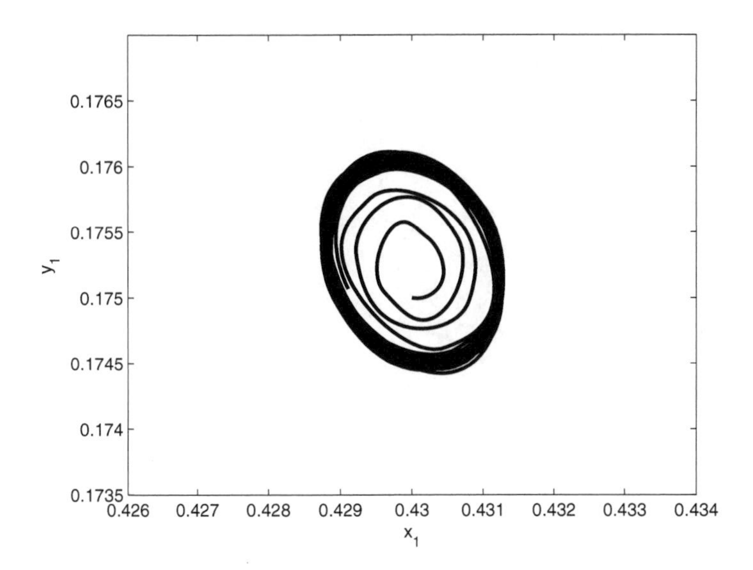

Fig. 9.7 When $\tau = 0.155$, bifurcation period solutions from equilibrium point Q_3 occur.

9.5 Conclusions

First, this chapter studies a delayed prey-predator singular bioeconomic model in the absence of diffusion. Transcritical bifurcation, singularity induced bifurcation, and Hopf bifurcation are obtained due to regarding the intrinsic mortality rate of predator, economic profit, and time delay as different bifurcation parameters, respectively.

Transcritical bifurcation implies that equilibrium point P_2 remains stable if the intrinsic mortality rate of predator d is larger than the critical point $\frac{1}{1+a}$; otherwise, the stability is lost and equilibrium point P_3 exists. In fact, stable P_2 means that the prey population will stay at a positive value and the predator population will be extinct, which is corresponding to the larger intrinsic mortality rate of predator d.

At the singularity induced bifurcation, one eigenvalue of system Jacobian A_1 at equilibrium point P_4 becomes unbounded and the other remains stable. From a biological point of view, singularity induced bifurcation shows that small positive economic profit causes impulse, i.e., rapid expansion of the biological population, which may induce ecosystem unbalance and even biological disaster. In order to avoid this phenomenon, necessary measures must be adopted to regulate the harvesting and economic profit so that the biological population stays at a steady state and the impulsive phenomenon can be eliminated.

It is well known that an individual prey killed does not contribute instantaneously to the predator population in most biological systems. Hence, this chapter studies the effect of gestation delay on dynamical behavior of the prey-predator singular bioeconomic system. It shows that time delay can alter the stability of the biological population and cause the population oscillations, i.e., time delay is responsible for Hopf bifurcation.

Moreover, a delayed prey-predator singular bioeconomic system in two identical patches linked by diffusion is investigated. When only diffusion is present, it is shown that diffusion can make the stability of the positive equilibrium point change and the larger dispersal rate of the predator is in favor of permanence and stability of biological populations. In addition, numerical simulations show that adjusting economic profit may restrain unstable or oscillational phenomena.

When both diffusion and time delay are present, a small amplitude periodic solution occurs at critical values of diffusion and time delay. Compared with the system without diffusion, the one with diffusion may create a new Hopf bifurcation at a different critical value of time delay due to the introduction of diffusion. It means that both time delay and diffusion are responsible for the dynamical behavior of biological populations.

Generally, gestation delay can be regarded as an inherence of biological populations and diffusion is related to their life habit. Thus, it is easier to adjust economic profit than gestation delay and diffusion in order to eliminate bifurcational phenomena and keep the population at a steady state. In fact, some measures adopted, such as adjusting revenue, drawing out favorable policy to encourage or improve fishery, and so on, can influence economic profit directly.

References

1. Ayasun, S., Nwankpa, C.O., Kwatny, H.G.: Computation of singular and singularity induced bifurcation points of differential-algebraic power system model. IEEE T. Circuits-I 51(8), 1525–1537 (2004)
2. Chen, F.D., Xie, X.D.: Permanence and extinction in nonlinear single and multiple species system with diffusion. Appl. Math. Comput. 177(1), 410–426 (2006)
3. Chen, L.N., Aihara, K.: Global searching ability of chaotic neural networks. IEEE T. Circuits-I 46(8), 974–993 (1999)
4. Chen, S.H., Wang, F., Young, T.: Positive periodic solution of two-species ratio-dependent predator-prey system with time delay in two-patch environment. Appl. Math. Comput. 150(3), 737–748 (2004)
5. Ding, X.Q., Wang, F.F.: Positive periodic solution for a semi-ratio-dependent predator-prey system with diffusion and time delays. Nonlinear Anal.-Real 9(2), 239–249 (2008)
6. Gakkhar, S., Sahani, S.K., Negi, K.: Effects of seasonal growth on delayed prey-predator model. Chaos Soliton. Fract. 39(1), 230–239 (2009)
7. Guckenheimer, J., Holmes, P.: Nonlinear Oscillations, Dynamical Systems and Bifurcations of Vector Field. Springer, New York (1983)
8. Li, F., Woo, P.Y.: Fault detection for linear analog IC-The method of short-circuit admittance parameter. IEEE T. Circuits-I 49(1), 105–108 (2002)
9. Marchenko, V.M., Poddubnaya, O.N., Zaczkiewicz, Z.: On the observability of linear differential-algebraic systems with delays. IEEE T. Automat. Contr. 51(8), 1387–1392 (2006)

10. Riaza, R.: Singularity-induced bifurcations in lumped circuits. IEEE T. Circuits-I 52(7), 1442–1450 (2005)
11. Somogyi, R., Sniegoski, C.: Modeling the complexity of genetic networks: Understanding multigenic and pleiotropic regulation. Complexity 1(6), 45–63 (1996)
12. Venkatasubramanian, V., Schaettler, H., Zaborszky, J.: Local bifurcations and feasibility regions in differential-algebraic systems. IEEE T. Automat. Contr. 40(12), 1992–2013 (1995)
13. Zhou, X.Y., Shi, X.Y., Song, X.Y.: Analysis of nonautonomous predator-prey model with nonlinear diffusion and time delay. Appl. Math. Comput. 196(1), 129–136 (2008)

Chapter 10
Bifurcation Analysis and Control of a Class of Hybrid Biological Economic Models

10.1 Introduction

For many real-world power systems with digital controllers, the dynamics of the generators as well as their continuous-time controllers and the load dynamics together define the ordinary differential equations while algebraic equalities are defined by the power balance equations of the transmission networks [4]. On the other hand, the difference equations govern the dynamics of the digital devices. These digital control devices require both continuous and discrete time descriptions that can be formulated as differential-difference-algebraic equations(or a hybrid dynamical system). Besides such industrial systems, models of a biological system such as neural networks [1, 2, 13] and genetic networks [11] can also be described as hybrid dynamical systems. For instance, chaotic dynamics of biological neurons [8] have been implemented by electronic circuit models both with discrete-time [7] and with continuous-time [10]. A hybrid system of chaotic neural networks obtained by combining these hardware neuron models can be described by DDA equations with electrical circuit constraints due to Kirchhoff's laws. Moreover, gene expression processes [5] and genome-proteome networks [9] are also typical models mixed with both continuous and discrete time sequences. Therefore, it is necessary to analyze nonlinear properties of such dynamical systems.

This chapter systematically studies a hybrid predator-prey economic model. First, combining the economic theory of a fishery resources [6] proposed by Gordon in 1954 with the classical Lotka-Volterra model, we can obtain a biological economic system expressed by the differential-algebraic equation

$$\begin{cases} \dfrac{\mathrm{d}\tilde{x}}{\mathrm{d}\tilde{t}} &= r\tilde{x}(1-\dfrac{\tilde{x}}{K}) - a\tilde{x}\tilde{y} - \tilde{E}\tilde{x}, \\ \dfrac{\mathrm{d}\tilde{y}}{\mathrm{d}\tilde{t}} &= -\tilde{d}\tilde{y} + \tilde{b}\tilde{x}\tilde{y} - \tilde{E}\tilde{y}, \\ 0 &= \tilde{E}(\tilde{p}_x\tilde{x} - \tilde{c}_x) + \tilde{E}(\tilde{p}_y\tilde{y} - \tilde{c}_y) - m, \end{cases} \tag{10.1}$$

Q. Zhang et al.: Complex., Analy. and Cntrl. of Singular Biological Systems, LNCIS 421, pp. 143–156.
springerlink.com

where $\tilde{x}$ and $\tilde{y}$ represent the prey density and predator density at time $\tilde{t}$, respectively. $r > 0$, $\tilde{d} > 0$ are the intrinsic growth rate of prey and the death rate of predator in the absence of food, respectively. $K > 0$ is the carrying capacity of prey. $a > 0$ and $\tilde{b} > 0$ measure the effect of the interaction of the two populations. $\tilde{E}$ represents harvesting effort. $\tilde{E}\tilde{x}$ and $\tilde{E}\tilde{y}$ indicate that the harvests for the prey and predator populations are proportional to their densities at time $\tilde{t}$. $\tilde{p}_x$ and $\tilde{p}_y$ are the harvesting rewards per unit weight of the prey and predator populations, respectively. $\tilde{c}_x$ and $\tilde{c}_y$ are the harvesting costs per unit harvesting effort for the prey and predator populations, respectively. m is the economic profit. $\tilde{p}_x, \tilde{p}_y, \tilde{c}_x, \tilde{c}_y, m$ are all positive constants.

We nondimensionalize system (10.1) with the following scaling:

$$x = \frac{\tilde{x}}{K}, y = \frac{a\tilde{y}}{r}, E = \frac{\tilde{E}}{r}, t = r\tilde{t},$$

and then obtain the form

$$\begin{cases} \dfrac{\mathrm{d}x}{\mathrm{d}t} = x(1-x) - xy - Ex, \\ \dfrac{\mathrm{d}y}{\mathrm{d}t} = -dy + bxy - Ey, \\ 0 = E(p_x x - c_x) + E(p_y y - c_y) - m, \end{cases} \tag{10.2}$$

where the non-dimensional parameters are defined as

$$d = \frac{\tilde{d}}{r}, b = \frac{\tilde{b}K}{r}, p_x = rK\tilde{p}_x, p_y = \frac{r^2}{a}\tilde{p}_y, c_x = r\tilde{c}_x, c_y = r\tilde{c}_y.$$

Suppose the predator population has no overlap between successive generations. Applying the Euler method to the predator equation of system (10.2) with integral step size τ, we have the following biological economic system expressed by differential-difference-algebraic equation

$$\begin{cases} \dfrac{\mathrm{d}x}{\mathrm{d}t} = x(1-x) - xy_n - Ex, \\ y_{n+1} = y_n + \tau[-dy_n + bx_n y_n - E_n y_n], \\ 0 = E(p_x x - c_x) + E_n(p_y y_n - c_y) - m, \end{cases} \tag{10.3}$$

where $x_n = x(n\tau)$, $E_n = E(n\tau)$, y_n are the value at the instant $n\tau$ and x_n, y_n, E_n are constantly held during $n\tau \le t < (n+1)\tau$ for the first and third equations of system (10.3).

Remark 10.1. System (10.3) above is called a hybrid dynamical system [3], which is formulated by differential-difference-algebraic equations.

For simplicity, let

$$\begin{pmatrix} f(x(t),x_n,y_n,E(t),E_n,m) \\ g(x_n,y_n,E_n,m) \\ h(x(t),x_n,y_n,E(t),E_n,m) \end{pmatrix} = \begin{pmatrix} x(1-x)-xy_n-Ex \\ y_n+\tau[-dy_n+bx_ny_n-E_ny_n] \\ E(p_xx-c_x)+E_n(p_yy_n-c_y)-m \end{pmatrix}.$$

Without confusion, t is occasionally dropped from the related variables, i.e., $x(t)$ and $E(t)$ are simply expressed by x and E, respectively.

In this chapter, we mainly discuss the effects of economic profit on the dynamics of system (10.3) in the following region: $R_+^5 = \{(x(t),x_n,y_n,E(t),E_n)|x(t)\geq 0, x_n\geq 0, y_n\geq 0, E(t)\geq 0, E_n\geq 0\}$.

10.2 Stability of Equilibrium and Dynamics of System (10.3)

10.2.1 Existence and Stability of Equilibrium

The equilibrium of system (10.3) satisfies the following equations:

$$\begin{cases} x(1-x-y_n-E)=0, \\ y_n+\tau(-dy_n+bx_ny_n-E_ny_n)=y_n, \\ E(p_xx-c_x)+E_n(p_yy_n-c_y)-m=0. \end{cases} \tag{10.4}$$

By the analysis of roots for system (10.4), we obtain the following theorem:

Theorem 10.1. *(1) If $p_x > c_x+c_y$ and $(p_x-c_x-c_y)^2\geq 4p_xm$, system (10.3) has an equilibrium $P_1(x_1,x_{n1},y_{n1},E_1,E_{n1})$, where x_1 satisfies the following equation,*

$$p_xx^2-(p_x+c_x+c_y)x+m+c_x+c_y=0,$$

and $x_1=x_{n1}$, $y_{n1}=0$, $E_1=E_{n1}=1-x_1$.
(2) If harvesting effort $0<E_2<(b-d)/(1+b)$ satisfies the following equation

$$(p_x-p_y-bp_y)E^2+[(p_x-p_y)d+(p_y-c_x-c_y)b]E-bm=0,$$

the system has a positive equilibrium $P_2(x_2,x_{n2},y_{n2},E_2,E_{n2})$, where $x_2=x_{n2}=(E_2+d)/b$, $y_{n2}=1-x_2-E_2$, and $E_{n2}=E_2$.

The Jacobian matrix of system (10.3) is

$$J=\begin{pmatrix} e^{\tau A}+(e^{\tau A}-1)A^{-1}B & (e^{\tau A}-1)A^{-1}C \\ \dfrac{\partial g_R}{\partial x_n} & \dfrac{\partial g_R}{\partial y_n} \end{pmatrix}, \tag{10.5}$$

where

$$
\begin{aligned}
A &= f_x - f_E h_E^{-1} h_x = 1 - 2x - y_n - E + \frac{p_x x E}{p_x x - c_x},\\
B &= f_{x_n} - f_E h_E^{-1}[h_{x_n} - h_{E_n}(h_E + h_{E_n})^{-1}(h_x + h_{x_n})] - f_{E_n}(h_E + h_{E_n})^{-1}(h_x + h_{x_n})\\
&= -\frac{p_x x E(p_y y_n - c_y)}{(p_x x - c_x + p_y y_n - c_y)(p_x x - c_x)},\\
C &= f_{y_n} - f_E h_E^{-1}[h_{y_n} - h_{E_n}(h_E + h_{E_n})^{-1} h_{y_n}] - f_{E_n}(h_E + h_{E_n})^{-1} h_{y_n}\\
&= -x + \frac{p_y E_n x}{p_x x - c_x + p_y y_n - c_y},\\
\frac{\partial g_R}{\partial x_n} &= g_{x_n} - g_{E_n}(h_E + h_{E_n})^{-1}(h_x + h_{x_n}) = b\tau y_n + \frac{\tau p_x E y_n}{p_x x - c_x + p_y y_n - c_y},\\
\frac{\partial g_R}{\partial y_n} &= g_{y_n} - g_{E_n}(h_E + h_{E_n})^{-1} h_{y_n}\\
&= 1 + \tau(-d + b x_n - E_n) + \frac{\tau p_y E_n y_n}{p_x x - c_x + p_y y_n - c_y}.
\end{aligned}
$$

By using the Jacobian matrix J, the local stability of an equilibrium for all instants is obtained by the following theorem:

Theorem 10.2. *Assume that $P^*(x^*, x^*, y^*, E^*, E^*)$ is an equilibrium of system (10.3) and satisfies $(p_x x^* - c_x)(p_x x^* - c_x + p_y y^* - c_y) \neq 0$, $1 - 2x^* - y^* - E^* + p_x x^* E^* / (p_x x^* - c_x) \neq 0$. If at P^*, the following conditions are satisfied,*

(1) $$\left| (e^{\tau A} + (e^{\tau A} - 1)A^{-1}B)\frac{\partial g_R}{\partial y_n} - (e^{\tau A} - 1)A^{-1}C\frac{\partial g_R}{\partial x_n} \right| < 1,$$

(2) $$\left| e^{\tau A} + (e^{\tau A} - 1)A^{-1}B + \frac{\partial g_R}{\partial y_n} \right| < 1 + (e^{\tau A} + (e^{\tau A} - 1)A^{-1}B)\frac{\partial g_R}{\partial y_n} - (e^{\tau A} - 1)A^{-1}C\frac{\partial g_R}{\partial x_n},$$

P^* is an asymptotically stable equilibrium.

10.2.2 Dynamics of System (10.3)

Assume that $P^*(x^*, x^*, y^*, E^*, E^*)$ is an equilibrium of system (10.3) and satisfies $(p_x x^* - c_x)(p_x x^* - c_x + p_y y^* - c_y) \neq 0$. Then, according to the implicit function theorem, there exists a unique C^∞ mapping in open neighborhoods of P^* such that at sampling instants,

$$E_n = \frac{m}{p_x x_n - c_x + p_y y_n - c_y},$$

and at the intersampling instants

$$E = \frac{m(p_x x_n - c_x)}{(p_x x - c_x)(p_x x_n - c_x + p_y y_n - c_y)}.$$

Then, we can obtain the following reduced system:

$$\begin{cases} \dfrac{dx}{dt} &= f_R(x, x_n, y_n) \\ &= x(1-x) - xy_n - \dfrac{mx(p_x x_n - c_x)}{(p_x x - c_x)(p_x x_n - c_x + p_y y_n - c_y)}, \\ y_{n+1} &= g_R(x_n, y_n) = y_n + \tau[-d + bx_n - \dfrac{m}{p_x x_n - c_x + p_y y_n - c_y}]y_n. \end{cases}$$

Therefore, the dynamics of system (10.3) are given as follows:
(1) When $t = 0$,

$$E(0) = E_0 = \frac{m}{p_x x_0 - c_x + p_y y_0 - c_y},$$

for the given x_0 and y_0.
(2) When $n\tau < t < (n+1)\tau$, for $n = 0, 1, 2, \cdots$,

$$\begin{cases} \dfrac{dx}{dt} = x(1-x) - xy_n - \dfrac{mx(p_x x_n - c_x)}{(p_x x - c_x)(p_x x_n - c_x + p_y y_n - c_y)}, \\ E(t) = \dfrac{m(p_x x_n - c_x)}{(p_x x - c_x)(p_x x_n - c_x + p_y y_n - c_y)}. \end{cases} \tag{10.6}$$

(3) When $t = (n+1)\tau$, for $n = 0, 1, 2, \cdots$,

$$\begin{cases} x_{n+1} &= \lim_{t \to (n+1)\tau} x(t), \\ y_{n+1} &= y_n + \tau[-d + bx_n - \dfrac{m}{p_x x_n + p_y y_n - c_x - c_y}]y_n, \\ E_{n+1} &= \dfrac{m}{(p_x x_{n+1} - c_x + p_y y_{n+1} - c_y)}. \end{cases} \tag{10.7}$$

10.3 Local Bifurcations at the Intersampling Instants

10.3.1 Saddle-Node Bifurcation

In the absence of a predator, system (10.3) reduces to the following differential-algebraic system:

$$\begin{cases} \dfrac{dx}{dt} = \tilde{f}(x, E) = x(1-x) - Ex, \\ 0 \quad = \tilde{g}(x, E) = E(p_x x - c_x) - m. \end{cases} \tag{10.8}$$

Theorem 10.3. *If $p_x > c_x$ and $m = m_{sn} = (p_x - c_x)^2/4p_x$, there exists a saddle-node bifurcation for the equilibrium $P_{sn}((p_x + c_x)/2p_x, (p_x - c_x)/2p_x)$ of system (10.8) at the intersampling instants.*

Proof. When $p_x > c_x$ and $m = m_{sn} = (p_x - c_x)^2/4p_x$, system (10.8) has an equilibrium $P_{sn}(p_x c_x/2p_x, (p_x - c_x)/2p_x)$, which satisfies

$$D_x\tilde{f}_R|_{(P_{sn},m_{sn})} = (D_x\tilde{f} - D_E\tilde{f}(D_E\tilde{g})^{-1}D_x\tilde{g})|_{(P_{sn},m_{sn})} = 0,$$

$$(\frac{\partial \tilde{f}_R}{\partial m})|_{(P_{sn},m_{sn})} = (D_m\tilde{f} - D_E\tilde{f}(D_E\tilde{g})^{-1}D_m\tilde{g})|_{(P_{sn},m_{sn})} = -\frac{p_x + c_x}{p_x(p_x - c_x)},$$

$$(\frac{\partial^2 \tilde{f}_R}{\partial x^2})|_{(P_{sn},m_{sn})} = (-2 - \frac{2mc_xp_x}{(p_xx - c_x)^3})|_{(P_{sn},m_{sn})} = -\frac{2(p_x + c_x)}{p_x - c_x}.$$

According to the theory of generic local bifurcation, system (10.8) undergoes saddle-node bifurcation at the equilibria P_{sn} for the intersampling instants. This completes the proof.

From the property of the saddle-node bifurcation, it is clear that the number of equilibrium becomes two from zero when m decreases through $m_{sn} = (p_x - c_x)^2/4p_x$, i.e., two positive equilibria for $m < m_{sn}$, one is stable,

$$P_{sn}^{(1)}(\frac{p_x + c_x + \sqrt{(p_x - c_x)^2 - 4p_xm}}{2p_x}, \frac{p_x - c_x - \sqrt{(p_x - c_x)^2 - 4p_xm}}{2p_x}),$$

and the other,

$$P_{sn}^{(2)}(\frac{p_x + c_x - \sqrt{(p_x - c_x)^2 - 4p_xm}}{2p_x}, \frac{p_x - c_x + \sqrt{(p_x - c_x)^2 - 4p_xm}}{2p_x})$$

is unstable; no equilibria when $m > m_{sn}$; and only one positive equilibrium when $m = m_{sn}$. Thus, if the economic profit $m < m_{sn}$, the prey population and the harvesting effort will stay at steady equilibrium $P_{sn}^{(1)}$ for the single population economic system (10.8), which is favorable to the breeding of the prey population and permanent and constant harvesting for fishermen.

10.3.2 Singularity Induced Bifurcation

When the economic profit $m = m_{sib} = [(p_x - c_x - bc_x + dp_x)p_y - c_yp_x](bc_x - dp_x)/p_x^2$, $P_{sib}(c_x/p_x, c_x/p_x, 1-(b+1)c_x/p_x + d, bc_x/p_x - d, bc_x/p_x - d)$ is the equilibrium of system (10.3) and satisfies

$$h_E|_{P_{sib}} = (p_xx - c_x)|_{P_{sib}} = 0.$$

The following theorem shows that system (10.3) undergoes singularity induced bifurcation at equilibrium P_{sib}.

Theorem 10.4. *If $0 < bc_x - dp_x < p_x - c_x$, system (10.3) undergoes singular induced bifurcation at equilibrium P_{sib} when bifurcation parameter m is increased through m_{sib} at the intersampling instants.*

Proof. Define

$$\delta = h_E = p_x x - c_x.$$

It is easy to calculate that system (10.3) at equilibrium P_{sib} satisfies

$$\delta|_{(P_{sib},m_{sib})} = 0,$$

$$b_E|_{(P_{sib},m_{sib})} = (-\mathrm{trace}(f_E \mathrm{adj}(h_E)h_x))|_{(P_{sib},m_{sib})} = c_x(\frac{bc_x}{p_x} - d),$$

$$\begin{aligned}\begin{vmatrix} f_x & f_E \\ h_x & h_E \end{vmatrix}_{(P_{sib},m_{sib})} &= \begin{vmatrix} 1-2x-y_n-E & -x \\ p_x E & p_x x - c_x \end{vmatrix}_{|(P_{sib},m_{sib})} \\ &= c_x(\frac{bc_x}{p_x} - d),\end{aligned}$$

$$\begin{aligned}\begin{vmatrix} f_x & f_E & f_m \\ h_x & h_E & h_m \\ \delta_x & \delta_E & \delta_m \end{vmatrix}_{(P_{sib},m_{sib})} &= \begin{vmatrix} 1-2x-y_n-E & -x & 0 \\ p_x E & p_x x - c_x & -1 \\ p_x & 0 & 0 \end{vmatrix}_{|(P_{sib},m_{sib})} \\ &= c_x,\end{aligned}$$

$$\begin{aligned}&c|_{(P_{sib},m_{sib})} \\ &= \left(\delta_m - (\delta_x, \delta_E)\begin{pmatrix} f_x & f_E \\ h_x & h_E \end{pmatrix}^{-1}\begin{pmatrix} f_m \\ h_m \end{pmatrix}\right)_{(P_{sib},m_{sib})} \\ &= \left(-(p_x, 0)\begin{pmatrix} 1-2x-y_n-E & -x \\ p_x E & p_x x - c_x \end{pmatrix}^{-1}\begin{pmatrix} 0 \\ -1 \end{pmatrix}\right)_{(P_{sib},m_{sib})} \\ &= \frac{p_x}{bc_x - dp_x}.\end{aligned}$$

Therefore, according to Theorem 3 in [12], system (10.3) undergoes singular induced bifurcation at positive equilibrium P_{sib} when bifurcation parameter m increases through m_{sib}. Because of $b_E/c > 0$, one eigenvalue of A moves from C^- to C^+ along the real axis by diverging through ∞ when m increases through m_{sib}. It brings impulse, i.e., rapid expansion of the population in terms of ecological implication. Therefore, system (10.3) becomes unstable at equilibrium P_{sib} for the intersampling instants. This completes the proof.

10.4 Local Bifurcations for the Sampling Instants

For simplicity, denote

$$J = \begin{pmatrix} a_{11} & a_{12} \\ a_{21} & a_{22} \end{pmatrix},$$

where
$a_{11} = e^{\tau A} + (e^{\tau A} - 1)A^{-1}B,$
$a_{12} = (e^{\tau A} - 1)A^{-1}C,$
$a_{21} = \partial g_R / \partial x_n,$
$a_{22} = \partial g_R / \partial y_n.$

According to Theorem 6 in [10], we obtain the following theorem:

Theorem 10.5. *Under the assumption that there exists an equilibrium of system (10.3) $P_{ns}(x_{ns}, x_{ns}, y_{ns}, E_{ns}, E_{ns})$ that satisfies $(a_{11}a_{22} - a_{12}a_{21})|_{P_{ns}} = 1$, $((a_{11} - a_{22})^2 + 4a_{12}a_{21})|_{P_{ns}} < 0$, $(p_x x_{ns} - c_x)(p_x x_{ns} - c_x + p_y y_{ns} - c_y) \neq 0$ and $1 - 2x_{ns} - y_{ns} - E_{ns} + p_x x_{ns} E_{ns}/(p_x x_{ns} - c_x) \neq 0$, there exists the Neimark-Sacker bifurcation that generates a unique closed invariant curve from equilibrium P_{ns}, if at P_{ns}*

$$\frac{d}{dm}(\sqrt{a_{11}a_{22} - a_{12}a_{21}}) \neq 0,$$

$$(\frac{1}{2}(a_{11} + a_{22}) + i\sqrt{4 - (a_{11} + a_{22})^2})^k \neq 1, \quad for \quad k = 1,2,3,4.$$

However, it is difficult to compute analytically the bifurcation value m_{ns}, hence, we will study Neimark-Sacker bifurcation in numerical simulation.

10.5 State Feedback Control

Unstable fluctuations, bifurcations, and impulsive phenomena have always been regarded as unfavorable ones from the point of view of ecological managers. In order to plan harvesting strategies and keep the sustainable development of a system, it is necessary to take action to stabilize the biological population. This section proposes following the state feedback control method

$$\begin{cases} \frac{dx}{dt} &= x(1-x) - xy_n - Ex, \\ y_{n+1} &= y_n + \tau[-dy_n + bx_n y_n - E_n y_n], \\ 0 &= E(p_x x - c_x) + E_n(p_y y_n - c_y) + L \cdot U_E - m, \end{cases} \tag{10.9}$$

where $L = (l_1, l_2)$ is the feedback gain, $U_E = (E - E^*, E_n - E_n^*)^T$, E^* and E_n^* are the coordinates of the equilibrium of system (10.3).

In fact, the third equation of the feedback control system above can be rewritten as

$$E(p_x x - \hat{c}_x) + E_n(p_y y_n - \hat{c}_y) - \hat{m} = 0,$$

where $\hat{c}_x = c_x - l_1$, $\hat{c}_y = c_y - l_2$, $\hat{m} = m + l_1 E^* + l_2 E_n^*$.

Therefore, the state feedback control method can be implemented by adjusting the harvesting costs and the economic profit so that the biological population can be controlled at steady states and those unfavorable phenomena will be eliminated. In our life, managers can take action, such as adjusting revenue, drawing out favorable policy to encourage fishery, abating pollution, and so on, in order to control the harvesting costs and economic profit.

The Jacobian matrix of the feedback control system is

$$\tilde{J} = \begin{pmatrix} e^{\tau\tilde{A}} + (e^{\tau\tilde{A}} - 1)\tilde{A}^{-1}\tilde{B} & (e^{\tau\tilde{A}} - 1)\tilde{A}^{-1}\tilde{C} \\ \dfrac{\partial \tilde{g}_R}{\partial x_n} & \dfrac{\partial \tilde{g}_R}{\partial y_n} \end{pmatrix}, \tag{10.10}$$

where

$$\tilde{A} = 1 - 2x - y_n - E + \frac{p_x x E}{p_x x - c_x + l_1},$$

$$\tilde{B} = -\frac{p_x x E(p_y y_n - c_y + l_2)}{(p_x x - c_x + p_y y_n - c_y + l_1 + l_2)(p_x x - c_x + l_1)},$$

$$\tilde{C} = -x + \frac{p_y E_n x}{p_x x - c_x + p_y y_n - c_y + l_1 + l_2},$$

$$\frac{\partial \tilde{g}_R}{\partial x_n} = b\tau y_n + \frac{\tau p_x E y_n}{p_x x - c_x + p_y y_n - c_y + l_1 + l_2},$$

$$\frac{\partial \tilde{g}_R}{\partial y_n} = 1 + \tau(-d + bx_n - E_n) + \frac{\tau p_y E_n y_n}{p_x x - c_x + p_y y_n - c_y + l_1 + l_2}.$$

Theorem 10.6. *If the gain matrix $L = (l_1, l_2)$ satisfies that the two eigenvalues of $\tilde{J}$ at equilibrium P^* for system (10.9) have moduli less than 1, P^* is an asymptotically stable equilibrium for all instants.*

10.6 Numerical Simulation

To illustrate the above results obtained, let us consider the following particular cases.
Case I: Saddle-node bifurcation at the intersampling instants.

For the parameter values $p_x = 3.7$, and $c_x = 1$, there exists a unique equilibrium $P_1^*(0.6351, 0.3649)$ for system (10.8) when the economic profit is $m_{sn} = 0.4926$. For $m < m_{sn}$, system (10.8) has two hyperbolic equilibria: take $m = 0.45$ for example, two equilibria are $P_1^{(1)}(0.7424, 0.2576)$ and $P_1^{(2)}(0.5279, 0.4721)$, whose eigenvalues are -0.3373 and 0.4396, respectively, i.e., $P_1^{(1)}$ is stable and $P_1^{(2)}$ is unstable. For $m > m_{sn}$, system (10.8) has no equilibria.
Case II: Singular induced bifurcation at the intersampling instants.

Choose $d = 0.5, b = 3, \tau = 1, p_x = 3.7, c_x = 1, p_y = 4.1, c_y = 1.2, m_{sib} = 0.1609$, then $P_2^*(0.2703, 0.2703, 0.4189, 0.3108, 0.3108)$ is the positive equilibrium of

system (10.3). Due to fixing of $x_n = 0.2703$, $y_n = 0.4189$, $E_n = 0.3108$, the dynamics of hybrid system (10.3) are reined by system (10.6). By simple calculation, we obtain that the equilibrium and eigenvalue of system (10.3) are as follows: $P_2^{(1)}(0.2701, 0.2703, 0.4189, 0.311, 0.3108)$ and -493.6009 for $m = 0.1607$ and then become $P_2^{(2)}(0.2705, 0.2703, 0.4189, 0.3106, 0.3108)$ and 365.4519 for $m = 0.1610$, respectively. It is obvious that the eigenvalue moves from C^- to C^+ along the real axis by diverging through ∞. From the point of biology, this means that economic profit may bring impulse at some critical value, i.e., rapid expansion of the density of the prey population. If the feedback gain is chosen as $L = (0.5, 1)^T$, the eigenvalues become 0.3523 and 0.4776 for $m = 0.1607$ and $m = 0.1610$, respectively. Therefore, the impulsive phenomenon is eliminated by controlling the harvesting effort. That is, the prey population can stay at a steady state.
Case III: Neimark-Sacker bifurcation for the sampling instants.

Take $d = 0.4, b = 2, \tau = 0.1, p_x = 5.7, c_x = 1.5, p_y = 3.5, c_y = 0.9, m_{ns} = 0.1$ for example, then system (10.3) has a unique positive equilibrium $P_3^*(0.2414, 0.2414, 0.6759, 0.0827, 0.0827)$ and eigenvalues are $\lambda_\pm = 0.9986 \pm 0.0522i$. For $m_{ns} = 0.1$, there are $|\lambda_\pm| = 1$ and $d|\lambda(m_{ns})|/dm = 0.1375$. Hence, from the analysis, there exists Neimark-Sacker bifurcation. Fig. 10.1 shows that there exists a unique closed invariant curve from equilibrium P_3^*. If the feedback gain is chosen as $L = (3, 5)^T$, the controlled system states stay at P_3^* after a short transient process, which is depicted in Fig. 10.2.

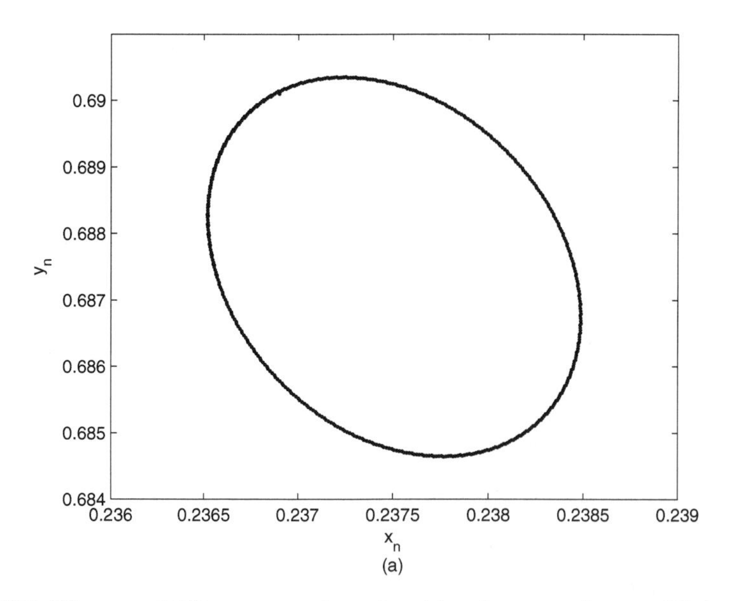

Fig. 10.1 When $m = 0.102 > m_{ns}$, a unique closed invariant curve from equilibrium P_3^* occurs, where (a) $x_n - y_n$ plane, (b) $n - x_n$ plane, (c) $n - y_n$ plane, (d) $n - E_n$ plane.

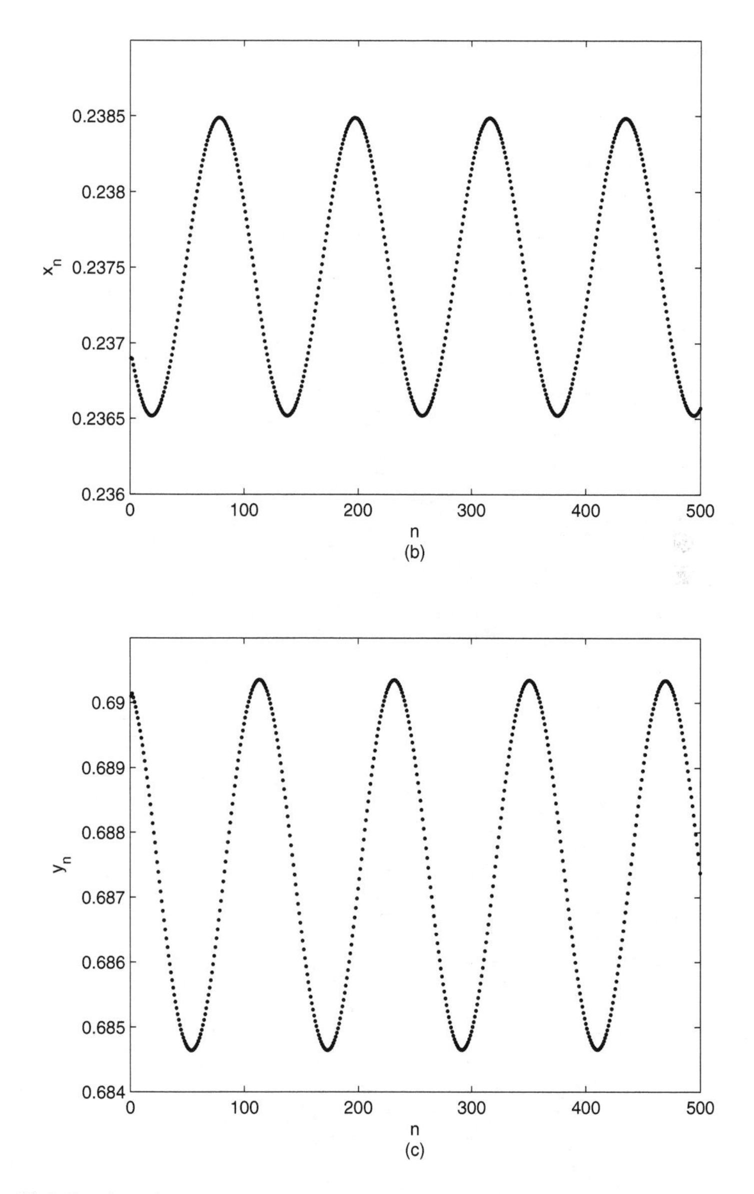

Fig. 10.1 *Continued*

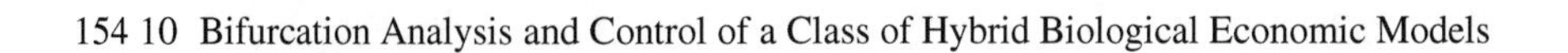

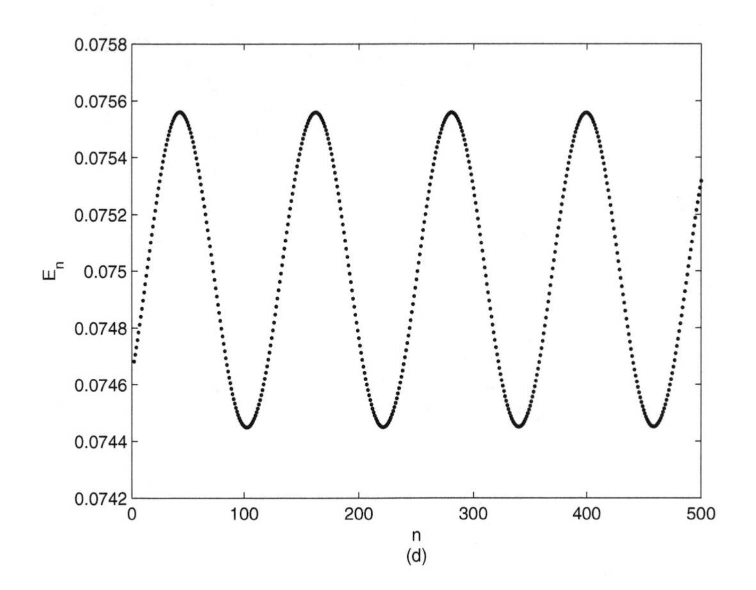

Fig. 10.1 *Continued*

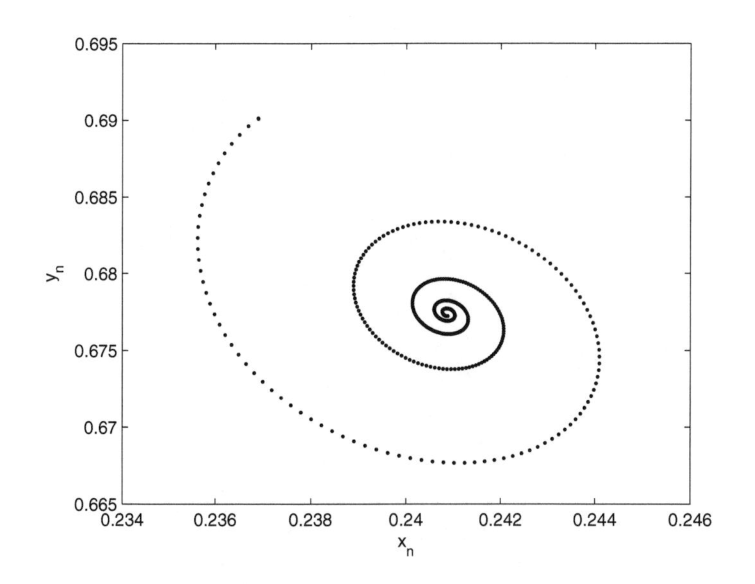

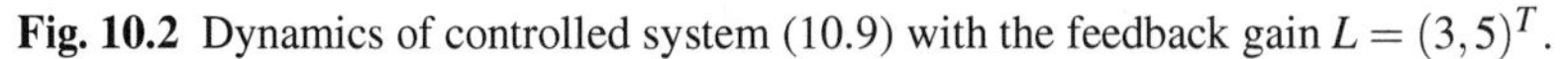
Fig. 10.2 Dynamics of controlled system (10.9) with the feedback gain $L = (3,5)^T$.

10.7 Conclusions

So far, considerable attention has been focused on continuous-time and discrete-time nonlinear systems in biology. However, to our best knowledge, the reports of differential-algebraic biological systems or differential-difference-algebraic biological systems are few. This chapter discusses a class of hybrid biological economic models, where the dynamics of the prey population are governed by a differential equation, the predator population governed by a difference equation, and economic theory by an algebraic equation. This chapter shows that this hybrid biological economic model exhibits the following three bifurcations:

(1) Saddle-node bifurcation at the intersampling instants: If the economic profit $m < m_{sn}$, the prey population and the harvesting effort will stay at steady states. Thus, it is favorable to the breeding of the prey population and permanent and constant harvesting for fishermen to keep the economic profit in some region.

(2) Singular induced bifurcation at the intersampling instants: When the economic profit increases through m_{sib}, the density of the population will expand rapidly, which induces ecosystem unbalance and even biological disaster. Applying the state feedback control method, which can be implemented by adjusting the harvesting costs and the economic profit, the impulsive phenomenon will be eliminated and the biological population can be controlled at steady states.

(3) Neimark-Sacker bifurcation for the sampling instants: For $m > m_{ns}$ and $|m - m_{ns}|$ small enough, the densities of the prey and predator population emerge as a periodical oscillation and a unique closed invariant curve occurs. After adjusting harvesting costs and the economic profit, two populations stay at steady states.

Therefore, the state feedback control method is efficient. In practice, we can regulate the harvesting costs and the economic profit, such as adjusting revenue, drawing out favorable policy to encourage or improve fishery, or abating pollution, so that the population can be driven to steady states.

References

1. Abeles, M., Bergman, H., Margalit, E., Vaadia, E.: Spatiotemporal firing patterns in the frontal cortex of behaving monkeys. J. Neuro-physiol. 70(4), 1629–1638 (1993)
2. Chen, L.N., Aihara, K.: Global searching ability of chaotic neural networks. IEEE T. Circuits-I 46, 974–993 (1999)
3. Chen, L.N., Aihara, K.: Stability and bifurcation analysis of differential-difference-algebraic equations. IEEE T. Circuits-I 48(3), 308–326 (2001)
4. Chen, L.N., Tada, Y., Okamoto, H., Tanabe, R., Ono, A.: Optimal operation solutions of power systems with transient stability constraints. IEEE T. Circuits-I 48, 327–339 (2001)
5. Gardner, T.S., Cantor, C.R., Collins, J.J.: Construction of a genetic toggle switch in Escherichia Coli. Nature 403, 339–342 (2000)
6. Gordon, H.S.: The economic theory of a common property resource: the fishery. J. Polit. Econ. 62(2), 124–142 (1954)

7. Herrera, R., Suyama, K., Horio, Y., Aihara, K.: IC implementation of a switched-current chaotic neuron. IEICE T. Fund. Electr. E82-A(9), 1776–17782 (1999)
8. Matsumoto, G., Aihara, K., Hanyu, Y., Takahashi, N., Yoshizawa, S., Nagumo, J.: Chaos and phase locking in normal squid axons. Phys. Lett. A. 123(4), 162–166 (1987)
9. Smolen, P., Baxter, D.A., Byrne, J.H.: Mathematical modeling of gene networks review. Neuron 26, 567–580 (2000)
10. Someya, H., Fujita, A., Sekine, Y., Aihara, K.: Hardware model of an active axon. Trans. Inst. Electron. Inf. Commun. Eng. 12, 655–661 (1999)
11. Somogyi, R., Sniegoski, C.: Modeling the complexity of genetic networks: Understanding multigenic and pleiotropic regulation. Complexity 1, 45–63 (1996)
12. Venkatasubramanian, V., Schaettler, H., Zaborszky, J.: Local bifurcations and feasibility regions in differential-algebraic systems. IEEE T. Automat. Contr. 40(12), 1992–2013 (1995)
13. Watanabe, M., Aihara, K., Kondo, S.: A dynamical neural network with temporal coding and functional connectivity. Biol. Cybern. 78, 87–93 (1998)

Chapter 11
Bifurcations of a Class of Discrete-Time Singular Biological Economic Models

11.1 Introduction

It is well known that the Lotka-Volterra prey-predator model is one of the fundamental population models. A predator-prey interaction was first described by two pioneers, Lotka [8] and Volterra [12], in two independent works. The research dealing with interspecific interactions has mainly focused on continuous prey-predator models of two variables. However, certain species, including many species of insects, have no overlap between successive generations and so their population evolves in discrete time-steps. Such populations can be modeled by difference equations, otherwise known as discrete dynamical systems or (iterative) maps. There is now a considerable literature on discrete ecosystem models, e.g., see [6, 9, 10] and references therein. Some works by Danca et al. [2], Jing and Yang [5], Liu and Xiao [7], and Agiza et al. [1] showed that, for the discrete-time prey-predator models, the dynamics can produce a much richer set of patterns than those observed in continuous-time models. Also, Summers et al. have examined four typical discrete-time ecosystem models under the effects of periodic forcing [11]. They found that a system that has simplistic behavior in its unforced state can assume chaotic behavior when subjected to periodic forcing, dependent on the values chosen for the controlling parameters.

Consider the following singular prey-predator economic system:

$$\begin{cases} \dfrac{\mathrm{d}x}{\mathrm{d}t} = x(r_1 + a_{11}x + a_{12}y), \\ \dfrac{\mathrm{d}y}{\mathrm{d}t} = y(r_2 + a_{21}x + a_{22}y) - Ey, \\ 0 = E(py - c) - m, \end{cases} \tag{11.1}$$

where x and y represent the prey density and predator density at time t, respectively. $r_1, r_2 > 0$ are the inherent net birth rates per unit of population per unit time of the prey and predator, respectively. $a_{12} < 0$ and $a_{21} > 0$ measure the effect of the

Q. Zhang et al.: Complex., Analy. and Cntrl. of Singular Biological Systems, LNCIS 421, pp. 157–170.
springerlink.com

interaction of the two species. $a_{ii} < 0 (i = 1,2)$ is the self-inhibition coefficient. E represents harvesting effort. Ey indicates that the harvest for predator population is proportional to its density at time t. p is the harvesting reward per unit weight of the predator population. c is the harvesting cost per unit harvesting effort for the predator population. m is the economic profit. And parameters p, c, m are all positive constants.

Suppose that the prey and predator population have no overlap between successive generations. Applying the Euler method to system (11.1) with integral step size δ, this section mainly studies the following discrete-time singular prey-predator economic system

$$\begin{cases} x_{n+1} = x_n + \delta x_n(r_1 + a_{11}x_n + a_{12}y_n), \\ y_{n+1} = y_n + \delta y_n(r_2 + a_{21}x_n + a_{22}y_n - E_n), \\ 0 = E_n(py_n - c) - m. \end{cases} \tag{11.2}$$

11.2 Existence and Stability of System (11.2)

The fixed points of system (11.2) satisfy the following equations:

$$\begin{cases} \delta x(r_1 + a_{11}x + a_{12}y) = 0, \\ \delta y(r_2 + a_{21}x + a_{22}y - E) = 0, \\ E(py - c) - m = 0. \end{cases}$$

The Jacobian matrix J of system (11.2) is given by

$$J = \begin{pmatrix} 1 + \delta(r_1 + 2a_{11}x + a_{12}y) & \delta a_{12}x \\ \delta a_{21}y & 1 + \delta(r_2 + a_{21}x + 2a_{22}y + \frac{c}{m}E^2) \end{pmatrix}. \tag{11.3}$$

The results of fixed point analysis are presented in the following theorem:

Theorem 11.1. *(1) System (11.2) has a fixed point $Q_1(0, y_1, E_1)$, where the coordinate $y_1 > c/p$ satisfies the following equation,*

$$a_{22}py^2 + (r_2p - a_{22}c)y - r_2c - m = 0,$$

and $E_1 = m/(py_1 - c)$. Moreover, if the following two conditions are satisfied,

$$-2 < \delta(r_1 + a_{12}y_1) < 0$$

and

$$-2m < \delta a_{22}my_1 + p\delta y_1E_1^2 < 0,$$

the fixed point Q_1 is asymptotically stable.
(2) System (11.2) has a positive fixed point $Q_2(x_2, y_2, E_2)$, where x_2 is the solution of the following equation,

$$[a_{22}(r_1+a_{11}x)-a_{12}(r_2+a_{21}x)][p(r_1+a_{11}x)+a_{12}c]-ma_{12}^2=0,$$

and $y_2=-(r_1+a_{11}x_2)/a_{12}$, $E_2=-ma_{12}/[p(r_1+a_{11}x_2)+ca_{12}]$. *Moreover, if*

$$m(a_{11}a_{22}-a_{12}a_{21})+pa_{11}E_2^2>0,$$

$$m[4+2\delta(a_{11}x_2+a_{22}y_2)+\delta^2x_2y_2(a_{11}a_{22}-a_{12}a_{21})]+\delta py_2E_2^2(2+\delta a_{11}x_2)>0$$

and

$$\left|(1+\delta a_{11}x_2)(m+\delta a_{22}my_2+p\delta y_2E_2^2)-\delta^2a_{12}a_{21}mx_2y_2\right|<m,$$

the fixed point Q_2 is asymptotically stable.

11.3 Bifurcation Behavior

Take the economic profit m as the bifurcation parameter in the analysis of bifurcation behavior.

(I) Flip bifurcation

The Jacobian matrix of system (11.2) at fixed point Q_2 is

$$J_{Q_2}=\begin{pmatrix}1+\delta a_{11}x_2 & \delta a_{12}x_2\\ \delta a_{21}y_2 & 1+\delta(a_{22}y_2+\frac{p}{m}y_2E_2^2)\end{pmatrix}, \tag{11.4}$$

and its characteristic equation is

$$\lambda^2+d_1(x_2,y_2,E_2)\lambda+d_2(x_2,y_2,E_2)=0,$$

where

$d_1(x_2,y_2,E_2)=-[2+\delta(a_{11}x_2+a_{22}y_2+py_2E_2^2/m)]$,

$d_2(x_2,y_2,E_2)=(1+\delta a_{11}x_2)(1+\delta a_{22}y_2+p\delta y_2E_2^2/m)-a_{12}a_{21}\delta^2x_2y_2$.

The eigenvalues of system (11.2) at $Q_2(x_2,y_2,E_2)$ are

$$\lambda_1=-1,\quad \lambda_2=1-d_1(x_2,y_2,E_2),$$

when $m=m_F$ satisfies the following equation:

$$m_F=-\frac{(py_2-c)^2}{\delta py_2(2+\delta a_{11}x_2)}[4+2\delta(a_{11}x_2+a_{22}y_2)+\delta^2x_2y_2(a_{11}a_{22}-a_{12}a_{21})]. \tag{11.5}$$

The condition $|\lambda_2|\neq 1$ means

$$\left|a_{12}a_{21}\delta^2x_2y_2-(1+\delta a_{11}x_2)(1+\delta a_{22}y_2+\frac{p\delta}{m_F}y_2E_2^2)\right|\neq 1. \tag{11.6}$$

Let $u = x - x_2$, $v = y - y_2$ and $\tilde{m} = m - m_F$. Take $\tilde{m}$ as a new bifurcation parameter, and then we obtain the following system:

$$\begin{cases} u_{n+1} = & p_{11}u_n + p_{12}v_n + \delta a_{11}u_n^2 + \delta a_{12}u_n v_n, \\ v_{n+1} = & p_{21}u_n + p_{22}v_n + p_{23}\tilde{m}_n + \dfrac{1}{2}b_{020}v_n^2 + b_{110}u_n v_n \\ & +b_{011}v_n\tilde{m}_n + \dfrac{1}{6}b_{030}v_n^3 + \dfrac{1}{2}b_{021}v_n^2\tilde{m}_n \\ & +o((\,|\,u_n\,|+|\,v_n\,|+|\,\tilde{m}_n\,|)^4), \\ \tilde{m}_{n+1} = \tilde{m}_n, \end{cases} \tag{11.7}$$

where
$p_{11} = 1 + \delta a_{11}x_2$,
$p_{12} = \delta a_{12}x_2$,
$p_{21} = \delta a_{21}y_2$,
$p_{22} = [1 + \delta(a_{22}y_2 + py_2E_2^2/m_F)]$,
$p_{23} = -\delta y_2 E_2/m_F$,
$b_{020} = 2\delta(a_{22} - pcE_2^3/m_F^2)$,
$b_{110} = a_{21}\delta$,
$b_{011} = \delta cE_2^2/m_F^2$,
$b_{030} = 6\delta cp^2E_2^4/m_F^3$,
$b_{021} = -2\delta pcE_2^3/m_F^3$.

Take the following transformation

$$\begin{pmatrix} u_n \\ v_n \\ \tilde{m}_n \end{pmatrix} = P_1 \begin{pmatrix} X_n \\ Y_n \\ M_n \end{pmatrix},$$

where $P_1 = \begin{pmatrix} p_{12} & p_{12} & p_{12}p_{23} \\ -(1+p_{11}) & \lambda_2 - p_{11} & (1-p_{11})p_{23} \\ 0 & 0 & 2(1-\lambda_2) \end{pmatrix}$, and then system (11.7) can be written as

$$\begin{pmatrix} X_{n+1} \\ Y_{n+1} \\ M_{n+1} \end{pmatrix} = \begin{pmatrix} -1 & 0 & 0 \\ 0 & \lambda_2 & 0 \\ 0 & 0 & 1 \end{pmatrix} \begin{pmatrix} X_n \\ Y_n \\ M_n \end{pmatrix} + \begin{pmatrix} \phi_1(X_n,Y_n,M_n) \\ \phi_2(X_n,Y_n,M_n) \\ 0 \end{pmatrix}, \tag{11.8}$$

where
$\phi_1(X_n,Y_n,M_n) = A_1u_n^2 + A_2u_nv_n + A_3v_n\tilde{m}_n + A_4v_n^2 + A_5v_n^2\tilde{m}_n + A_6v_n^3$
$\quad +o((\,|\,X_n\,|+|\,Y_n\,|+|\,\tilde{M}_n\,|)^4)$,
$\phi_2(X_n,Y_n,M_n) = B_1u_n^2 + B_2u_nv_n + B_3v_n\tilde{m}_n + B_4v_n^2 + B_5v_n^2\tilde{m}_n + B_6v_n^3$
$\quad +o((\,|\,X_n\,|+|\,Y_n\,|+|\,\tilde{M}_n\,|)^4)$,
$A_1 = \delta a_{11}(\lambda_2 - p_{11})/p_{12}(1+\lambda_2)$,
$A_2 = [\delta a_{12}(\lambda_2 - p_{11}) - p_{12}b_{110}]/p_{12}(1+\lambda_2)$,
$A_3 = -b_{011}/(1+\lambda_2)$,
$A_4 = -b_{020}/2(1+\lambda_2)$,

$A_5 = -b_{021}/2(1+\lambda_2)$,
$A_6 = -b_{030}/6(1+\lambda_2)$,
$B_1 = \delta a_{11}(1+p_{11})/p_{12}(1+\lambda_2)$,
$B_2 = [\delta a_{12}(1+p_{11})+p_{12}b_{110}]/p_{12}(1+\lambda_2)$,
$B_3 = b_{011}/(1+\lambda_2)$,
$B_4 = b_{020}/2(1+\lambda_2)$,
$B_5 = b_{021}/2(1+\lambda_2)$,
$B_6 = b_{030}/6(1+\lambda_2)$,
$u_n = p_{12}X_n + p_{12}Y_n + p_{12}p_{23}M_n$,
$v_n = -(1+p_{11})X_n + (\lambda_2 - p_{11})Y_n + (1-p_{11})p_{23}M_n$,
$\tilde{m}_n = 2(1-\lambda_2)M_n$.

Applying the center manifold theorem, the center manifold for system (11.8) can be represented as follows:

$$W^c(0,0,0) = \{ (X_n,Y_n,M_n)|Y_n = k_1X_n^2 + k_2X_nM_n + k_3M_n^2 + o((|X_n| + |M_n|)^3)\}.$$

By simple computation for the center manifold, we obtain

$$k_1 = \frac{1}{1-\lambda_2}[B_1p_{12}^2 - B_2p_{12}(1+p_{11}) + B_4(1+p_{11})^2],$$

$$k_2 = \frac{1}{1+\lambda_2}[-2B_1p_{12}^2p_{23} + 2B_2p_{11}p_{12}p_{23} + 2B_3(1+p_{11})(1-\lambda_2) + 2B_4p_{23}(1-p_{11}^2)],$$

$$k_3 = \frac{1}{1-\lambda_2}[B_1p_{12}^2p_{23}^2 + B_2(1-p_{11})p_{12}p_{23}^2 + 2B_3p_{23}(\lambda_2-1) \cdot (p_{11}-1) + B_4p_{23}^2(1-p_{11})^2].$$

Hence, the system restricted to the center manifold is given by

$$\begin{aligned} X_{n+1} &= F(X_n,M_n) \\ &= -X_n + C_1X_n^2 + C_2X_nM_n + C_3M_n^2 + C_4X_n^3 + C_5X_n^2M_n \\ &\quad + C_6X_nM_n^2 + C_7M_n^3 + o((|X_n| + |M_n|)^4), \end{aligned}$$

where
$C_1 = A_1 p_{12}^2 - A_2 p_{12}(1+p_{11}) + A_4(1+p_{11})^2,$
$C_2 = 2A_1 p_{12}^2 p_{23} - 2A_2 p_{11} p_{12} p_{23} - 2A_3(1+p_{11})(1-\lambda_2) - 2A_4(1-p_{11}^2)p_{23},$
$C_3 = A_1 p_{12}^2 p_{23}^2 + A_2 p_{12} p_{23}^2 (1-p_{11}) + 2A_3(1-p_{11})p_{23}(1-\lambda_2)$
$\quad +A_4(1-p_{11})^2 p_{23}^2,$
$C_4 = 2A_1 k_1 p_{12}^2 + A_2 k_1 p_{12}(\lambda_2 - 1 - 2p_{11}) - 2A_4 k_1(1+p_{11})(\lambda_2 - p_{11})$
$\quad -A_6(1+p_{11})^3,$
$C_5 = 2A_1 p_{12}^2(k_1 p_{23} + k_2) + A_2 p_{12}[k_1 p_{23}(1+\lambda_2 - 2p_{11}) + k_2(\lambda_2 - 1 - 2p_{11})]$
$\quad +2A_3 k_1(1-\lambda_2)(\lambda_2 - p_{11}) + 2A_4(\lambda_2 - p_{11})[k_1 p_{23} - k_2 - p_{11}(k_1 p_{23} + k_2)]$
$\quad +2A_5(1+p_{11})^2(1-\lambda_2) + 3A_6(1+p_{11})^2(1-p_{11})p_{23},$
$C_6 = 2A_1 p_{12}^2(k_2 p_{23} + k_3) + A_2 p_{12}[k_2 p_{23}(1+\lambda_2 - 2p_{11}) + k_3(\lambda_2 - 1 - 2p_{11})]$
$\quad +2A_3 k_2(1-\lambda_2)(\lambda_2 - p_{11}) + 2A_4(\lambda_2 - p_{11})[k_2 p_{23} - k_3 - p_{11}(k_2 p_{23} + k_3)]$
$\quad -4A_5(1-p_{11}^2)p_{23}(1-\lambda_2) - 3A_6(1-p_{11})^2(1+p_{11})p_{23}^2,$
$C_7 = 2A_1 k_3 p_{12}^2 p_{23} + A_2 k_3 p_{12} p_{23}(1+\lambda_2 - 2p_{11}) + 2A_3 k_3(\lambda_2 - p_{11})(1-\lambda_2)$
$\quad +2A_4 p_{23} k_3(\lambda_2 - p_{11})(1-p_{11}) + 2A_5 p_{23}^2(1-p_{11})^2(1-\lambda_2)$
$\quad +A_6 p_{23}^3(1-p_{11})^3.$

Since

$$\sigma_1 = \left(\frac{\partial F}{\partial M_n}\frac{\partial^2 F}{\partial X_n^2} + 2\frac{\partial^2 F}{\partial X_n \partial M_n}\right)_{(0,0)} = 2C_2,$$

$$\sigma_2 = \left(\frac{1}{2}\left(\frac{\partial^2 F}{\partial X_n^2}\right)^2 + \frac{1}{3}\frac{\partial^3 F}{\partial X_n^3}\right)_{(0,0)} = 2C_1^2 + 2C_4,$$

according to the above analysis and flip bifurcation theorem of Guckenheimer and Holmes [3], we have the following theorem:

Theorem 11.2. *If $\sigma_1\sigma_2 \neq 0$, system (11.2) undergoes a flip bifurcation at the fixed point $Q_2(x_2, y_2, E_2)$ when bifurcation parameter m varies in the small neighborhood of m_F satisfying (11.5) and (11.6). Moreover, the sign of σ_2 determines the stability and direction of the bifurcation orbits of period 2, i.e., if σ_2 is positive, the orbits are stable; if σ_2 is negative, they are unstable.*

(II) Hopf bifurcation

If the bifurcation parameter $m = m_H$ satisfies

$$\begin{aligned}&(a_{11}m_H x_2 + a_{22}m_H y_2 + p y_2 E_2^2)\\ \times\ &(\delta a_{11}m_H x_2 + \delta a_{22}m_H y_2 + \delta p y_2 E_2^2 + 4m_H) < 0,\end{aligned} \tag{11.9}$$

$$\begin{aligned}&a_{11}m_H x_2 + a_{22}m_H y_2 + p y_2 E_2^2 + \delta x_2 y_2\\ \times\ &(a_{11}a_{22}m_H - a_{12}a_{21}m_H + a_{11}pE_2^2) = 0,\end{aligned} \tag{11.10}$$

the eigenvalues of system (11.2) are complex conjugate with modulus 1, which is written as

$$\lambda, \bar{\lambda} = \alpha_1 \pm i\alpha_2, \tag{11.11}$$

where $\alpha_1 = (p_{11} + p_{22})/2, \alpha_2 = \sqrt{-4p_{12}p_{21} - (p_{11} - p_{22})^2}/2.$

When the bifurcation parameter $m = m_H$, let $u = x - x_2$ and $v = y - y_2$, and then system (11.2) translates the fixed point $Q_2(x_2, y_2, E_2)$ into the origin

$$\begin{cases} u_{n+1} = p_{11}u_n + p_{12}v_n + \delta a_{11}u_n^2 + \delta a_{12}u_n v_n, \\ v_{n+1} = p_{21}u_n + p_{22}v_n + \frac{1}{2}b_{02}v_n^2 + b_{11}u_n v_n + \frac{1}{6}b_{03}v_n^3 + o((|u_n| + |v_n|)^4), \end{cases} \tag{11.12}$$

where $b_{02} = 2\delta(a_{22} - pcm_H)/(py_2 - c)^3$, $b_{11} = \delta a_{21}$, $b_{03} = 6\delta p^2 cm_H/(py_2 - c)^4$. Construct an invertible matrix

$$P_2 = \begin{pmatrix} p_{12} & 0 \\ \alpha_1 - p_{11} & -\alpha_2 \end{pmatrix},$$

and use

$$\begin{pmatrix} u_n \\ v_n \end{pmatrix} = P_2 \begin{pmatrix} X_n \\ Y_n \end{pmatrix},$$

then system (11.12) becomes

$$\begin{pmatrix} X_{n+1} \\ Y_{n+1} \end{pmatrix} = \begin{pmatrix} \alpha_1 & -\alpha_2 \\ \alpha_2 & \alpha_1 \end{pmatrix} \begin{pmatrix} X_n \\ Y_n \end{pmatrix} + \begin{pmatrix} \varphi_1(X_n, Y_n) \\ \varphi_2(X_n, Y_n) \end{pmatrix},$$

where

$\varphi_1(X_n,Y_n) = q_1 X_n^2 + q_2 X_n Y_n,$

$\varphi_2(X_n,Y_n) = s_1 X_n^2 + s_2 X_n Y_n + s_3 Y_n^2 + s_4 X_n^3 + s_5 X_n^2 Y_n + s_6 X_n Y_n^2 + s_7 Y_n^3 + o((|X_n| + |Y_n|)^4),$

$q_1 = \delta[a_{11}p_{12} + a_{12}(\alpha_1 - p_{11})],$

$q_2 = -\delta a_{12}\alpha_2,$

$s_1 = -(\alpha_1 - p_{11})[2p_{12}(b_{11} - \delta a_{11}) + (\alpha_1 - p_{11})(b_{02} - 2\delta a_{12})]/2\alpha_2,$

$s_2 = (\alpha_1 - p_{11})(b_{02} - \delta a_{12}) + p_{12}b_{11},$

$s_3 = -b_{02}\alpha_2/2,$

$s_4 = -b_{03}(\alpha_1 - p_{11})^3/6\alpha_2,$

$s_5 = b_{03}(\alpha_1 - p_{11})^2/2,$

$s_6 = -b_{03}(\alpha_1 - p_{11})\alpha_2/2,$

$s_7 = b_{03}\alpha_2^2/6.$

According to the Hopf bifurcation theory, the discriminatory quantity a is given by

$$a = -\mathrm{Re}[\frac{(1-2\lambda)\bar{\lambda}^2}{1-\lambda}\xi_{11}\xi_{20}] - \frac{1}{2} \mid \xi_{11} \mid^2 - \mid \xi_{02} \mid^2 + \mathrm{Re}(\bar{\lambda}\xi_{21}), \tag{11.13}$$

where

$$\begin{aligned} \xi_{20} &= [(\varphi_{1XX} - \varphi_{1YY} + 2\varphi_{2XY}) + i(\varphi_{2XX} - \varphi_{2YY} - 2\varphi_{1XY})]/8 \\ &= \{(q_1 + s_2 + 2s_5X_n + 2s_6Y_n) + i[(s_1 - s_3 - q_2) + (3s_4 - s_6)X_n \\ &\quad +(s_5 - 3s_7)Y_n]\}/4, \\ \xi_{11} &= [(\varphi_{1XX} + \varphi_{1YY}) + i(\varphi_{2XX} + \varphi_{2YY})]/4 \\ &= \{q_1 + i[(s_1 + s_3) + (3s_4 + s_6)X_n + (s_5 + 3s_7)Y_n]\}/2, \\ \xi_{02} &= [(\varphi_{1XX} - \varphi_{1YY} - 2\varphi_{1XY}) + i(\varphi_{2XX} - \varphi_{2YY} + 2\varphi_{1XY})]/8 \\ &= \{(q_1 - s_2 - 2s_5X_n - 2s_6Y_n) + i[(s_1 - s_3 + q_2) + (3s_4 - s_6)X_n + (s_5 - 3s_7)Y_n]\}/4, \end{aligned}$$

$$\xi_{21} = [(\varphi_{1XXX} + \varphi_{1XYY} + \varphi_{2XXY} + \varphi_{2YYY}) + i(\varphi_{2XXX} + \varphi_{2XYY} - \varphi_{1XXY} - \varphi_{1YYY})]/16$$
$$= [(s_5 + s_6) + i(3s_4 + s_6)]/8.$$

The Hopf bifurcation theorem [4] requires $\lambda^k(m_H) \neq 1$, for $k = 1,2,3,4$, i.e.,

$$\left(\frac{1}{2}(p_{11} + p_{22}) + \frac{1}{2}i\sqrt{-4p_{12}p_{21} - (p_{11} - p_{22})^2}\right)^k_{m_H} \neq 1, \tag{11.14}$$

where $k = 1,2,3,4$ and another condition

$$\zeta = \left(\frac{\mathrm{d}}{\mathrm{d}m}(|\lambda(m)|)\right)_{m=m_H} \neq 0,$$

that is

$$\zeta = \left(\frac{\partial}{\partial x}(|\lambda(x)|)\frac{\partial x}{\partial m} + \frac{\partial}{\partial y}(|\lambda(y)|)\frac{\partial y}{\partial m}\right)_{m_H} \neq 0. \tag{11.15}$$

Summarize the above analysis into the following theorem:

Theorem 11.3. *If conditions (11.14) and (11.15) hold, system (11.2) undergoes a Hopf bifurcation at $Q_2(x_2, y_2, E_2)$ when bifurcation parameter m varies in the small neighborhood of m_H satisfying (11.9) and (11.10). Moreover, if $a < 0$ and $\zeta > 0$, the Hopf bifurcation is supercritical and an attracting invariant closed curve exists, surrounding Q_2 for $m > m_H$ and $\left| m - m_H \right|$ is small enough.*

11.4 State Delayed Feedback Control

Unstable fluctuations, bifurcations, and chaos have always been regarded as unfavorable phenomena in biology, so they are harmful for the breeding of the biological population. Naturally, we need to take action to stabilize the biological population. This section proposes the following state delayed feedback control method:

$$\begin{cases} x_{n+1} = x_n + \delta x_n(r_1 + a_{11}x_n + a_{12}y_n), \\ y_{n+1} = y_n + \delta y_n(r_2 + a_{21}x_n + a_{22}y_n - E_n) + K(y_n - y_{n-1}), \\ 0 \quad = E_n(py_n - c) - m, \end{cases} \tag{11.16}$$

where K is a feedback gain.

Denote $\omega_n = y_n - y_{n-1}$, system (11.16) becomes the following equations:

$$\begin{cases} x_{n+1} = x_n + \delta x_n(r_1 + a_{11}x_n + a_{12}y_n), \\ y_{n+1} = y_n + \delta y_n(r_2 + a_{21}x_n + a_{22}y_n - E_n) + K\omega_n, \\ \omega_{n+1} = \delta y_n(r_2 + a_{21}x_n + a_{22}y_n - E_n) + K\omega_n, \\ 0 \quad = E_n(py_n - c) - m. \end{cases}$$

By the stability criterion of the discrete system, we obtain the following results:

Theorem 11.4. *Under the conditions of Theorem 11.2, the fixed point* $Q_2(x_2,y_2,E_2)$ *is locally asymptotically stable if the following conditions are satisfied:*
(*i*) $m(2+\delta a_{11}x_2+\delta a_{22}y_2)+\delta p y_2 E_2^2<0$,
(*ii*) $K(2+\delta a_{11}x_2)>0$,
(*iii*) $|d_2(x_2,y_2,E_2)+K[1+(1+\delta a_{11}x_2)(1+d_1(x_2,y_2,E_2)-K)]|$
$<1-K^2(1+\delta a_{11}x_2)^2$.

Theorem 11.5. *Under the conditions of Theorem 11.3, the fixed point* $Q_2(x_2,y_2,E_2)$ *is locally asymptotically stable if the following conditions are satisfied:*
(*i*) $m(a_{11}x_2+a_{22}y_2)+py_2E_2^2<0$,
(*ii*) $m(4+\delta a_{11}x_2+\delta a_{22}y_2+4K+2\delta a_{11}Kx_2)+\delta p y_2 E_2^2>0$,
(*iii*) $|1+K[1+(1+\delta a_{11}x_2)(1+d_1(x_2,y_2,E_2)-K)]|<1-K^2(1+\delta a_{11}x_2)^2$.

In fact, the first equation of system (11.25) can be rewritten as

$$y_{n+1}=y_n+\delta y_n(r_2+a_{21}x_n+a_{22}y_n-\hat{E}_n), \tag{11.17}$$

where $\hat{E}_n=E_n-K(y_n-y_{n-1})/\delta y_n$.

Remark 11.1. From Eq. (11.17), we note that the state delayed feedback control method can be implemented by adjusting the harvesting effort for the predator population, which depends on the present predator density and the past one. From the view of application, managers can control the harvesting effort by adopting some methods, such as adjusting revenue, drawing out favorable policy to encourage fishery, abating pollution, and so on, so that those unfavorable phenomena can be eliminated and the biological population will stay at steady states.

11.5 Numerical Simulation

To illustrate the results obtained, let us consider the following particular cases.
Case I. Choose parameter values

$$r_1=2,a_{11}=-2,a_{12}=-0.6,r_2=-0.5,a_{21}=0.6,$$
$$a_{22}=0,\delta=1.2537,p=6,c=0.5,m=m_F=0.02,$$

then system (11.2) has a positive fixed point $P_2^F(0.8463,0.5124,0.0078)$, and its eigenvalues are $\lambda_1=-1$ and $\lambda_2=0.8897$. The bifurcation diagrams of system (11.2) in $(m-x)$ and $(m-y)$ planes are given by Fig.11.1 and Fig.11.2. We can see that any orbit with respectively initial values of prey and predator 0.845 and 0.512 approaches to a positive fixed point for $m\in(0.01,0.02)$ and there exists the period-2 orbit for the region of $m\in(0.02,0.023)$. Choose the feedback gain $K=-0.3$, and then Fig. 11.3 and Fig. 11.4 show that the period-2 orbit is stabilized at the positive fixed point P_2^F when economic profit $m=0.021$.

Case II. For the parameter values $r_1 = 0.7, a_{11} = -3.2696, a_{12} = -1, r_2 = -0.5$, $a_{21} = 6$, $a_{22} = 0$, $\delta = 1.2537$, $p = 6$, $c = 0.5$, and $m = m_H = 0.02$, system (11.2) has a fixed point $P_2^H(0.0850, 0.4222, 0.0098)$ and complex conjugate eigenvalues $\lambda_\pm = 0.8335 \pm i0.5526$. Fig. 11.5 shows that an attracting invariant closed curve from P_2^H occurs. If the feedback gain is chosen as $K = -0.2$, the controlled system states stay at the fixed point P_2^H after a short transient process, which is depicted in Fig. 11.6 and Fig. 11.7.

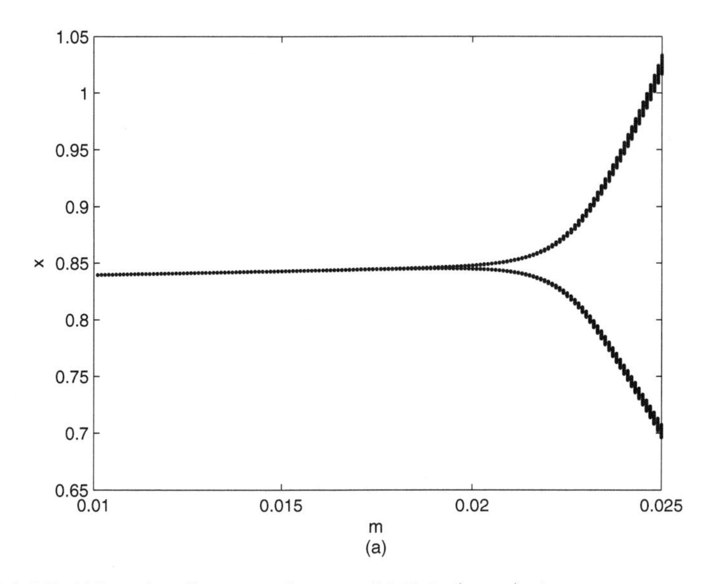

Fig. 11.1 Flip bifurcation diagrams of system (11.2) in $(m-x)$ plane.

11.6 Conclusions

In this section, we investigated the behaviors of a class of discrete-time singular prey-predator economic systems. By using the center manifold theorem and bifurcation theory, the system undergoes flip bifurcation and Hopf bifurcation when economic profits pass some critical values. Moreover, the proposed state delay feedback control method eliminates these bifurcational phenomena and drives the discrete-time singular prey-predator economic system to a steady state. Analysis shows that this control method can be implemented only by adjusting the harvesting effort for the predator population after observing the present predator density and the past one. In fact, managers can draw out some policies to encourage or restrain the harvesting efforts in application, such as adjusting revenue and market price, making an allowance to fishermen, abating pollution, and so on, so that the biological population can stay at steady states.

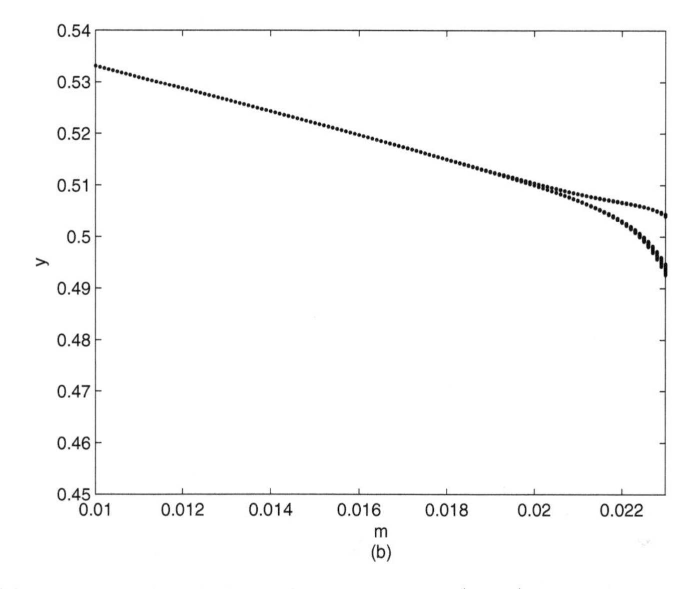

Fig. 11.2 Flip bifurcation diagrams of system (11.2) in $(m-y)$ plane.

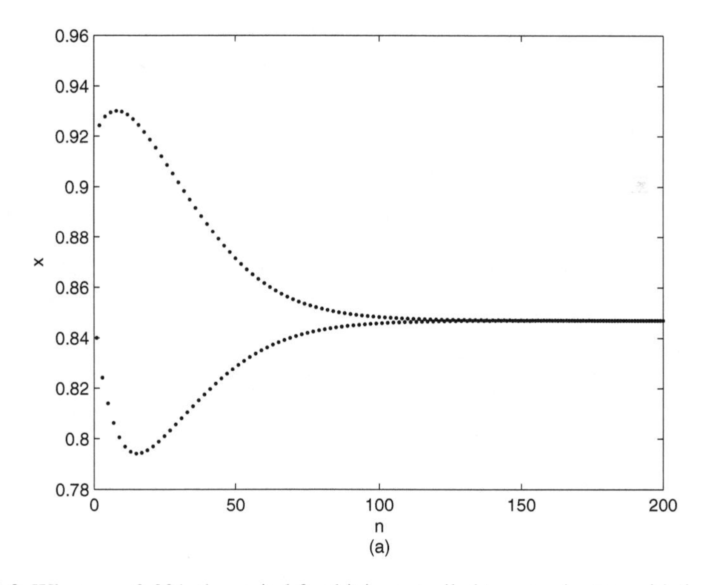

Fig. 11.3 When $m = 0.021$, the period-2 orbit is controlled to a steady state with the feedback gain $K = -0.3$ in $(n-x)$ plane.

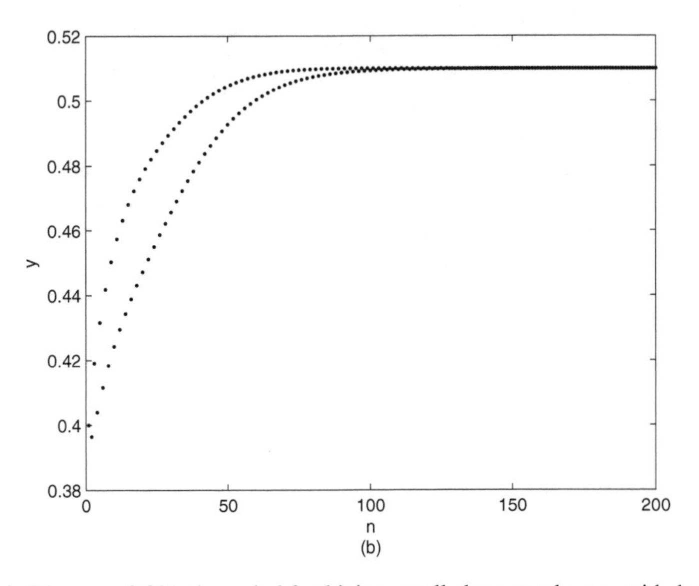

Fig. 11.4 When $m = 0.021$, the period-2 orbit is controlled to a steady state with the feedback gain $K = -0.3$ in $(n-y)$ plane.

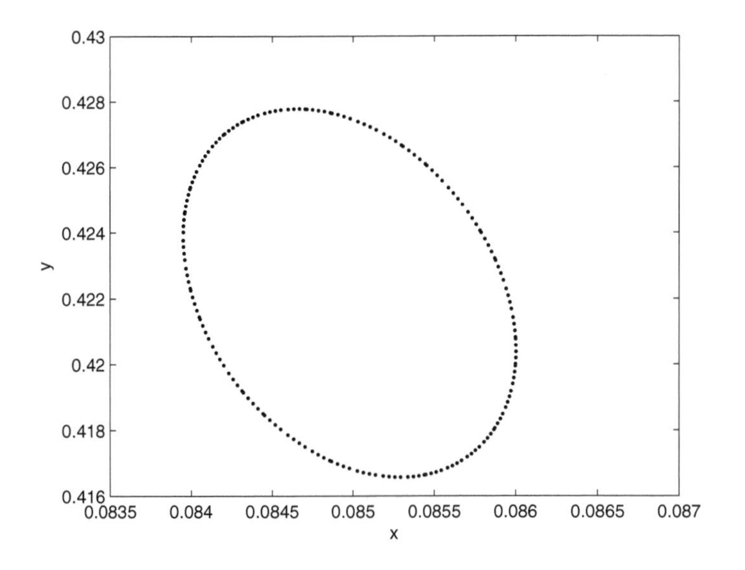

Fig. 11.5 An attracting bifurcating period solution from P_2^H occurs.

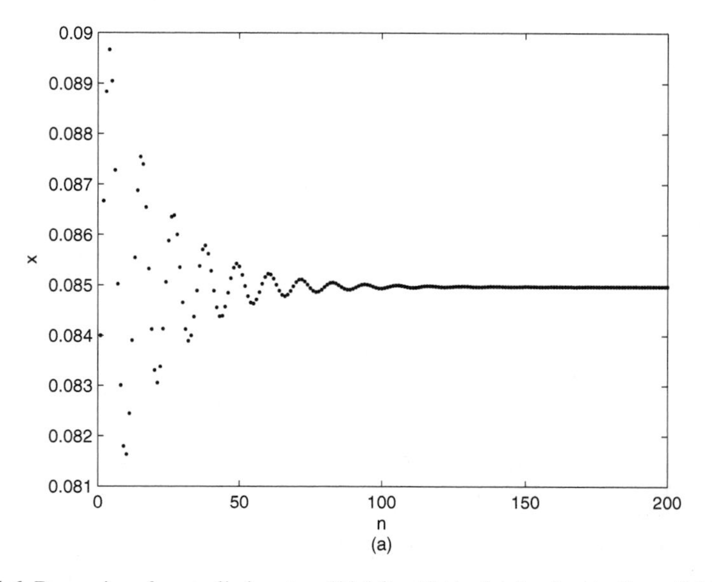

Fig. 11.6 Dynamics of controlled system (11.16) with the feedback gain $K = -0.2$ in $(n-x)$ plane.

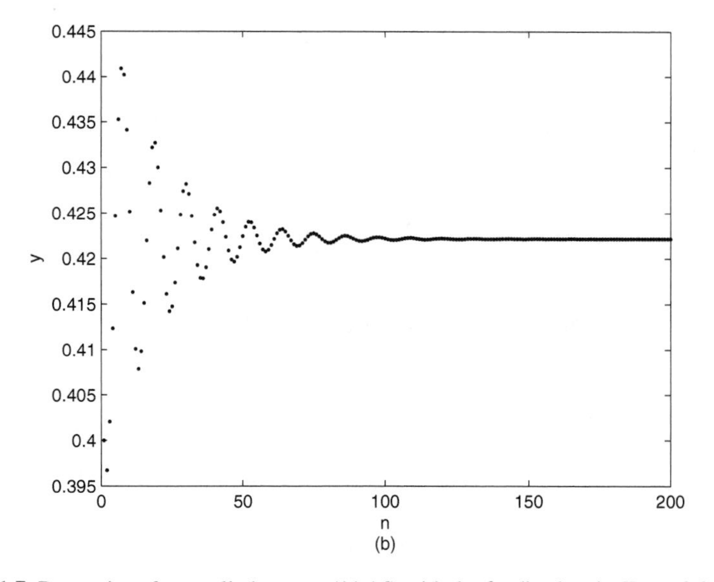

Fig. 11.7 Dynamics of controlled system (11.16) with the feedback gain $K = -0.2$ in $(n-y)$ plane.

References

1. Agiza, H.N., Elabbasy, E.M., El-Metwally, H., Elsadany, A.A.: Chaotic dynamics of a discrete prey-predator with Holling type II. Nonlinear Anal-Real. 10(1), 116–129 (2009)
2. Danca, M., Codreanu, S., Bako, B.: Detailed analysis of a nonlinear prey-predator model. J. Biol. Phys. 23, 11–20 (1997)
3. Guckenheimer, J., Holmes, P.: Nonlinear Oscillations, Dynamical Systems and Bifurcations of Vector Field. Springer, New York (1983)
4. Hale, J.K.: Theory of Functional Differential Equations. Springer, New York (1997)
5. Jing, Z.J., Yang, J.P.: Bifurcation and chaos in discrete-time predator-prey system. Chaos Soliton. Fract. 27(1), 259–277 (2006)
6. Liao, X.Y., Ouyang, Z.G., Zhou, S.F.: Permanence and stability of equilibrium for a two-prey one-predator discrete model. Appl. Math. Comput. 186(1), 93–100 (2007)
7. Liu, X., Xiao, D.: Complex dynamic behaviors of a discrete-time predator-prey system. Chaos Soliton. Fract. 32, 80–94 (2007)
8. Lotka, A.J.: Elements of Mathematical Biology. Dover, New York (1962)
9. May, R.M.: Simple mathematical models with very complicated dynamics. Nature 261, 459–467 (1976)
10. Murray, J.D.: Mathematical biology. Springer, New York (1989)
11. Summers, D., Justian, C., Brian, H.: Chaos in periodically forced discrete-time ecosystem models. Chaos Soliton. Fract. 11, 2331–2342 (2000)
12. Volterra, V.: Opere matematiche: mmemorie e note. Acc. Naz. dei Lincei, Roma, Cremon 4, 1914–1925 (1962)

Chapter 12
Mechanics and Control in Singular Physiological Complex Systems

12.1 Passivity and Optimal Control of Singular Biological Complex Systems

Passivity and passive control research for singular systems has been effectively developed in [22, 6, 9, 20, 2, 14, 13, 3, 18, 19]. However, there are few references involving any applications of these passive theorems for singular systems to biological complex systems. Therefore, passivity and its associated control research for biological complex systems are of great interest.

12.1.1 Model Formulation

Consider the physiological process of the endocrine disruptor Diethylstibestrol (DES) moving in human body. The model of this process can be developed in the following Fig. 12.1.

Let x_1,x_2,x_3, and x_4 be the quantity of the endocrine disruptor DES in different chambers (organs of human body). Let k_{ij} represent the rate of DES transferring from one chamber to another (from i to j, $i,j = 1,2,3,4$). Suppose that In denotes the quantity of intake and that the process of transferring is described by first-order reactions. Based on the above analysis, the differential equation, which follows the conceptual model, is established as follows:

$$\begin{cases} \dot{x}_1(t) = In_1(t) + k_{21}x_2 + k_{31}x_3 + k_{41}x_4 - k_{12}x_1 - k_{13}x_1 - k_{14}x_1, \\ \dot{x}_2(t) = k_{12}x_1 - k_{21}x_2, \\ \dot{x}_3(t) = In_3(t) + k_{13}x_1 - k_{31}x_3, \\ \dot{x}_4(t) = k_{14}x_1 - k_{41}x_4 - k_{45}x_4. \end{cases} \quad (12.1)$$

Q. Zhang et al.: Complex., Analy. and Cntrl. of Singular Biological Systems, LNCIS 421, pp. 171–183.
springerlink.com

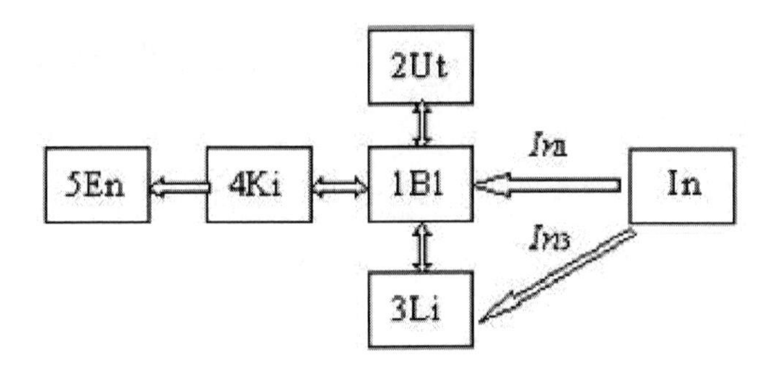

Fig. 12.1 Poly-chambers conceptual model of the process of endocrine disruptor DES transferring in the human body (1 B1: circulation systems and blood; 2 Ut: reproduction system and uterus; 3 Li: digestive system and liver; 4 Ki: excretion system and kidney; 5 En: environment outside body; In: ingesting and injection).

A medicinal filler to deliver the DES is designed to simultaneously contain two types of drugs. The initial values at different time points are given, and they satisfy the following formulas: $In_1(t) = (1/360)t, In_3(t) = (1/120)t, t = (0,1,2,\cdots,360)$, hence $In_3(t) = 3In(t)$. Suppose that $u(t) = In_1(t) = (1/360)t$, then model system (12.1) can be rewritten as

$$\begin{pmatrix} \dot{x}_1 \\ \dot{x}_2 \\ \dot{x}_3 \\ \dot{x}_4 \end{pmatrix} = \begin{pmatrix} -k_{12}-k_{31}-k_{41} & k_{21} & k_{31} & k_{41} \\ k_{12} & -k_{21} & 0 & 0 \\ k_{13} & 0 & -k_{31} & 0 \\ k_{14} & 0 & 0 & -k_{45}-k_{41} \end{pmatrix} \times \begin{pmatrix} x_1 \\ x_2 \\ x_3 \\ x_4 \end{pmatrix} + \begin{pmatrix} 1 \\ 0 \\ 3 \\ 0 \end{pmatrix} u(t). \tag{12.2}$$

For convenience, let

$$A = \begin{pmatrix} -k_{12}-k_{31}-k_{41} & k_{21} & k_{31} & k_{41} \\ k_{12} & -k_{21} & 0 & 0 \\ k_{13} & 0 & -k_{31} & 0 \\ k_{14} & 0 & 0 & -k_{45}-k_{41} \end{pmatrix}, B = \begin{pmatrix} 1 \\ 0 \\ 3 \\ 0 \end{pmatrix},$$

and $x = \begin{pmatrix} x_1 & x_2 & x_3 & x_4 \end{pmatrix}^T$.

According to differential equation models (12.1) and (12.2), the state equation and output equation can be expressed in the simple form,

$$\begin{cases} \dot{x}(t) = Ax(t) + Bu(t), \\ y(t) = Cx(t), \end{cases} \tag{12.3}$$

where $C = \begin{pmatrix} 0 & 0 & 0 & k_{45} \end{pmatrix}$.

According to singular system theory [5], if the dimension of the output variable is identical to that of the input variable, then model system (12.3) can take the following form:

$$\begin{cases} E\dot{x}(t) = A_1x(t) + B_1u(t) + B_2w(t), \\ y(t) = C_1x(t) + D_1u(t) + D_2w(t). \end{cases} \tag{12.4}$$

where $x(t), u(t)$, and $w(t)$ are the state variable, control input variable, and exogenous input variable, respectively.

It should be noted that the biological interpretations of the above variables are also obvious, which are given as follows: the quantity of endocrine disruptor medicine DES in different chambers, the intake quantity of endocrine disruptor medicine DES, and the influence of intake DES fluctuation on the quantity of DES in different chambers

$$E = \begin{pmatrix} 1\,0\,0\,0\,0 \\ 0\,1\,0\,0\,0 \\ 0\,0\,1\,0\,0 \\ 0\,0\,0\,1\,0 \\ 0\,0\,0\,0\,0 \end{pmatrix}.$$

A_1, B_1, C_1 can be obtained by calculating A and B, which are given as follows:

$$A_1 = \begin{pmatrix} -k_{12} - k_{31} - k_{41} & k_{21} & k_{31} & k_{41} & 0 \\ k_{12} & -k_{21} & 0 & 0 & 0 \\ k_{13} & 0 & -k_{31} & 0 & 0 \\ k_{14} & 0 & 0 & -k_{45} - k_{41} & 0 \\ 1 & 1 & 1 & 1 & 1 \end{pmatrix}, B_1 = \begin{pmatrix} 1 \\ 0 \\ 3 \\ 0 \\ -4 \end{pmatrix},$$

$$C_1 = \begin{pmatrix} 0\,0\,0\,k_{45}\,0 \end{pmatrix}.$$

B_2, D_1, and D_2 can be evaluated according to the physiological process rules of DES moving in a human body:

$$B_2 = \begin{pmatrix} 0.5 \\ 0 \\ 1.5 \\ 0 \\ -2 \end{pmatrix}, D_1 = 0, D_2 = 0.8.$$

12.1.2 Passive Description of Singular Biological Complex Systems

For model system (12.4), the following definition is given:

Definition 12.1. [7] A state space is strictly passive if there exist two functions $V(x)$ and $W(x)$ that are half positive definitive and positive definitive, respectively, and satisfy

$$\int_0^T w^T y\mathrm{d}t \geq V(x(\tau)) - V(x(0)) + \int_0^T W(x)\mathrm{d}t, \tag{12.5}$$

for all initial values $x(0)$ and times τ.

To determine the passivity, we differentiate system (12.5) from both sides with respect to τ. It is easy to show that if

$$w^T y \geq \dot{V}(x) + W(x) \tag{12.6}$$

is true for any initial value $x(0)$ and time τ, then the state space is strictly passive.

It is easy to determine a Lyapunov function for system (12.1) that is nonnegative. We can suppose $V = x^T E^T P x$ where P is to be determined and satisfies $E^T P = P^T E \geq 0$.

To obtain a strictly passive feedback controller, we first analyze the strictly passive features of free singular systems.

Theorem 12.1. *For a free system ($u(t) = 0$), if $D_2 + D_2^T > 0$ and $E^T P = P^T E \geq 0$, then there exists a feasible solution to the linear matrix inequality*

$$\begin{pmatrix} A_1^T P + P^T A_1 + \delta I & P^T B_2 - \frac{1}{2}C_1^T \\ B_2^T P - \frac{1}{2}C_1 & -\frac{1}{2}D_2 - \frac{1}{2}D_2^T \end{pmatrix}, \tag{12.7}$$

where $\delta > 0$, then the free singular system is strictly passive.

Proof. Note that

$$\begin{aligned} \dot{V} &= \dot{x}^T E^T P x + x^T P^T E \dot{x} \\ &= (A_1 x + B_2 w)^T P x + x^T P^T (A_1 x + B_2 w), \end{aligned}$$

where $\delta > 0$, and let $W(x) = \delta x^T x$, then it can be obtained that

$$\begin{aligned} &\dot{V} + W(x) - w^T y \\ &= (A_1 x + B_2 w)^T P x + x^T P^T (A_1 x + B_2 w) + \delta x^T x - w^T (C_1 x + D_2 w) \\ &= \begin{pmatrix} x \\ w \end{pmatrix}^T \begin{pmatrix} A_1^T P + P^T A_1 + \delta I & P^T B_2 - \frac{1}{2}C_1^T \\ B_2^T P - \frac{1}{2}C_1 & -\frac{1}{2}D_2 - \frac{1}{2}D_2^T \end{pmatrix} \begin{pmatrix} x \\ w \end{pmatrix} \leq 0. \end{aligned}$$

The proof is completed if there exists a feasible solution to linear matrix inequality (12.7), then free singular system (12.4) is strictly passive.

12.1.3 Strictly Passive Feedback Controller Design of a Biological Complex System

Consider closed-loop system (12.4).

Theorem 12.2. *For a singular system, if $\delta > 0, D_2 + D_2^T > 0$ and $E^T P = P^T E \geq 0$, then there exists a feasible solution to linear matrix inequality (12.8) ,*

$$\begin{pmatrix} (A_1X+B_1W)^T+(A_1X+B_1W) & B_2-\frac{1}{2}(C_1X+D_1W)^T & X^T \\ B_2^T-\frac{1}{2}(C_1X+D_1W) & -(\frac{1}{2}D_2+\frac{1}{2}D_2^T) & 0 \\ X & 0 & -\delta^{-1}I \end{pmatrix} < 0, \quad (12.8)$$

then the closed-loop system is strictly passive. Furthermore, if W^*, X^* *is one of the feasible solutions of (12.8), then the strictly passive feedback controller is given by* $u = Kx = W^*(X^*)^{-1}x$.

Proof. According to Theorem 12.1, for a closed-loop system such that $D_2 + D_2^T > 0$ and $E^T P = P^T E \geq 0$, if there exists a feasible solution,

$$\begin{pmatrix} \bar{A_1}^T + P^T\bar{A_1}+\delta I & P^TB_2-\frac{1}{2}\bar{C}^T \\ B_2^TP-\frac{1}{2}\bar{C} & -\frac{1}{2}D_2-\frac{1}{2}D_2^T \end{pmatrix} \leq 0, \quad (12.9)$$

then the closed-loop system is strictly passive, where $\bar{A}_1 = A_1 + B_1K$ and $\bar{C} = C_1 + D_1K$.

Premultiply inequality (12.9) by diag$(P^{-1})^T, I$ and postmultiply inequality (12.9) by diagP^{-1}, I. Let $X = P^{-1}$ and $W = KX$, and it can be obtained that inequality (12.9) is equivalent to (12.8) based on the Schur complementary theorem [3]. Hence, if W^*, X^* is a feasible solution of (12.8), the strictly passive feedback controller is given by $u = Kx = W^*(X^*)^{-1}x$.

12.1.4 Numerical Simulation for Strictly Passive State Feedback Controller

Numerical simulations are carried out to show the effectiveness of feedback controllers that are utilized for singular system (12.4).

Since k_{ij} represents the rate of DES transferring from one chamber to another, k_{ij} can be set based on the physiological process rules of DES moving in a human body, which are given as follows: $k_{21} = 0.2, k_{12} = 0.5, k_{31} = 0.5, k_{13} = 0.6, k_{41} = 0.2, k_{14} = 0.5$, and $k_{45} = 0.3$.

$$A = \begin{pmatrix} -1.6 & 0.2 & 0.5 & 0.2 & 0 \\ 0.5 & -0.2 & 0 & 0 & 0 \\ 0.6 & 0 & -0.5 & 0 & 0 \\ 0.5 & 0 & 0 & -0.5 & 0 \\ 1 & 1 & 1 & 1 & 1 \end{pmatrix},$$

$$C_1 = \begin{pmatrix} 0 & 0 & 0 & 0.3 & 0 \end{pmatrix}, E = \begin{pmatrix} 1 & 0 & 0 & 0 & 0 \\ 0 & 1 & 0 & 0 & 0 \\ 0 & 0 & 1 & 0 & 0 \\ 0 & 0 & 0 & 1 & 0 \\ 0 & 0 & 0 & 0 & 0 \end{pmatrix},$$

$$B_1 = \begin{pmatrix} 1 \\ 0 \\ 3 \\ 0 \\ -4 \end{pmatrix}, B_2 = \begin{pmatrix} 0.5 \\ 0 \\ 1.5 \\ 0 \\ -2 \end{pmatrix}, D_1 = 0, D_2 = 0.8.$$

According to Theorem 12.2, it is easy to obtain that $\delta = 0.0604$:

$$A = \begin{pmatrix} 3.4696 & -1.3604 & -0.8482 & -1.1303 & 0 \\ -1.3604 & 3.6110 & -1.6229 & -0.9016 & 0 \\ -0.8482 & -1.6229 & 4.7215 & -2.2589 & 0 \\ -1.1309 & -0.9616 & -2.2589 & 5.4964 & 0 \\ 0.3730 & -0.0564 & 1.5837 & -0.1952 & -4.2448 \end{pmatrix},$$

$$W = \begin{pmatrix} -0.474 & -0.0917 & -0.6733 & 0.0251 & 0.0563 \end{pmatrix}.$$

Therefore, the strictly passive state feedback controller is given by

$$u = Kx = \begin{pmatrix} -0.5157 & -1.5795 & -1.5242 & -1.1931 & -0.0133 \end{pmatrix} x.$$

12.2 Singular Model of Hepatitis B and Quarantine and Vaccination Control

Hepatitis B is a disease caused by HBV (hepatitis B virus), which infects the liver of hominoidae, including humans, and causes an inflammation called hepatitis. Hepatitis B is a potentially life-threatening liver infection caused by the hepatitis B virus. It is a major global health problem and the most serious type of viral hepatitis. It can cause chronic liver disease and puts people at high risk of death from cirrhosis of the liver and liver cancer.

Worldwide, an estimated two billion people have been infected with HBV, and more than 350 million have chronic (long-term) liver infections. A vaccine against hepatitis B has been available since 1982. The hepatitis B vaccine is ninety five percent effective in preventing HBV infection and its chronic consequences and is the first vaccine against a major human cancer. In the past decades, many mathematical models have been established to express the infective mechanism of HBV [23, 8, 15, 17, 1, 16, 21]. However, there are few references concentrating on carrying out associated strategies to control the disease.

12.2.1 Model Formulation

Generally, the transmission process of HBV is as follows: The hepatitis B virus is transmitted between people by contact with the blood or other body fluids (i.e., semen and vaginal fluid) of an infected person. Modes of transmission are the same for the human immunodeficiency virus (HIV), but HBV is 50 to 100 times more infectious. Unlike HIV, HBV can survive outside the body for at least 7 days. During that time, the virus can still cause infection if it enters the body of a person who is not infected.

According to the infectious mechanism of HBV, people can be divided into five groups: susceptible individuals (S); infected individuals (Y); asymptomatic individuals who have been exposed to the virus but have not yet developed clinical symptoms of HBV (C); symptomatic individuals (H); and recovered individuals (M). The transformation relationship among them is in the following Fig. 12.2, where μ represents the birth rate and death rate, total number of population can be maintained when birth rate is equal to death rate; λ represents infective rate; $\alpha, \beta, \gamma, k, \phi, \theta, \rho$ represents the transformation rate among different state, respectively. v represents the proportion of HBV carrier gravida, whose baby will become HBV carrier if there is not any quarantine control strategies.

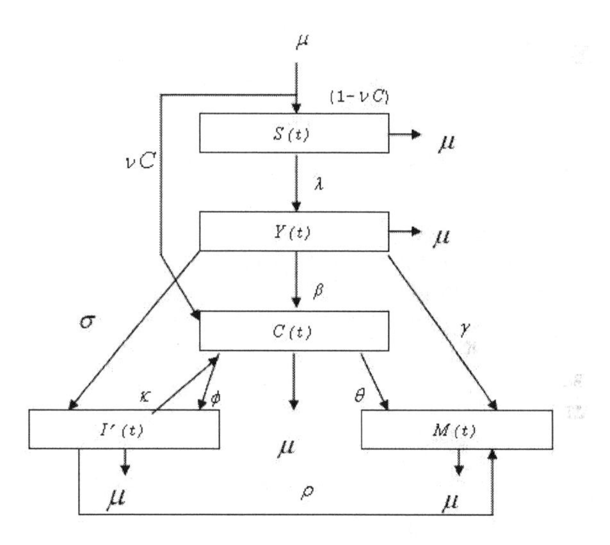

Fig. 12.2 State transformations among $S(t), I(t), C(t), Y(t)$ and $M(t)$.

Since $H(t)$ and $C(t)$ are all infectious and can be transformed with each other, these two groups can be generally summarized as a whole group that are represented by $I(t)$.

The transformation relationship among $S(t), I(t), Y(t)$ and $M(t)$ can be shown in the following Fig. 12.3, where μ, λ share the same physiological interpretations shown in Fig. 12.2; β, θ, γ represents the transformation rate among different states, respectively; k represents the proportion of HBV carrier gravidas in the whole infective population.

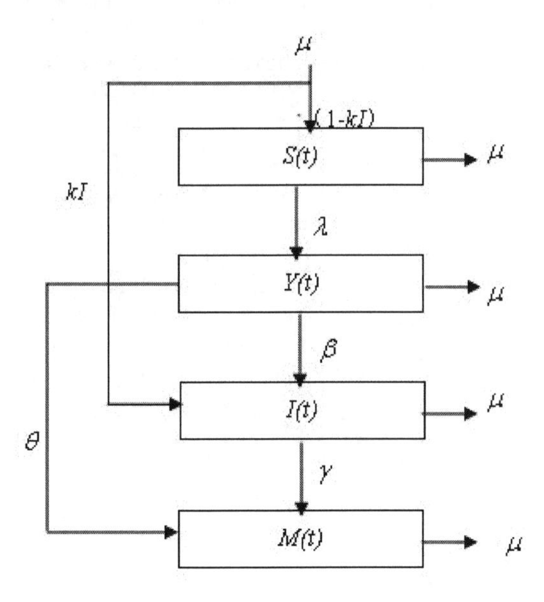

Fig. 12.3 State transformations among $S(t), I(t), Y(t)$, and $M(t)$.

Suppose the total number $N(t)$ of the population is constant at time t. People can be divided into four different groups: $S(t), Y(t), I(t)$, and $M(t)$, who all have natural death rate and birth rate μ, $S(t)+I(t)+Y(t)+M(t)=1$.

According to the above assumptions, a singular model can be established as follows:

$$\begin{cases} \dfrac{\mathrm{d}S}{\mathrm{d}t} = \lambda SI - \mu S - \mu kI + \mu, \\ \dfrac{\mathrm{d}Y}{\mathrm{d}t} = \lambda SI - (\mu+\beta+\theta)Y, \\ \dfrac{\mathrm{d}I}{\mathrm{d}t} = \beta Y - (\mu+\gamma-\mu k)I, \\ 0 = 1 - S - I - Y - M. \end{cases} \tag{12.10}$$

The initial values are given as follows: $S(0)=S_0>0, I(0)=I_0>0, Y(0)=Y_0>0, M_0 \geq 0$.

12.2.2 Quarantine and Vaccination Control for HBV

Let $B = \{(S,Y,I,M) : S \geq 0, Y \geq 0, I \geq 0, M \geq 0, S+Y+I+M \leq 1\}$. It is easy to show that B is a positive invariant set of model system (12.10). For model system (12.10), there are two equilibria: $G(1,0,0,0)$ and $F(S^*,I^*,Y^*,M^*)$, where $S^* = \frac{(\mu+\gamma-k\mu)(\mu+\beta+\theta)}{\lambda\beta}, I^* = \frac{\mu(1-S^*)}{\lambda S^*+k\mu}, Y^* = \frac{\mu+\gamma-k\mu}{\beta}I^*, M^* = 1-S^*-I^*-Y^*$.

Theorem 12.3. *The sufficient and necessary condition for the existence of the interior equilibrium of model system (12.10) is $S^* < 1$, which follows that*

$$\beta\lambda-(\mu+\gamma-k\mu)(\mu+\beta+\theta)>0.$$

Theorem 12.4. *If the interior equilibrium of model system (12.10) exists, then interior equilibrium $G(1,0,0,0)$ is unstable.*

Proof. The characteristic equation of the model system evaluated around interior equilibrium $G(1,0,0,0)$ is as follows:

$$(x+\mu)[x^2+(2\mu+\beta+\theta+\gamma-k\mu)x+(\mu+\gamma-k\mu)(\mu+\beta+\theta)-\beta\lambda]=0. \quad (12.11)$$

It follows from Theorem 12.3 that there exist two positive roots and one negative root for Eq. (12.11). Hence, model system (12.10) is not stable around interior equilibrium $G(1,0,0,0)$. With the purpose of stabilizing the model system around the interior equilibrium, some associated controlling strategies are designed in the following part.

Remark 12.1. Generally, HBV infection is inversely proportional to age, and fetuses are the group that can be mostly exposed to infection. Hence, it is necessary for HBV carrier gravidas to take corresponding vaccination strategies, and then infants can avoid contracting HBV. Among these strategies, a combination of HBV bacterin and immunoglobulin is considered as a constructive way, whose efficiency is up to ninety-seven percent [12, 11]. In the meanwhile, some immunity control measures can be applied, which are beneficial to eliminating HBV from the infant body.

Consider the quarantine and vaccination control strategies, model system (12.10) can be rewritten as follows:

$$\begin{cases} \frac{\mathrm{d}S}{\mathrm{d}t} = \lambda(1-\omega)SI-\mu S+(1-\omega)\mu, \\ \frac{\mathrm{d}Y}{\mathrm{d}t} = \lambda(1-\omega)SI-(\mu+\beta+\theta)Y, \\ \frac{\mathrm{d}I}{\mathrm{d}t} = \beta Y-(\mu+\gamma)I, \\ 0=1-S-I-Y-M. \end{cases} \quad (12.12)$$

where ω represents immunity rate. The interior equilibria of model system (12.12) can be obtained as follows: one is $\acute{E}(1-\omega,0,0,\omega)$, another is $\acute{F}(\acute{S}^*,\acute{I}^*,\acute{Y}^*,\acute{M}^*)$, where $\acute{S}^* = \frac{(\mu+\gamma-k\mu)(\mu+\beta+\theta)}{\lambda\beta(1-\omega)}$, $\acute{I}^* = \frac{\mu(1-\omega-\acute{S}^*)}{\lambda\acute{S}^*}$, $\acute{Y}^* = \frac{\mu+\gamma}{\beta}\acute{I}^*$, and $\acute{M}^* = 1-\acute{S}^*-\acute{S}^*-\acute{I}^*-\acute{Y}^*$.

By using quarantine and vaccination control strategies, the interior equilibrium $\acute{E}(1-\omega,0,0,\omega)$ is expected to be a globally stable equilibrium, which biologically interprets the elimination of HBV.

Let $\bar{S}=S-(1-\omega),\bar{Y}=Y,\bar{I}=I,\bar{M}=M-\omega$, and then $\omega-1\leq\bar{S}\leq\omega$, $-\omega\leq\tilde{M}\leq1-\omega$; model system (12.12) is transformed as follows:

$$\begin{cases}\dfrac{\mathrm{d}\bar{S}}{\mathrm{d}t}=\lambda(1-\omega)(\bar{S}+1-\omega)\bar{I}-\mu\bar{S}+(1-\omega)\mu,\\ \dfrac{\mathrm{d}\bar{Y}}{\mathrm{d}t}=\lambda(1-\omega)(\bar{S}+1-\omega)\bar{I}-(\mu+\beta+\theta)\bar{Y},\\ \dfrac{\mathrm{d}\bar{I}}{\mathrm{d}t}=\beta\bar{Y}-(\mu+\gamma)\bar{I},\\ 0=\bar{S}+\bar{I}+\bar{Y}+\bar{M}.\end{cases}\tag{12.13}$$

It is obvious that $C(0,0,0,0)$ is the interior equilibrium of model system (12.13).

Theorem 12.5. *If* $\beta\lambda-(\mu+\gamma-k\mu)(\mu+k+\theta)>0$, *there must exist* $0<\omega\leq1$, *which satisfies that* $1-\dfrac{(\mu+\gamma)(\mu+\beta+\theta)}{\lambda\beta}\leq\omega\leq1$.

Proof. If $\beta\lambda-(\mu+\gamma-k\mu)(\mu+k+\theta)>0$, there must exist $0<\omega\leq1$, which satisfies that $(1-\omega)\beta\lambda-(\mu+\gamma)(\mu+\beta+\theta-k\mu)<0$.

Since $(1-\omega)\beta\lambda-(\mu+\gamma)(\mu+\beta+\theta)<(1-\omega)\beta\lambda-(\mu+\gamma)(\mu+\beta+\theta-k\mu)$, it can be obtained that $(1-\omega)\beta\lambda-(\mu+\gamma)(\mu+\beta+\theta)<0$, which follows that $(1-\omega)\beta\lambda-(\mu+\gamma)(\mu+\beta+\theta-k\mu)<0$.

Theorem 12.6. *For model system (12.13), if* $1-\dfrac{(\mu+\gamma)(\mu+\beta+\theta)}{\lambda\beta}<\omega\leq1$, *then interior equilibrium* $C(0,0,0,0)$ *is globally asymptotically stable.*

Proof. Constructing the following Lyapunov function, $V=\bar{Y}+\dfrac{\mu+\beta+\theta}{\beta}\bar{I}$.

By differentiating V with respect to t, it derives that, $\dfrac{\mathrm{d}V}{\mathrm{d}t}=\lambda(1-\omega)\bar{I}(\bar{S}-\omega)+\dfrac{\lambda(1-\omega)\beta-(\mu+\gamma)(\mu+\beta+\theta)}{\beta}\bar{I}$.

It can be derived that $V>0$ due to the positivity of $\bar{Y}$ and $\bar{I}$. According to $0<\omega\leq1$ and $\bar{S}\leq\omega$, it is easy to show $\dfrac{\mathrm{d}V}{\mathrm{d}t}<0$, then model system (12.13) is globally asymptotically stable around $C(0,0,0,0)$, which derives that model system (12.12) is globally asymptotically stable around $E(1-\omega,0,0,\omega)$.

12.2.3 Numerical Simulation

In 1985, a clinical HBV survey was carried out in four provinces (Hebei province, Hunan province, Heilongjiang province, Henan province) of P.R. China, and the age range was from 0 to 70. It showed that 6,051 out of 10,484 were infected by HBV. After a 2-year blood serum transformation survey, the proportion of HBV carrier was up to around 10.1[4]. Suppose that the infected were transformed as virus carriers with the rate ε and the average interval $\frac{1}{\varepsilon}$. It follows that

$$\theta = \frac{1-\beta-\mu}{\varepsilon}.$$

Here, we take $\mu = \frac{1}{75} = 0.013, k = 0.001, \gamma = 0.001, \beta = 0.2506$, and then $\theta = 1.4278, \lambda = 0.3809$. Fig. 12.4 shows dynamical behavior of model system (12.10) before control, and Fig. 12.5 shows dynamical behavior of model system (12.10) after control. Based on Fig. 12.5, it can be concluded that the control strategies designed in this chapter are useful to control the infection of HBV and total number of infected tends to zero.

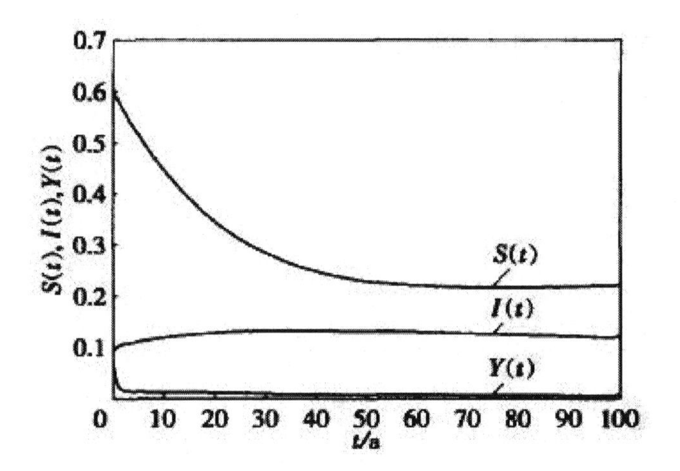

Fig. 12.4 Dynamical behavior of model system (12.10).

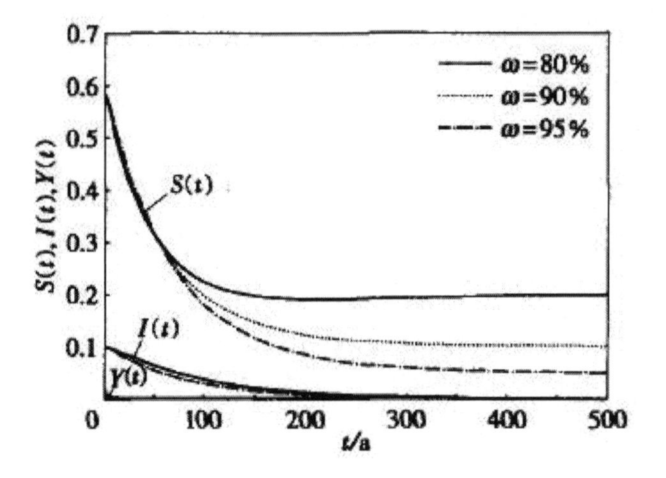

Fig. 12.5 Dynamical behavior of model system (12.10) with control.

References

1. Alfonseca, M., Bravo, M.T.M., Torrea, J.L.: Mathematical model for the analysis of hepatitis B and AIDS epidemics. Simu. 74(4), 219–226 (2000)
2. Belura, M.N., Trentelmanb, H.L.: The strict dissipativity synthesis problems and the rank of the coupling QDF. Syst. Cont. Lett. 51, 247–258 (2004)
3. Boyd, S., Ghaoui, E., Feron, E., Balakrishnan, V.: Linear Matrix Inequalities in Systems and Control Theory. SIAM, Philadelphia (1994)
4. Cao, H.L.: Prevalent situation of hepatitis virus B in China. Chi. J. Vac. Immu. 2(2), 88–90 (1996)
5. Dai, L.: Singular Control System. Springer, New York (1989)
6. Dong, X.Z., Zhang, Q.L.: Robust passive control for singular systems with time-varying uncertainties. Cont. Theo. Appl. 21, 517–520 (2004)
7. Dong, X.Z., Zhang, Q.L., Zhao, L.C.: Passive control of linear singular systems. J. Biomath. 19, 185–187 (2004)
8. Gohary, A.E., Bukhari, F.A.: Optimal stabilization of steady states of the genital herpes epidemic during infinite and finite time intervals. Appl. Mathe. Compu. 137(1), 33–47 (2003)
9. Fridman, E., Shaked, U.: On delay-dependent passivity. IEEE Trans. Auto. Cont. 47, 664–669 (2002)
10. Harrison, P.T.: Reproductive health in humans and wildlife: Are adverse trends associated with environmental chemical exposure. Sci. Tot. Envi. 205, 97–106 (1997)
11. Hung, J.H., Chu, C.J., Sung, P.L., Chen, C.Y., Chao, K.C., Yang, M.J., Huang, S.C.: Lamivudine therapy in the treatment of chronic hepatitis B with acute exacerbation during pregnancy. J. Chi. Med. Asso. 71(3), 155–158 (2008)
12. Ingardia, C.J., Kelley, L., Steinfeld, J.D., Wax, J.R.: Hepatitis B vaccination in pregnancy: Factors influencing efficacy. Obs. Gyn. 93(6), 983–986 (1999)
13. Jimenez, B.: Environmental effects of endocrine disruptors and current methodologies for assessing wildlife health effects. Tre. Anal. Che. 16, 596–606 (1997)

14. Li, G.F., Li, H.Y.: Observer-based passive control for uncertain linear systems with delay in state and control input. Chn. Phys. 14, 2379–2386 (2005)
15. Matveev, A.S., Savkin, A.V.: Application of optimal control theory to analysis of cancer chemotherapy regimens. Syst. Cont. Lett. 46(5), 311–321 (2002)
16. Medley, G.F., Lindop, N.A., Edmunds, W.J.: Hepatitis B virus endemicity: Heterogeneity catastrophic dynamics and control. Nat. Med. 7(5), 619–624 (2001)
17. Ogren, P., Martin, C.F.: Vaccination strategeies for epidemics in highly mobile populations. Appl. Mathe. Compu. 137(1), 33–47 (2003)
18. Tsuguchika, K.: Modeling of signaling pathways for endocrine disruptors. Bios. 55, 23–31 (1999)
19. Warren, G.F.: Human development exposure to endocrine active compounds. Envi. Toxi. Pha. 12, 75–81 (2002)
20. Xie, S.L., Xie, L.H.: Robust dissipative control for linear systems with dissipative uncertainty and nonlinear perturbation. Syst. Cont. Lett. 29, 255–268 (1997)
21. Xu, J., Zhou, Y.C.: Hepatitis epidemics models with proportional and pulse vaccination. J. Biomath. 19(2), 149–155 (2004)
22. Yu, L., Pan, H.T.: Robust passive control of linear systems with time-varying uncertain parameters. Acta Auto. Sini. 24, 368–372 (1998)
23. Zhao, S.J., Xu, Z.Y., Lu, Y.: Mathematical model of hepatitis B virus transmission and its application for vaccination strategy in China. Int. J. Epide. 29(4), 744–752 (2000)

Chapter 13
Complex Dynamical Behavior in Bio-economic Prey-Predator Models with Competition for Prey

13.1 Introduction

In recent years, harvesting in the prey-predator ecosystem is one of the most important fields of interest. Much research effort [1, 3, 2, 14, 11, 16, 5, 6, 8] has been put into investigating the interaction and coexistence mechanism of the harvested prey-predator ecosystem. A numerical analysis of a harvested prey-predator model is performed in [1], and the model can be expressed as follows:

$$\begin{cases} \dot{x}_1(t) = x_1(t)(\varepsilon_1 - a_{11}x_1(t) - a_{12}x_2(t) - a_{13}y(t)), \\ \dot{x}_2(t) = x_2(t)(\varepsilon_2 - a_{21}x_1(t) - a_{22}x_2(t) - a_{23}y(t)), \\ \dot{y}(t) = y(t)(-\varepsilon_3 + a_{31}x_1(t) + a_{32}x_2(t)) - H(t), \end{cases} \tag{13.1}$$

where $x_1(t)$ and $x_2(t)$ denote populations of the prey 1 and prey 2, respectively; $y(t)$ represents population of the predator; $\varepsilon_i (i = 1,2,3)$ are intrinsic rates of growth and decay of the three species; $a_{ij}(i, j = 1,2,3)$ with $i \neq j$ are the inter-species and $a_{ii}(i = 1,2)$ are the intra-species coefficients of competitive interactions; a_{13} and a_{23} are the coefficients for the loss of prey 1 and prey 2, respectively; a_{31} and a_{32} are the coefficients for growth in the predator as a result of consumption of prey by them; and $H(t)$ is a harvest function, which represents harvesting strategies applied to the predator.

In [1], two different harvesting strategies for the predator, namely, constant harvest quota $H(t) = H_q$ (H_q is a constant) and constant harvest effort $H(t) = hy(t)$ ($H(t)$ is proportional to the predator population $y(t)$, where h is a constant) are applied, and the temporal behavior of the population dynamics for different harvest intensities is investigated, which reveals stability of the entire system on introduction of the predator to the prey species system. The model proposed in [1] is further investigated in [10] by using Hopf bifurcation theory. The authors there study an interacting one-predator two-prey model with harvesting of the predator at a constant rate, where harvest rate is used as a control parameter. It is found that when a

Q. Zhang et al.: Complex., Analy. and Cntrl. of Singular Biological Systems, LNCIS 421, pp. 185–201.
springerlink.com

particular species is harvested, the equilibrium state is drastically changed. Furthermore, the stable equilibrium state becomes unstable and bifurcates to the periodic solution. Along with this line of research, the work done in [10] is extended in [13]. By assuming that the reproduction of the predator after seizing the prey is not instantaneous but mediates by some discrete time lag required for gestation of the predator, some discrete time delays are incorporated into the model proposed in [13]. By choosing the time delay as a bifurcation parameter, it is shown that Hopf bifurcation can occur as the time delay crosses some critical values.

It is well known that harvesting has a strong impact on the dynamic evolution of a population. Nowadays, the biological resources in the prey-predator ecosystem are commercially harvested and sold with the aim of achieving economic interest. By considering the economic interest of harvest effort on the predator, a differential-algebraic model is proposed in this chapter. It is utilized to investigate the dynamical behavior of the prey-predator ecosystem due to the variation of economic interest of harvesting, which extends the previous work done in [1, 10, 13] from an economic perspective. We aim to obtain some results that are theoretically beneficial to maintaining the sustainable development of the ecosystem as well as keeping the persistent prosperity of harvesting.

The rest of this chapter is organized as follows. In the second section, a differential-algebraic model, which investigates a prey-predator system with harvest effort on predator, is established. The third section deals with the local stability analysis of the proposed model system around the interior equilibrium. In order to stabilize the proposed model system around the interior equilibrium, a state feedback controller is designed and numerical simulations are given in the fourth section. Finally, this chapter ends with a conclusion.

13.2 Model Formulation

In 1954, Gordon proposed the economic theory of a common-property resource in [7], which studies the effect of the harvest effort on the ecosystem from an economic perspective. In [7], an equation is proposed to investigate the economic interest of the yield of the harvest effort, which takes form as follows:

Net Economic Revenue (NER) = Total Revenue (TR)łTotal Cost (TC). (13.2)

Let $E(t)$ and $Y(t)$ represent the harvest effort and the density of the harvested population, respectively, and then $TR = wE(t)Y(t)$, $TC = cE(t)$, w represents unit price constant, and c is a constant representing the unit cost of harvest effort. Associated with model system (13.1), an algebraic equation, which considers the economic interest v of the harvest effort $E(t)$ on the predator $y(t)$, is established as follows:

$$E(t)(wy(t) - c) = v. \tag{13.3}$$

Recently, a class of differential-algebraic biological economic models is proposed in [17, 12, 18, 19], which is established by several differential equations and an algebraic equation. Especially, the algebraic equations are established based on the economic theory proposed in [7]. Compared with the conventional harvested models proposed in the previous related references (see [1, 3, 2, 14, 11, 16, 5, 6, 10, 13] and the references therein), the models proposed in [17, 12, 18, 19] not only investigate the interaction mechanism of the harvested prey-predator ecosystem, but also offer a simpler way to explore dynamical behavior due to the variation of economic interest of harvesting.

Inspired by the work done in [17, 12, 18, 19], a differential-algebraic model of three species (two prey populations and one predator population) is established. Based on (13.1) and (13.3), $H(t)$ takes the form of $H(t) = E(t)y(t)$ and a differential-algebraic model that consists of three differential equations and an algebraic equation can be established as follows:

$$\begin{cases} \dot{x}_1(t) = x_1(t)(\varepsilon_1 - a_{11}x_1(t) - a_{12}x_2(t) - a_{13}y(t)), \\ \dot{x}_2(t) = x_2(t)(\varepsilon_2 - a_{21}x_1(t) - a_{22}x_2(t) - a_{23}y(t)), \\ \dot{y}(t) = y(t)(-\varepsilon_3 + a_{31}x_1(t) + a_{32}x_2(t)) - E(t)y(t), \\ 0 = E(t)(wy(t) - c) - v, \end{cases} \tag{13.4}$$

where $x_1(t), x_2(t), y(t), \varepsilon_i(i = 1,2,3), a_{ij}(i,j = 1,2,3), a_{ii}(i = 1,2), w, c,$ and v have the same biological interpretations as mentioned in (13.1) and (13.3).

Differential-algebraic model system (13.4) can be expressed in the following form:

$$A(t)\dot{X}(t) = G(x_1(t), x_2(t), y(t), E(t)), \tag{13.5}$$

where $X(t) = (x_1(t), x_2(t), y(t), E(t))^T$, $A(t) = \begin{bmatrix} 1 & 0 & 0 & 0 \\ 0 & 1 & 0 & 0 \\ 0 & 0 & 1 & 0 \\ 0 & 0 & 0 & 0 \end{bmatrix}$,

and $G(x_1(t), x_2(t), y(t), E(t)) = \begin{bmatrix} x_1(t)(\varepsilon_1 - a_{11}x_1(t) - a_{12}x_2(t) - a_{13}y(t)) \\ x_2(t)(\varepsilon_2 - a_{21}x_1(t) - a_{22}x_2(t) - a_{23}y(t)) \\ y(t)(-\varepsilon_3 + a_{31}x_1(t) + a_{32}x_2(t)) - E(t)y(t) \\ E(t)(wy(t) - c) - v \end{bmatrix}$.

Remark 13.1. The algebraic equation in differential-algebraic model system (13.5) contains no differentiated variables, hence, the leading matrix $A(t)$ in (13.5) has a corresponding zero row.

Remark 13.2. The models proposed in [1, 10, 13], which consist of differential equations, only discuss the interaction and coexistence mechanism of prey-predator. Compared with the models proposed in [1, 10, 13], an algebraic equation is introduced into the model established in this chapter, which focuses on the economic interest of harvesting. Hence, the established model investigates not only interaction and coexistence mechanism of the harvested prey-predator system, but also the dynamical behavior due to the variation of economic interest of harvesting. The theorem in the following section investigates the dynamical behavior of differential-algebraic model system (13.4) around the interior equilibrium due to variation of economic interest of the harvesting.

Remark 13.3. It should be noted that the differential-algebraic models presented in [17, 12, 18, 19] investigate only some biological-economic systems with a single population or population with stage structure. Compared with these works, the model proposed in this chapter investigates the dynamical behavior of three more interacting species (two prey and one harvested predator), which means the work done here has some new and positive features.

13.3 Local Stability Analysis

In this section, dynamical behavior of differential-algebraic system (13.4) is investigated by using the differential-algebraic system theory and bifurcation theory. By performing the local stability analysis of the model system around the interior equilibrium, the bifurcation phenomenon that can reveal the instability mechanism of the model system is investigated, and the biological interpretations of the bifurcation phenomenon are also discussed. In this chapter, we concentrate only on the interior equilibrium of model system (13.4), since the biological interpretations of interior equilibrium imply that prey, predators, and harvest effort on predators all exist, which is relevant to our study.

Theorem 13.1. *Differential-algebraic model system (13.4) has a singularity induced bifurcation around the interior equilibrium, $v = 0$ is a bifurcation value, and model system (13.4) is unstable around the interior equilibrium as v increases through* 0.

Proof. According to the economic theory of a common-property resource in [7], there is a phenomenon of bio-economic equilibrium when the economic interest of harvesting is zero ($v = 0$). For model system (13.4), an interior equilibrium can be obtained in the case of the phenomenon of bio-economic equilibrium, i.e., $P^*(x_1^*, x_2^*, y^*, E^*)$, where

$$x_1^* = \frac{a_{12}(w\varepsilon_2 - ca_{23}) - a_{22}(w\varepsilon_1 - ca_{13})}{w(a_{12}a_{21} - a_{11}a_{22})},$$

$$x_2^* = \frac{a_{21}(w\varepsilon_1 - ca_{13}) - a_{11}(w\varepsilon_2 - ca_{23})}{w(a_{12}a_{21} - a_{11}a_{22})}, y^* = \frac{c}{w},$$

$$E^* = \frac{(a_{12}a_{31} - a_{11}a_{32})(w\varepsilon_2 - ca_{23}) + (a_{21}a_{32} - a_{22}a_{31})(w\varepsilon_1 - ca_{13})}{w(a_{12}a_{21} - a_{11}a_{22})} - \frac{w\varepsilon_3(a_{21}a_{12} - a_{11}a_{22})}{w(a_{12}a_{21} - a_{11}a_{22})}.$$

Like Azar et al. [1], Kumar et al. [10], and Liu et al. [13], values of parameters used in our investigation are also chosen as follows:

$\varepsilon_1 = \varepsilon_2 = \varepsilon_3 = a_{11} = a_{12} = a_{22} = a_{23} = 1$,

$a_{21} = 1.5$, $a_{32} = 0.5$ and $a_{13} = 2a_{31} = a$,

where a is a positive constant.

Since $a_{12} < a_{21}$, the first prey has a competitive advantage, i.e., $x_1(t)$ is the dominant prey and $x_2(t)$ is the sub-dominant prey, and the relation $a_{12}a_{21} - a_{11}a_{22} > 0$ implies that the system is unstable without predation.

By virtue of these parameter values, $P^*(x_1^*, x_2^*, y^*, E^*)$ can be rewritten as follows: $x_1^* = \frac{2(a-1)c}{w}, x_2^* = \frac{w+2c-3ac}{w}, E^* = \frac{2(a-1)(w-c)+(w-ac)(3-2a)-2w}{2w}$, $y^* = \frac{c}{w}$,. Based on the biological interpretations of the interior equilibrium, the following expressions are satisfied:

$$\begin{cases} a > 1, \\ w + 2c - 3ac > 0, \\ 2(a-1)(w-c) + (w-ac)(3-2a) - 2w > 0. \end{cases} \tag{13.6}$$

Let v be a bifurcation parameter, $U(t) = (x_1(t), x_2(t), y(t))^T$, and

$$f(U(t), E(t), v) = \begin{bmatrix} x_1(t)(\varepsilon_1 - a_{11}x_1(t) - a_{12}x_2(t) - a_{13}y(t)) \\ x_2(t)(\varepsilon_2 - a_{21}x_1(t) - a_{22}x_2(t) - a_{23}y(t)) \\ y(t)(-\varepsilon_3 + a_{31}x_1(t) + a_{32}x_2(t)) - E(t)y(t) \end{bmatrix},$$

$$g(U(t), E(t), v) = E(t)(wy(t) - c) - v.$$

By simple computing, $g(U(t), E(t), v)|_{P^*} = \begin{cases} 0, & v=0; \\ \text{nonzero}, & v \neq 0. \end{cases}$, which implies

$$\operatorname{dimker}(D_E g(U(t), E(t), v)|_{P^*}) = 1, \tag{13.7}$$

$$\operatorname{trace}[D_E f \operatorname{adj}(D_E g) D_U g]|_{P^*} = \frac{c[2w - 2(a-1)(w-c) - (w-ac)(3-2a)]}{2w}.$$

It follows from (13.6) that

$$\operatorname{trace}[D_E f \operatorname{adj}(D_E g) D_U g]|_{P^*} \neq 0. \tag{13.8}$$

It can be calculated that

$$\begin{vmatrix} D_U f & D_E f \\ D_U g & D_E g \end{vmatrix}|_{P^*} = \frac{c^2(1-a)(w+2c-3ac)[2(a-1)(w-c)+(w-ac)(3-2a)-2w]}{2w^3}.$$

By virtue of (13.6), it follows that

$$\begin{vmatrix} D_U f & D_E f \\ D_U g & D_E g \end{vmatrix}|_{P^*} \neq 0. \tag{13.9}$$

According to Part A of Section IV in [15], $\Delta(U(t),E(t),v)$ can be defined as follows: $\Delta(U(t),E(t),v) = D_E g(U(t),E(t),v) = wy(t) - c$.

By simple computing,

$$\begin{vmatrix} D_U f & D_E f & D_v f \\ D_U g & D_E g & D_v g \\ D_U \Delta & D_E \Delta & D_v \Delta \end{vmatrix} |_{P^*} = \frac{2c^2(1-a)(w+2c-3ac)}{2w^2}.$$

By virtue of (13.6) it gives that

$$\begin{vmatrix} D_U f & D_E f & D_v f \\ D_U g & D_E g & D_v g \\ D_U \Delta & D_E \Delta & D_v \Delta \end{vmatrix} |_{P^*} \neq 0. \tag{13.10}$$

Based on the above analysis, four items can be obtained as follows:

(i) It can be computed that $f(U(0),E(0),0) = 0$, $g(U(0),E(0),0) = 0$, and $D_E g(U(t),E(t),v)|_{P^*}$ has an algebraically simple zero eigenvalue. Furthermore, it follows from (13.8) that $\text{trace}[D_E f \text{adj}(D_E g) D_U g]|_{P^*} \neq 0$.

(ii) From (13.9), it can be shown that $\begin{bmatrix} D_U f & D_E f \\ D_U g & D_E g \end{bmatrix}$ is nonsingular at P^*.

(iii) By virtue of (13.10), it can be shown that $\begin{bmatrix} D_U f & D_E f & D_v f \\ D_U g & D_E g & D_v g \\ D_U \Delta & D_E \Delta & D_v \Delta \end{bmatrix}$ is nonsingular at P^*, so rank $\begin{bmatrix} D_U f & D_E f & D_v f \\ D_U g & D_E g & D_v g \\ D_U \Delta & D_E \Delta & D_v \Delta \end{bmatrix} = 5$.

(iv) Since $\text{rank}(f(U(t),E(t),v)) = 3$ and $\text{rank}(g(U(t),E(t),v)) = 1$, it can be concluded that
rank $\begin{bmatrix} D_U f & D_E f & D_v f \\ D_U g & D_E g & D_v g \\ D_U \Delta & D_E \Delta & D_v \Delta \end{bmatrix}$
$= \text{rank}(f(U(t),E(t),v)) + \text{rank}(g(U(t),E(t),v)) + 1$.

The condition for occurrence of singularity induced bifurcation is introduced in Theorem 3 in [15]. According to the four above items (i-iv), it follows that the condition for occurrence of singularity induced bifurcation is satisfied, hence, differential-algebraic model system (13.4) has a singularity induced bifurcation around the interior equilibrium P^* and the bifurcation value is $v = 0$.

Along with the line of the above proof, for differential-algebraic model system (13.4), by simple computing,

$$M = -\text{trace}[D_E f \text{adj}(D_E g) D_U g]|_{P^*} = \frac{c[2(a-1)(w-c) + (w-ac)(3-2a) - 2w]}{2w},$$

$$N = [D_v\Delta - (D_U\Delta, D_E\Delta)\begin{bmatrix} D_U f & D_E f \\ D_U g & D_E g \end{bmatrix}^{-1}\begin{pmatrix} D_v f \\ D_v g \end{pmatrix}]|_{P^*}$$
$$= \frac{2w}{2(a-1)(w-c)+(w-ac)(3-2a)-2w},$$

and then it can be obtained that ,

$$\frac{M}{N} = \frac{c[2(a-1)(w-c)+(w-ac)(3-2a)-2w]^2}{4w^2} > 0. \tag{13.11}$$

The inequality (13.11) satisfies Theorem 3 in [15]. According to Theorem 3 in [15], when v increases through 0, one eigenvalue of differential-algebraic model system (13.4) moves from C^- to C^+ along the real axis by diverging through ∞. According to the stability theory in [9], it can be concluded that model system (13.4) is unstable around the interior equilibrium as v increases through 0.

Remark 13.4. As stated in Theorem 3.1, model system (13.4) is unstable in the case of the positive economic interest of harvesting. Concerning the economic interest of harvesting, it should be restricted within an interval from the viewpoint of mathematical perspective, which guarantees the existence of all species in the harvested ecosystem. Let v^* represent a positive economic interest of harvesting, then the interior equilibrium in the case of positive economic interest v^* is obtained, i.e., $\tilde{P}^*(\tilde{x}_1^*, \tilde{x}_2^*, \tilde{y}^*, \tilde{E}^*)$, where $\tilde{x}_1^* = 2(a-1)\tilde{y}^*$, $\tilde{x}_2^* = 1+(2-3a)\tilde{y}^*$, $\tilde{E}^* = \frac{v^*}{w\tilde{y}^*-c}$, and $\tilde{y}^*$ satisfy the following equation:

$$w(2a^2-5a+2)\tilde{y}^{*2} - (w+c(2a^2-5a+2))\tilde{y}^* + (c-2v^*) = 0. \tag{13.12}$$

According to the biological interpretations of interior equilibrium $\tilde{P}^*$, the following inequalities should also be satisfied:

$$\begin{cases} \tilde{y}^* > 0, \\ a > 1, \\ 1+(2-3a)\tilde{y}^* > 0, \\ w\tilde{y}^* - c > 0. \end{cases} \tag{13.13}$$

Based on the above analysis, the cases for solutions of Eq. (13.12) due to the variation of positive economic interest v^* are obtained as follows:

Case 1. If $2a^2-5a+2=0$, then Eq. (13.12) has a unique solution, i.e., $\tilde{y}^* = \frac{c-2v^*}{w}$. By simple calculation, $w\tilde{y}^* - c = -2v^* < 0$, which contradicts (13.13).

Case 2. If $2a^2-5a+2 \neq 0$ and $v^* > \frac{c}{2} - \frac{[w+c(2a^2-5a+2)]^2}{8w(2a^2-5a+2)}$, then Eq. (13.12) has two solutions, which are denoted as $\tilde{y}_1^*$ and $\tilde{y}_2^*$ ($\tilde{y}_1^* > \tilde{y}_2^*$).

(a) If $2a^2-5a+2 > 0$ and $\frac{c}{2} - \frac{[w+c(2a^2-5a+2)]^2}{8w(2a^2-5a+2)} < v^* < \frac{c}{2}$, then it can be obtained that $\frac{w+c(2a^2-5a+2)}{w(2a^2-5a+2)} > 0$ and $\frac{(c-2v^*)}{w(2a^2-5a+2)} > 0$. By using the Vitem theorem, it gives that $\tilde{y}_1^* > \tilde{y}_2^* > 0$. Furthermore, based on (13.6), v^* should also satisfy the following inequalities:

$$\begin{cases} 1+(2-3a)\tilde{y}_1^* > 0, \\ w\tilde{y}_2^* - c > 0. \end{cases} \Rightarrow \begin{cases} v^* < \frac{c}{2} - \frac{[w+c(2a^2-5a+2)]^2 - [w+c(2a^2-5a+2) - \frac{2w(2a^2-5a+2)}{3a-2}]^2}{8w(2a^2-5a+2)}, \\ v^* < \frac{c}{2}. \end{cases}$$

(b) If $2a^2-5a+2<0$, then $v^* > \frac{c}{2} - \frac{[w+c(2a^2-5a+2)]^2}{8w(2a^2-5a+2)} > \frac{c}{2}$, i.e., $v^* > \frac{c}{2}$, which follows that $\frac{(c-2v^*)}{w(2a^2-5a+2)} > 0$.

- If $w+c(2a^2-5a+2)<0$, then $\frac{w+c(2a^2-5a+2)}{w(2a^2-5a+2)} > 0$ and $\frac{(c-2v^*)}{w(2a^2-5a+2)} > 0$; it follows from the Vitem theorem that $\tilde{y}_1^* > \tilde{y}_2^* > 0$. Based on (13.6), v^* should also satisfy the following inequalities:

$$\begin{cases} 1+(2-3a)\tilde{y}_1^* > 0, \\ w\tilde{y}_2^* - c > 0. \end{cases} \Rightarrow \begin{cases} v^* < \frac{c}{2} - \frac{[w+c(2a^2-5a+2)]^2 - [w+c(2a^2-5a+2) - \frac{2w(2a^2-5a+2)}{3a-2}]^2}{8w(2a^2-5a+2)}, \\ v^* < \frac{c}{2}. \end{cases}$$

  By solving the above inequality, it can be derived that $v^* < \frac{c}{2}$, which contradicts the assumption $v^* > \frac{c}{2}$ presented in Case 2(b).
- If $w+c(2a^2-5a+2)>0$, then $\frac{w+c(2a^2-5a+2)}{w(2a^2-5a+2)} < 0$; it follows from the Vitem theorem that $\tilde{y}_2^* < \tilde{y}_1^* < 0$, which obviously contradicts (13.13).

Case 3. If $2a^2-5a+2 \neq 0$ and $v^* = \frac{c}{2} - \frac{[w+c(2a^2-5a+2)]^2}{8w(2a^2-5a+2)}$, then Equation (13.12) has a unique solution, i.e., $\tilde{y}_1^* = \tilde{y}_2^* = \frac{w+c(2a^2-5a+2)}{2w(2a^2-5a+2)}$. Based on (13.13), the following inequalities should be satisfied:

$$\begin{cases} w(2-3a)+(2a^2-5a+2)[2w+c(2-3a)] > 0, \\ \frac{w-c(2a^2-5a+2)}{(2a^2-5a+2)} > 0. \end{cases}$$

Case 4. if $2a^2-5a+2 \neq 0$ and $v^* < \frac{c}{2} - \frac{[w+c(2a^2-5a+2)]^2}{8w(2a^2-5a+2)}$, then Eq. (13.12) does not have a solution.

Remark 13.5. The above research results show that there is a phenomenon of singularity induced bifurcation around the interior equilibrium in the case of zero economic interest of harvesting, and the model system becomes unstable when the economic interest of harvesting is positives. From the biological and economic perspective, the harvested prey-predator ecosystem can be influenced in the following two aspects.

- The singularity induced bifurcation can result in an impulse phenomenon, which may lead to the collapse of the model system. In the prey-predator ecosystem, the impulse phenomenon of the ecosystem is usually connected with the rapid growth of the species population. If this phenomenon lasts for a long duration, the species population will be beyond the carrying capacity of the environment and the prey-predator ecosystem will be out of balance, which is disastrous for the sustainable development of the prey-predator ecosystem.
- People always try to make a profit of harvesting, so they are usually interested in the case of positive economic interest of harvesting. However, Theorem 13.1 shows that the dynamical behavior of differential-algebraic model system (13.4)

is unstable in the case of positive economic interest of harvesting, which hampers persistent prosperity of harvesting.

With the purpose of maintaining the sustainable development of the biological resource as well as economic interest of harvest effort at an ideal level, some related measures should be applied to eliminate the impulse phenomenon caused by singularity induced bifurcation and stabilize differential-algebraic model system (13.3) around the interior equilibrium.

13.4 State Feedback Control for Model System

13.4.1 Design of State Feedback Controller

13.4.1.1 Case of Zero Economic Interest $v = 0$

For model system (13.4) with zero economic interest $v = 0$, the Jacobian of the differential-algebraic model system evaluated around the corresponding interior equilibrium P^* takes the form as follows:

$$J_{P^*} = \begin{bmatrix} -a_{11}x_1^* & -a_{12}x_1^* & -a_{13}x_1^* & 0 \\ -a_{21}x_2^* & -a_{22}x_2^* & -a_{23}x_2^* & 0 \\ a_{31}y^* & a_{32}y^* & 0 & -y^* \\ 0 & 0 & wE^* & 0 \end{bmatrix}.$$

According to the leading matrix A in model system (13.5) and J_{P^*}, it can be calculated that $\text{rank}(J_{P^*}, AJ_{P^*}, A^2J_{P^*}, A^3J_{P^*}) = 4$. By using the Theorem 2-2.1 in [4], it is easy to show differential-algebraic model system (13.4) is locally controllable around the interior equilibrium. Consequently, a state feedback controller can be applied to stabilize differential-algebraic model system (13.4) around the interior equilibrium.

By using Theorem 3-1.2 in [4], a state feedback controller $u(t) = k(E(t) - E^*)$ (k is a feedback gain and E^* is a component of the interior equilibrium P^*) can be applied to stabilize differential-algebraic model system (13.4) around the interior equilibrium.

Apply the controller $u(t) = k(E(t) - E^*)$ into differential-algebraic model system (13.4) in the case of $v = 0$, and then a controlled differential-algebraic model system is as follows:

$$\begin{cases} \dot{x}_1(t) = x_1(t)(\varepsilon_1 - a_{11}x_1(t) - a_{12}x_2(t) - a_{13}y(t)), \\ \dot{x}_2(t) = x_2(t)(\varepsilon_2 - a_{21}x_1(t) - a_{22}x_2(t) - a_{23}y(t)), \\ \dot{y}(t) = y(t)(-\varepsilon_3 + a_{31}x_1(t) + a_{32}x_2(t)) - E(t)y(t), \\ 0 = E(t)(wy(t) - c) + k(E(t) - E^*). \end{cases} \tag{13.14}$$

The controller $u(t) = k(E(t) - E^*)$ will be designed in the following theorem.

Theorem 13.2. *If the feedback gain k satisfies the following inequality,*

$$k > \max\{\frac{cE^*}{x_1^* + x_2^*}, \frac{2cE^*(x_1^* + x_2^*)}{a^2 x_1^* y^* + x_2^* y^* - x_1^* x_2^*}, \frac{2cE^*}{(5a - 2a^2 - 2)y^*}\},$$

then differential-algebraic model system (13.14) is stable around P^*.

Proof. The Jacobian of differential-algebraic model system (13.4) with controller evaluated around P^* takes the form

$$\tilde{J}_{P^*} = \begin{bmatrix} -a_{11}x_1^* & -a_{12}x_1^* & -a_{13}x_1^* & 0 \\ -a_{21}x_2^* & -a_{22}x_2^* & -a_{23}x_2^* & 0 \\ a_{31}y^* & a_{32}y^* & 0 & -y^* \\ 0 & 0 & wE^* & k \end{bmatrix}.$$

According to the leading matrix A in model system (13.5) and $\tilde{J}_{P^*}$, we obtain the characteristic equation of differential-algebraic model system (13.4) with controller around P^* is $\det(\lambda A - \tilde{J}_{P^*}) = 0$, which can be expressed as follows:

$$\lambda^3 + \tilde{\Delta}_1 \lambda^2 + \tilde{\Delta}_2 \lambda + \tilde{\Delta}_3 = 0,$$

where

$$\tilde{\Delta}_1 = a_{11}x_1^* + a_{22}x_2^* - \frac{cE^*}{k} = x_1^* + x_2^* - \frac{cE^*}{k},$$

$$\begin{aligned} \tilde{\Delta}_2 &= a_{11}a_{22}x_1^*x_2^* + a_{13}a_{31}x_1^*y^* + a_{23}a_{32}x_2^*y^* - a_{12}a_{21}x_1^*x_2^* - \frac{cE^*}{k}(a_{11}x_1^* + a_{22}x_2^*) \\ &= \frac{a^2 x_1^* y^* + x_2^* y^* - x_1^* x_2^*}{2} - \frac{cE^*}{k}(x_1^* + x_2^*), \end{aligned}$$

$$\begin{aligned} \tilde{\Delta}_3 &= (a_{13}a_{31}a_{22} + a_{23}a_{32}a_{11} - a_{13}a_{21}a_{32} - a_{12}a_{23}a_{31})x_1^*x_2^*y^* - \frac{cE^*(a_{11}a_{22} - a_{12}a_{21})x_1^*x_2^*}{k} \\ &= (\frac{2a^2 - 5a + 2}{4})x_1^*x_2^*y^* + \frac{cE^*}{2k}x_1^*x_2^*. \end{aligned}$$

By using the Routh-Hurwitz criteria (see [9]), the sufficient and necessary condition for the stability of differential-algebraic model system (13.4) with controller around P^* is the feedback gain k satisfying

$$k > \max\{\frac{cE^*}{x_1^* + x_2^*}, \frac{2cE^*(x_1^* + x_2^*)}{a^2 x_1^* y^* + x_2^* y^* - x_1^* x_2^*}, \frac{2cE^*}{(5a - 2a^2 - 2)y^*}\}.$$

Consequently, if the feedback gain k satisfies the above inequality, then differential-algebraic model system (13.4) with controller is stable around the interior equilibrium P^*.

13.4.1.2 Case of Positive Economic Interest $v > 0$

For model system (13.4) with controller in the case of positive economic interest $v^* > 0$, the Jacobian of the differential-algebraic model system evaluated around the corresponding interior equilibrium $\tilde{P}^*$ takes the form as follows:

$$\hat{J}_{\tilde{P}^*} = \begin{bmatrix} -a_{11}\tilde{x}_1^* & -a_{12}\tilde{x}_1^* & -a_{13}\tilde{x}_1^* & 0 \\ -a_{21}\tilde{x}_2^* & -a_{22}\tilde{x}_2^* & -a_{23}\tilde{x}_2^* & 0 \\ a_{31}\tilde{y}^* & a_{32}\tilde{y}^* & 0 & -\tilde{y}^* \\ 0 & 0 & w\tilde{E}^* & 0 \end{bmatrix}.$$

According to the leading matrix A in model system (13.5) and $\hat{J}_{\tilde{P}^*}$, it can be calculated that $\text{rank}(\hat{J}_{\tilde{P}^*}, A\hat{J}_{\tilde{P}^*}, A^2\hat{J}_{\tilde{P}^*}, A^3\hat{J}_{\tilde{P}^*}) = 4$. By using Theorem 3-1.2 in [4], a state feedback controller $u(t) = k(E(t) - \tilde{E}^*)$ can be applied to stabilize differential-algebraic model system (13.4) around the interior equilibrium $\tilde{P}^*$, where k is a feedback gain and $\tilde{E}^*$ is a component of the interior equilibrium $\tilde{P}^*$. Apply the controller $u(t) = k(E(t) - \tilde{E}^*)$ into differential algebraic model system (13.4), and then a controlled differential-algebraic model system is as follows:

$$\begin{cases} \dot{x}_1(t) = x_1(t)(\varepsilon_1 - a_{11}x_1(t) - a_{12}x_2(t) - a_{13}y(t)), \\ \dot{x}_2(t) = x_2(t)(\varepsilon_2 - a_{21}x_1(t) - a_{22}x_2(t) - a_{23}y(t)), \\ \dot{y}(t) = y(t)(-\varepsilon_3 + a_{31}x_1(t) + a_{32}x_2(t)) - E(t)y(t), \\ 0 = E(t)(wy(t) - c) - v^* + k(E(t) - E^*). \end{cases} \tag{13.15}$$

Theorem 13.3. *If the feedback gain k satisfies the following inequality,*

$$k > \max\{\frac{c\tilde{E}^*}{\tilde{x}_1^* + \tilde{x}_2^*}, \frac{2c\tilde{E}^*(\tilde{x}_1^* + \tilde{x}_2^*)}{a^2\tilde{x}_1^*\tilde{y}^* + \tilde{x}_2^*\tilde{y}^* - \tilde{x}_1^*\tilde{x}_2^*}, \frac{2c\tilde{E}^*}{(5a - 2a^2 - 2)\tilde{y}^*}\},$$

then model system (13.2) is stable around $\tilde{P}^*$.

Proof. By using the similar way of proof in Theorem 13.2, the proof of Theorem 13.3 is obvious.

Remark 13.6. After applying the feedback controller into model system (13.4), the model system can be stabilized around the interior equilibrium and the impulse phenomenon caused by the singularity induced bifurcation is also eliminated, and the prey-predator ecosystem is restored to ecological balance. Furthermore, according to the design of the state feedback controller, the biological interpretation of the feedback controller is also obvious. It means that by enhancing the harvest effort on the predator, both sustainable development of the prey-predator ecosystem and the ideal economic interest of harvesting can be obtained, as is in line with intuition.

13.4.2 Numerical Simulation

Simulation work with a hypothetical set of parameters is performed to illustrate the effectiveness of the designed state feedback controller with the help of MATLAB. $w = 20$, $c = 1$, and $a_{13} = 2a_{31} = a = 6$ are set in appropriate units. Like Azar et al. [1], Kumar et al. [10], and Liu et al. [13], other parameter values are also set in appropriate units, which were given in Section 13.1, i.e., $\varepsilon_1 = \varepsilon_2 = \varepsilon_3 = a_{11} = a_{12} = a_{22} = a_{23} = 1$, $a_{21} = 1.5$, and $a_{32} = 0.5$.

13.4.2.1 Simulation for the Case $v = 0$

In the case of zero economic interest, interior equilibrium $P^*(x_1^*, x_2^*, y^*, E^*) = (0.5, 0.2, 0.05, 0.6)$ can be obtained with the given values of parameter values and (13.6) is also satisfied. Differential-algebraic model system (13.4) takes the following form by virtue of the given parameter values:

$$\begin{cases} \dot{x}_1(t) = x_1(t)(1 - x_1(t) - x_2(t) - 6y(t)), \\ \dot{x}_2(t) = x_2(t)(1 - 1.5x_1(t) - x_2(t) - y(t)), \\ \dot{y}(t) = y(t)(-1 + 3x_1(t) + 0.5x_2(t)) - E(t)y(t), \\ 0 = E(t)(20y(t) - 1). \end{cases} \tag{13.16}$$

It follows from Theorem 13.1 that differential-algebraic model system (13.4) has a singularity induced bifurcation around $P^*(0.5, 0.2, 0.05, 0.6)$. Based on the analysis in Section 13.3, a state feedback controller $u(t) = k(E(t) - 0.6)$ can be applied to stabilize differential-algebraic model system (13.4) around P^*, and then model system (13.4) with the state feedback controller takes the form as follows:

$$\begin{cases} \dot{x}_1(t) = x_1(t)(1 - x_1(t) - x_2(t) - 6y(t)), \\ \dot{x}_2(t) = x_2(t)(1 - 1.5x_1(t) - x_2(t) - y(t)), \\ \dot{y}(t) = y(t)(-1 + 3x_1(t) + 0.5x_2(t)) - E(t)y(t), \\ 0 = E(t)(20y(t) - 1) + k(E(t) - 0.6). \end{cases} \tag{13.17}$$

By using Theorem 13.2, if feedback gain k satisfies $k > 1.037$, then singularity induced bifurcation is eliminated and differential-algebraic model system (13.4) is stable around P^*, whose dynamical responses of differential-algebraic model system (13.4) can be shown in Fig. 13.1.

13.4.2.2 Simulation for the Case $v > 0$

By virtue of these parameter values, it can be computed that

$$\frac{c}{2} - \frac{[w + c(2a^2 - 5a + 2)]^2}{8w(2a^2 - 5a + 2)} = -0.1808.$$

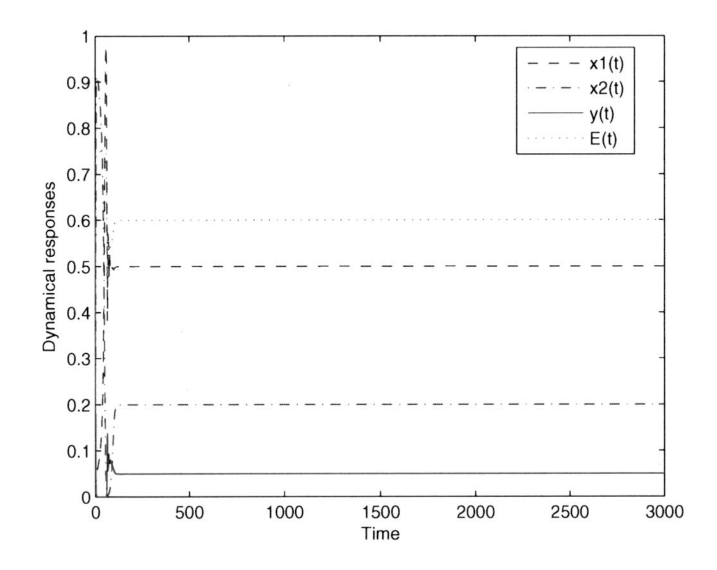

Fig. 13.1 Dynamical responses of model system (13.4) with controller in the case of zero economic interest.

Since $v^* > 0$ represents a positive economic interest of harvesting, it is obvious that $v^* > \frac{c}{2} - \frac{[w + c(2a^2 - 5a + 2)]^2}{8w(2a^2 - 5a + 2)}$.

If $v^* < \frac{c}{2} = 0.5$, then Case 2(a) for the solution of Eq. (13.12) is satisfied, and it follows that $\frac{w + c(2a^2 - 5a + 2)}{w(2a^2 - 5a + 2)} > 0$ and $\frac{(c - 2v^*)}{w(2a^2 - 5a + 2)} > 0$. By using the Vitem theorem, it can be shown that Eq. (13.12) has two positive solutions $\tilde{y}_1^* > \tilde{y}_2^* > 0$. Actually, by solving the corresponding inequalities in Case 2(a), it can be obtained that $0 < v^* < 0.2188$, which is in accordance with the assumption $0 < v^* < \frac{c}{2} = 0.5$.

According to the above analysis, the positive economic interest is set as $v^* = 0.12$ in the appropriate unit, which is arbitrarily selected within the interval $(0, 0.2188)$ and is enough to merit the theoretical analysis obtained in the above analysis. Based on the selected $v^* = 0.12$, there are two roots of Eq. (13.12), i.e., $\tilde{y}_1^* = 0.0578$ and $\tilde{y}_2^* = 0.0149$. However, further calculations show that $20\tilde{y}_1^* - 1 > 0$ and $20\tilde{y}_2^* - 1 < 0$. Hence, $\tilde{y}_2^*$ is neglected due to its contradiction to (13.13), and the interior equilibrium with the corresponding economic interest $v^* = 0.12$ is $\tilde{P}^*(0.5778, 0.0752, 0.0578, 0.7692)$. Fig. 13.2 shows the variation of the population density in the prey-predator ecosystem with the change of economic interest of harvesting. It illustrates that the economic interest of harvesting should be restricted within some interval, which may guarantee the survival of all species in the harvested ecosystem.

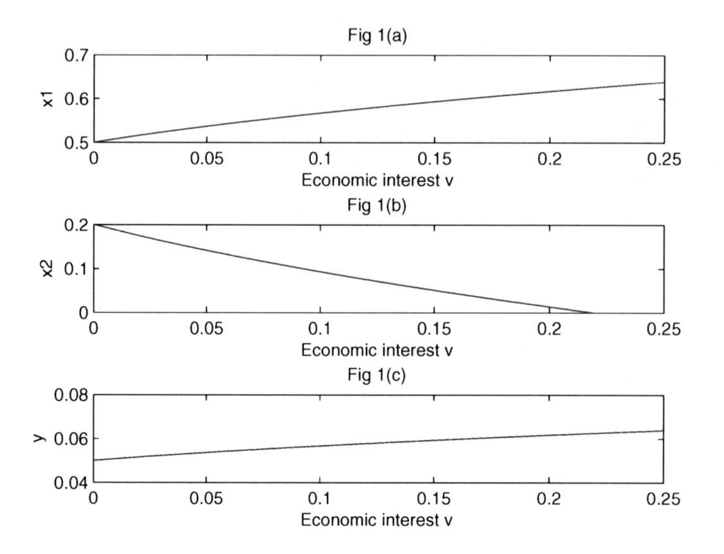

Fig. 13.2 Population density of dominant prey, sub-dominant prey, and predator with the change of the economic interest v of harvesting. As shown in Fig. 13.2(a) and Fig. 13.2(c), the population density of the dominant prey and predator increases with the economic interest of harvesting. However, the population density of the sub-dominant prey decreases with the economic interest of harvesting, which can be shown in Fig. 13.2(b). The biological interpretation is also obvious. It implies that the sub-dominant prey comes close to extinction when the economic interest of harvesting increases through a critical value ($v^* = 0.2188$).

Differential-algebraic model system (13.4) in the case of positive economic interest takes the following form with the given parameter values:

$$\begin{cases} \dot{x}_1(t) = x_1(t)(1 - x_1(t) - x_2(t) - 6y(t)), \\ \dot{x}_2(t) = x_2(t)(1 - 1.5x_1(t) - x_2(t) - y(t)), \\ \dot{y}(t) = y(t)(-1 + 3x_1(t) + 0.5x_2(t)) - E(t)y(t), \\ 0 = E(t)(20y(t) - 1) - 0.12. \end{cases} \tag{13.18}$$

By using Theorem 13.1, differential-algebraic model system (13.4) is unstable around $\tilde{P}^* = (0.5778, 0.0752, 0.0578, 0.7692)$ in the case of $v^* = 0.12$. Based on the analysis in Section 13.4, a state feedback controller $u(t) = k(E(t) - 0.7692)$ can be applied to stabilize differential-algebraic model system (13.4) around $\tilde{P}^*$, and then model system (13.4) with the state feedback controller takes the form as follows:

$$\begin{cases} \dot{x}_1(t) = x_1(t)(1 - x_1(t) - x_2(t) - 6y(t)), \\ \dot{x}_2(t) = x_2(t)(1 - 1.5x_1(t) - x_2(t) - y(t)), \\ \dot{y}(t) = y(t)(-1 + 3x_1(t) + 0.5x_2(t)) - E(t)y(t), \\ 0 = E(t)(20y(t) - 1) - 0.12 + k(E(t) - 0.7692). \end{cases} \tag{13.19}$$

By using Theorem 13.3 of this chapter, if feedback gain k satisfies $k > 1.1779$, then differential-algebraic model system (13.4) with controller is stable around $\tilde{P}^*$, and dynamical responses of differential-algebraic model system (13.4) can be shown in Fig. 13.3.

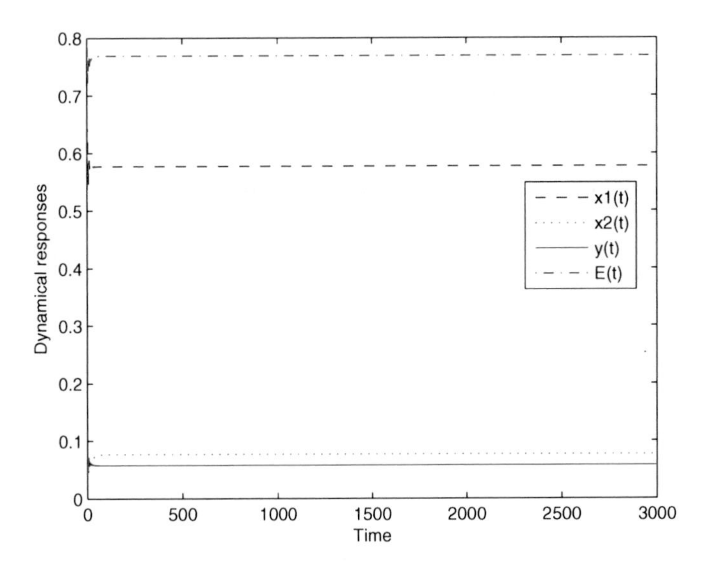

Fig. 13.3 Dynamical responses of model system (13.4) with controller in the case of positive economic interest.

13.5 Conclusion

It is well known that harvesting has a strong impact on the dynamic evolution of a population. Nowadays, the biological resources in the prey-predator ecosystem are mostly harvested with the aim of achieving economic interest, which motivates the introduction of harvesting in the prey-predator models. Compared with the previous related work done in [1, 3, 2, 14, 11, 16, 5, 6, 10, 13], a harvested differential-algebraic prey-predator model is proposed, which is governed by three-dimension differential equations and an algebraic equation. The differential equations investigate the dynamics of dominant prey, subdominant prey, and predator, and the algebraic equation studies the harvest effort on predator from an economic perspective; further analysis focuses on the variation of population density with the change of economic interest of harvesting, which extends previous work [1, 10, 13]. Moreover, compared with the differential models proposed in [1, 10, 13], the advantages of the proposed differential-algebraic model can be found in Remark 13.3.

By analyzing the local stability and bifurcation phenomenon around the interior equilibrium of the differential-algebraic system, the mechanism of the sustainable development of the prey-predator system and harvesting effects on the prey-predator system are studied. Furthermore, a state feedback controller is designed to stabilize the differential-algebraic model system around the interior equilibrium, whose bio-economic interpretations can be found in Remark 13.4. Numerical simulations are carried out to illustrate the effectiveness of the state feedback controller, which can be shown in Fig. 13.1 and Fig. 13.3. On the other hand, numerical simulation also shows that the harvest effort significantly affects the population density of sub-dominant prey, which may face extinction when the economic interest of harvesting increases to a certain level, as can be vividly illustrated in Fig. 13.2.

Theorem 13.2 and Theorem 13.3 provide a bio-economic way to maintain the sustainable development of the harvested prey-predator ecosystem in the case of zero and positive economic interest of harvesting, respectively. Here, we take fishery for example. When there is a negative growth trend and less economic interest than usual. From a long view of maintaining the ecological balance and persistent prosperity of fishery, it is desirable for the government to provide subsidies and formulate special policies for fishermen to encourage them to maintain the amount of capture effort on the predator fish. On the other hand, related polices should also be formulated to regulate the capture effort within a certain range, so that some biological resources, especially those in the sub-dominant position of eco-competition, can survive over-harvesting.

References

1. Azar, C., Holmberg, J., Lindgren, K.: Stability analysis of harvesting in a predator-prey model. J. Theo. Bio. 174, 13–19 (1995)
2. Costa, M.I.S., Kaszkurewicz, E., Bhaya, A., Hsu, L.: Achieving global convergence to an equilibrium population in predator-prey systems by the use of a discontinuous harvesting policy. Ecol. Model 128, 89–99 (2000)
3. Dai, G.R., Tang, M.X.: Coexistence region and global dynamics of a harvested predator-prey system. SIAM J. Appl. Math. 58(1), 193–210 (1998)
4. Dai, L.: Singular Control System. Springer, New York (1989)
5. Edwards, H.J., Dytham, C., Pitchford, J.W., Righton, D.: Prey selection, vertical migrations and the impacts of harvesting upon the population dynamics of a predator-prey system. Bull. Mathe. Biol. 69, 1827–1846 (2007)
6. Gakkhar, S., Singh, B.: The dynamics of a food web consisting of two preys and a harvesting predator. Chaos Soli. Frac. 34(4), 1346–1356 (2007)
7. Gordon, H.S.: The economic theory of a common property resource: The fishery. J. Polit. Econ. 62(2), 124–142 (1954)
8. Kar, T.K., Matsuda, H.: Controllability of a harvested prey-predator with time delay. J. Biol. Syst. 14(2), 243–254 (2006)
9. Kot, M.: Elements of Mathematical Biology. Cambridge Press, Cambridge (2001)
10. Kumar, S., Srivastava, S.K., Chingakham, P.: Hopf bifurcation and stability analysis in a harvested one-predator-two-prey model. Appl. Math. Comp. 129, 107–118 (2002)

11. Liu, B., Zhang, Y.J., Chen, L.S.: Dynamic complexities in a Lotka-Volterra predator-prey model concerning impulsive control strategy. Int. J. Bifur. Chao. 15(2), 517–531 (2005)
12. Liu, C., Zhang, Q.L., Zhang, Y., Duan, X.D.: Bifurcation and control in a differential-algebraic harvested prey-predator model with stage structure for predator. Int. J. Bifur. Chaos. 18(10), 3159–3168 (2008)
13. Liu, Z., Yuan, R.: Stability and bifurcation in a harvested one-predator-two-prey model with delays. Chaos Soli. Frac. 27, 1395–1407 (2006)
14. Martin, A., Ruan, S.G.: Predator-prey models with delay and prey harvesting. J. Math. Biol. 43(3), 247–267 (2001)
15. Venkatasubramanian, V., Schaettler, H., Zaborszky, J.: Local bifurcations and feasibility regions in differential-algebraic systems. IEEE Trans. Auto. Contr. 40(12), 1992–2013 (1995)
16. Zhang, H., Georgescu, P., Chen, L.S.: An impulsive predator-prey system with Beddington-DeAngelis functional response and time delays. Int. J. Bio. 1(1), 1–17 (2008)
17. Zhang, X., Zhang, Q.L., Zhang, Y.: Bifurcations of a class of singular biological economic models. Chaos Soli. Frac. 42(3), 1485–1494 (2009)
18. Zhang, Y., Zhang, Q.L.: Chaotic control based on descriptor bioeconomic systems. Contr. Deci. 22(4), 445–452 (2007) (in Chinese)
19. Zhang, Y., Zhang, Q.L., Zhao, L.C.: Bifurcations and control in singular biological economical model with stage structure. J. Syst. Engi. 22(3), 232–238 (2007) (in Chinese)

Chapter 14
Analysis and Control of an SEIR Epidemic System with Nonlinear Transmission Rate

14.1 Introduction

Mathematical models describing the population dynamics of infectious diseases have been playing an important role in better understanding epidemiological patterns and disease control for a long time. In order to predict the spread of infectious disease among regions, many epidemic models have been proposed and analyzed in recent years (see [7, 4, 22, 25, 36, 19, 20]). However, most of the literature researched on epidemic systems (see [36, 24, 8, 13]) assumes that the disease incubation is negligible that, once infected, each susceptible individual (in class S) becomes infectious instantaneously (in class I) and later recovers (in class R) with a permanent or temporary acquired immunity. The model based on these assumptions is customarily called an SIR (susceptible-infectious-recovered) or SIRS (susceptible-infectious-recovered-susceptible) system (see [10, 9]). Many diseases such as measles, severe acute respiratory syndromes (SARS), and so on, however, incubate inside the hosts for a period of time before the hosts become infectious. So the systems that are more general than SIR or SIRS types need to be studied to investigate the role of incubation in disease transmission. We may assume that a susceptible individual first goes through a latent period (and is said to become exposed or in the class E) after infection before becoming infectious. Thus, the resulting models are of SEIR (susceptible-exposed-infectious-recovered) or SEIRS (susceptible-exposed-infectious-recovered-susceptible) types, respectively, depending on whether the acquired immunity is permanent or not. Many researchers have studied the stability, bifurcation, or chaos behavior of SEIR or SEIRS epidemic systems (see [26, 21, 27, 30, 32, 23]). Y. L. Michael et al. [26] studied the global stability of an SEIR epidemic system in the interior of the feasible region. D. Greenhalgh [11] discussed Hopf bifurcation in models of SEIRS type with density dependent contact and death rates. In addition, some literature on the SEIR-type

Q. Zhang et al.: Complex., Analy. and Cntrl. of Singular Biological Systems, LNCIS 421, pp. 203–225.
springerlink.com

age-independent epidemic systems was investigated by many authors (see [32, 5]) and their threshold theorems are well obtained.

Many authors find that most practical systems are more exactly described by differential and algebraic equations, which appear in engineering systems such as power systems, aerospace engineering, biological systems, economic systems, and so on (see [31, 29, 38, 6]). Although many epidemic systems can be described by differential and algebraic equations (see [39, 40]), they are studied by reducing the dimension of epidemic models to differential systems and the dynamical behaviors of the whole system are not better described. By reducing the dimension of an SEIR epidemic system via substituted algebraic constraint into differential equations and using the methods of reconstructed phase and correlation dimension, L. F. Olsen and W. M. Schaffer [27] studied the system described by differential equations that is chaos with the degree of seasonality $\beta_1 = 0.28$. However, we can find more complex dynamical behaviors if the SEIR epidemic system is described by differential and algebraic equations via analysis of the whole system. The systemic parameters in this chapter are the same as [27]. In particular, the system is hyperchaos when systemic parameter $\beta_1 = 0.28$ in this chapter. Some authors study biologic systems based on seasonal forcing. M. Kamo and A. Sasaki [18] discuss dynamical behaviors of a multi-strain SIR epidemiological model with seasonal forcing in transmission rate. H. Broer et al. [1] study dynamics of a predator-prey model with seasonal forcing.

Differential and algebraic systems are also referred to as descriptor systems, singular systems, generalized state space systems, and so on. Differential and algebraic systems are governed by the so-called singular differential equations, which endow the systems with many special features that are not found in classical systems. Among these are impulse terms and input derivatives in the state response, nonproperness of transfer matrix, noncausality between input and state (or output), consistent initial conditions, and so on. Research on nonlinear differential and algebraic systems has focused on systems with the following description:

$$\begin{cases} \Xi(t)\dot{x}(t) = H(x(t),u(t),t), \\ y(t) = J(x(t),u(t),t), \end{cases} \tag{14.1}$$

where $\Xi(t) \in R^n \times R^n$ is singular; H and J are appropriate dimensional vector functions in $x(t)$, $u(t)$, and t; $x(t)$, $u(t)$, and $y(t)$ are appropriate dimensional state, input and output vectors, respectively; and t is the time variable. In particular, systems (14.1) is normal if rank$\Xi = n$. Some authors have discussed chaotic dynamical behavior and chaotic control based on differential and algebraic systems. Y. Zhang et al. [39, 40] discussed chaos and their control of singular biological economy systems by the theory of differential and algebraic systems.

The literature mentioned above is concerned about low-dimensional chaotic systems with one positive Lyapunov exponent. The attractor of chaotic systems that may have two or more positive Lyapunov exponents is called hyperchaos. However, many researchers have investigated hyperchaotic systems, which are the classical hyperchaotic systems, such as hyperchaotic Chen systems, hyperchaotic Rossler systems, hyperchaotic Lorenz systems, and so on (see [17, 15, 34, 35]). They are all based on hyperchaos synchronization and hyperchaos control. Up to now, a wide variety of approaches have been used to control hyperchaotic systems, for example, the sliding mode control, state feedback control, adaptive control, tracking control, and so on. (see [16, 28, 41, 37, 14, 33]). However, no literature discusses hyperchaos and its control based on differential and algebraic systems.

To the best of our knowledge, hyperchaos first appears in differential and algebraic systems based on this chapter. The contribution of this chapter can be divided into three main parts. In the first part, an SEIR epidemic system with seasonal forcing in transmission rate, which is a new form of differential and algebraic system, is modeled. We discuss the cases of only one varying parameter, two varying parameters, and three varying parameters, respectively. For the case of one varying parameter, the periodic, chaotic, and hyperchaotic dynamical behaviors of the system are analyzed via the bifurcation diagrams, Lyapunov exponents spectrum diagram and Poincaré section. For the cases of two and three varying parameters, the dynamical behaviors of the system are investigated by using Lyapunov diagrams. In the second part, for the hyperchaotic dynamical behavior of the system, we design a tracking controller such that the infectious trajectory of the system tracks an ideal state $i_d = 0$. In the last part, the case for the degree of seasonality $\beta_1 = 0$ is studied. Taking isolation control, we reach the aim of elimination of the disease, and it is easy to implement in real life.

This chapter is organized as follows. In Section 14.2, some preliminaries for the differential and algebraic systems are introduced and the SEIR model is described by differential and algebraic equations. In Section 14.3, the dynamical behaviors of the model are analyzed and a tracking controller is designed for the hyperchaos system, such that the infected gradually disappears. In particular, the case for the degree of seasonality $\beta_1 = 0$ is studied. Taking isolation control, the aim of elimination of the disease can be reached. Simulation results are presented to demonstrate the validity of the controller. Some concluding remarks are given in Section 14.4.

14.2 Preliminaries and Description of Model System

In this section, we describe the SEIR epidemic model and introduce some correlative definitions about differential and algebraic systems. We describe an SEIR epidemic model with nonlinear transmission rate as follows. The population of size $N(t)$ is divided into classes containing susceptible, exposed (infected but not yet infectious), infectious, and recovers. At time t, there are susceptible $S(t)$, exposed $E(t)$, infectious $I(t)$, and recovers $R(t)$. The host total population is $N(t) = S(t)+I(t)+R(t)+E(t)$ at time t. And we assume that immunity is permanent and recovered individuals do not revert to the susceptible class. It is assumed that all newborns are susceptible (no vertical transmission) and there is a uniform birth rate. The dynamical transfer of the population is depicted in Fig. 14.1.

The parameter $b > 0$ is the rate for natural birth and $d > 0$ is the rate for natural death. The parameter $\alpha > 0$ is the rate at which the exposed individuals become infectious, so $1/\alpha$ means the latent period and $\gamma > 0$ is the rate of recovery. The force of infection is $\frac{\beta I}{N}$, where $\beta > 0$ is effective per capita contact rate of infectious individuals and the incidence rate is $\frac{\beta IS}{N}$.

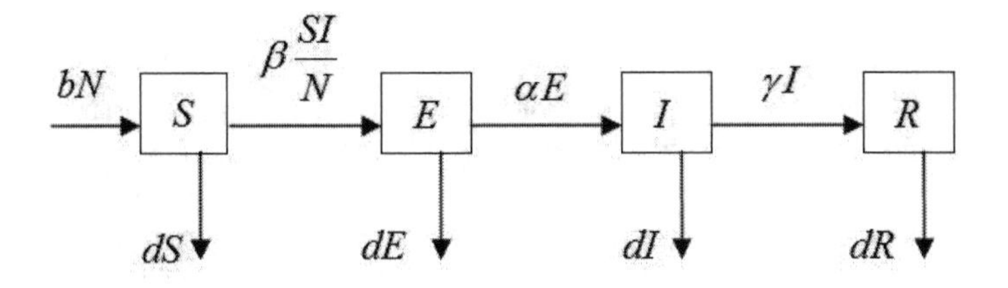

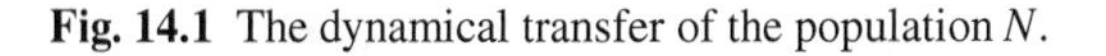

Fig. 14.1 The dynamical transfer of the population N.

The following differential algebraic system is derived based on the basic assumptions and using the transfer diagram:

$$\begin{cases} \dot{S}(t) = bN(t) - dS(t) - \beta \frac{S(t)I(t)}{N(t)}, \\ \dot{E}(t) = \beta \frac{S(t)I(t)}{N(t)} - (\alpha + d)E(t), \\ \dot{I}(t) = \alpha E(t) - (\gamma + d)I(t), \\ \dot{R}(t) = \gamma I(t) - dR(t), \\ 0 = S(t) + E(t) + I(t) + R(t) - N(t). \end{cases} \tag{14.2}$$

Remark 14.1. Model system (14.2) is a classical epidemiological one (see [4]) when the population size $N(t)$ is assumed to be a constant and normalized to 1.

Remark 14.2. The rate of removal α of individuals from the exposed class is assumed to be a constant so that $1/\alpha$ can be regarded as the mean latent period. In the limiting case, when $\alpha \to \infty$ and, the latent period $1/\alpha \to 0$, the SEIR model becomes an SIR model (see [18]).

The first to fourth differential equations of system (14.2) describe dynamical behaviors of every dynamic element for the whole epidemic system (14.2) and the last algebraic equation describes restriction of every dynamic element of system (14.2). That is, differential algebraic system (14.2) can describe the whole behavior of certain epidemic spreads in a certain area.

We consider transmission rate with seasonal forcing in this chapter as follows:

$$\beta = \beta_0(1 + \beta_1 \cos(2\pi t)),$$

where β_0 is the base transmission rate, and β_1 $(0 \leq \beta_1 \leq 1)$ measures the degree of seasonality.

Some transformation is as follows,

$$s = \frac{S}{N}, e = \frac{E}{N}, i = \frac{I}{N}, r = \frac{R}{N}.$$

By utilizing the above transformations, the differential-algebraic system can be obtained in the following:

$$\begin{cases} \dot{s}(t) = b - bs(t) - \beta s(t)i(t), \\ \dot{e}(t) = \beta s(t)i(t) - (\alpha + b)e(t), \\ \dot{i}(t) = \alpha e(t) - (\gamma + b)i(t), \\ \dot{r}(t) = \gamma i(t) - br(t), \\ 0 = s(t) + e(t) + i(t) + r(t) - 1, \end{cases} \tag{14.3}$$

where $s(t), e(t), i(t), r(t)$ denotes the proportions of susceptible, exposed, infectious and recovers, respectively. Note that the total population size $N(t)$ does not appear in system (14.3); this is a direct result of the homogeneity of system (14.2). Also observe that the variable $r(t)$ is described by differential equation $\dot{r}(t) = \gamma i(t) - br(t)$ as well as algebraic equation $r(t) = 1 - s(t) - e(t) - i(t)$, but there is no the variable $r(t)$ in the first to third equations of system (14.3). This allows us to attack system (14.3) by studying the subsystem

$$\begin{cases} \dot{s}(t) = b - bs(t) - \beta s(t)i(t), \\ \dot{e}(t) = \beta s(t)i(t) - (\alpha + b)e(t), \\ \dot{i}(t) = \alpha e(t) - (\gamma + b)i(t), \\ 0 = s(t) + e(t) + i(t) + r(t) - 1. \end{cases} \tag{14.4}$$

Model system (14.4) is also a differential and algebraic system. The dynamical transfer of the epidemic model such as measles, smallpox, chicken-pox, and so on. accords with the description of system (14.4).

From biological considerations, we study system (14.4) in the closed set:

$$\Gamma = \{(s(t), e(t), i(t), r(t)) \in R_+^4 | s(t) + e(t) + i(t) + r(t) = 1\},$$

where R_+^4 denotes the nonnegative cone of R^4.

We consider the following differential and algebraic system [22]:

$$\begin{cases} \dot{X}(t) = f(X(t), Y(t)) + g(X(t), Y(t), t)u(t), \\ 0 = p(X(t), Y(t)), \end{cases} \tag{14.5}$$

where $X(t) = (X_i(t))^T, i = 1, 2, \cdots, n$ and $u(t) \in R$ are the n dimensional state variable, m dimensional constraint variable, and control input, respectively. $f : R^n \times R^m \to R^n$; $g : R^n \times R^m \to R^n$, and $p : R^n \times R^m \to R^n$ are smooth vector fields, and

$$\text{rank}(\frac{\partial p(X, Y)}{\partial Y}) = m, \forall (X(t), Y(t)) \in \Sigma \subset R^n \times R^m,$$

where Ω is an open connectible set.

14.3 Qualitative Analysis of Model System

14.3.1 Dynamical Behavior Analysis

In this subsection, we not only consider the case of only one varying parameter β_1, but also discuss the cases of two and three varying parameters β_1. For the case of only one varying parameter, the dynamical behaviors of system (14.4) are analyzed by using the bifurcation diagrams, Lyapunov exponents spectrum diagram and Poincaré section. In particular, there is hyperchaotic dynamical behavior for system (14.4) with $\beta_1 = 0.28$, i.e., system (14.4) has two positive Lyapunov exponents. For the cases of two and three varying parameters, the dynamical behaviors of system (14.4) are analyzed by using Lyapunov diagrams.

14.3.1.1 Only One Varying Parameter

Let β_1 be a varying parameter of system (14.4), and the remaining parameters are as follows: $b = 0.02, \alpha = 35.84, \gamma = 100$, and$\beta_0 = 1800$, respectively (see [27]). The bifurcation diagrams of systemic parameter β_1 and every variable of system (14.4) are shown in Fig. 14.2.

From Fig. 14.2 (a), (b), (c) and (d), we can easily see that there are complicated dynamical behaviors for system (14.4) with parameter β_1 in some areas. The corresponding Lyapunov exponents spectrum diagram is given in Fig. 14.3. Fig. 14.4, and Fig. 14.5 show how the dynamics of system (14.4) change with the increasing value of the parameter β_1. We can observe that the Lyapunov exponents spectrum gives results completely consistent with the bifurcation diagram. In particular, Fig. 14.5 shows that there are two positive Lyapunov exponents with the parameter $\beta_1 = 0.28$, which implies that system (14.4) with $\beta_1 = 0.28$ is hyperchaotic.

Assume that $\lambda_i, i = 1,2,3,4,5$ are Lyapunov exponents of system (14.4), satisfying the condition $\lambda_i > \lambda_{i+1}, i = 1,2,3,4$. The dynamical behaviors of system (14.4) based on the Lyapunov exponents are given in Table 14.1.

Table 14.1 Attractor type of system (14.4) based on the Lyapunov exponents.

Lyapunov exponents	Attractor type
$\lambda_1 > 0, \lambda_2 > 0, \lambda_3 = 0, \lambda < 0, \lambda_5 < 0$	hyperchaotic attractor
$\lambda_1 > 0, \lambda_2 = 0, \lambda_3 < 0, \lambda < 0, \lambda_5 < 0$	chaotic attractor
$\lambda_1 = 0, \lambda_2 < 0, \lambda_3 < 0, \lambda < 0, \lambda_5 < 0$	periodical attractor

Hyperchaotic dynamical behavior is analyzed via phase plots as follows. The projection of the hyperchaotic attractor on the phase plan of system (14.4) with $\beta_1 = 0.28$ is given in Fig. 14.6.

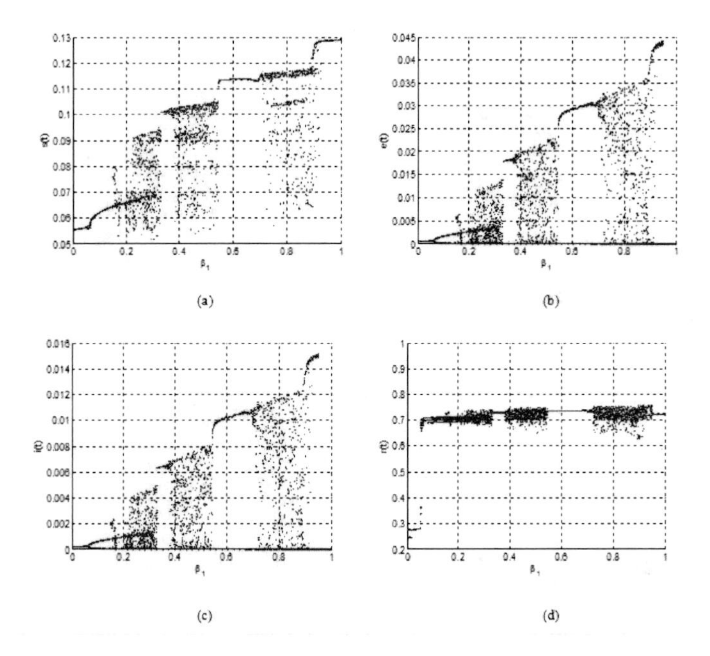

Fig. 14.2 Bifurcation diagrams of parameter $\beta_1 (0 \le \beta_1 \le 1)$ and every variable of system (14.4). (a) $\beta_1 - s(t)$; (b) $\beta_1 - e(t)$; (c) $\beta_1 - i(t)$; (d) $\beta_1 - r(t)$.

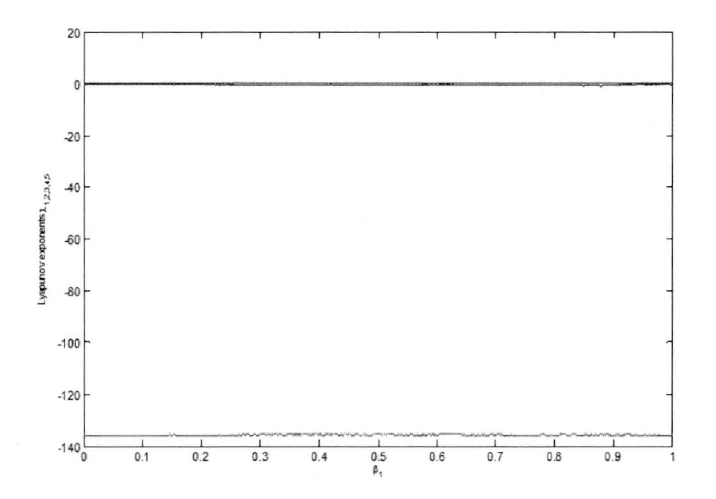

Fig. 14.3 Corresponding Lyapunov exponents of system (14.3) versus parameter β_1.

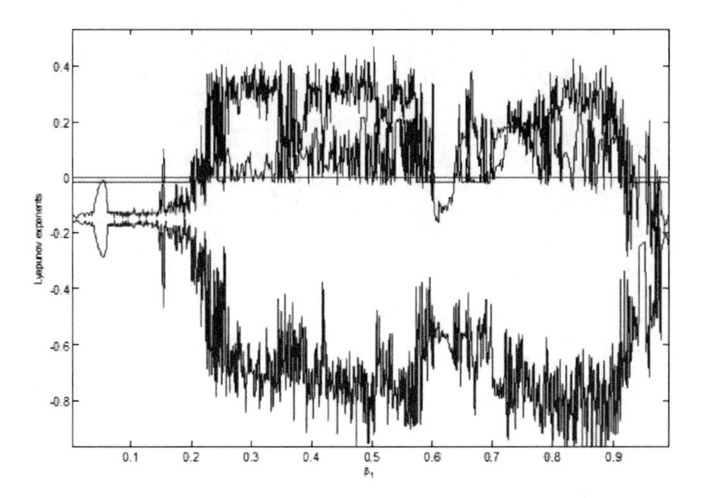

Fig. 14.4 Local amplification of Fig. 14.3 for Lyapunov exponents values in (-1,0.5).

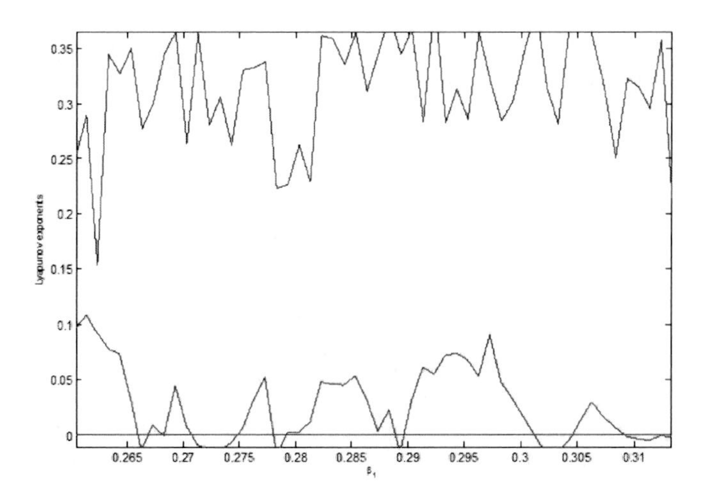

Fig. 14.5 Local amplification of Fig. 14.4 for neighborhood $\beta_1 = 0.28$.

The hyperchaotic attractor of system (14.4) with $\beta_1 = 0.28$ is shown in Fig. 14.7 (a), (b), (c), and (d). The Poincaré section of system (14.4) with $\beta_1 = 0.28$ is given in Fig. 14.8.

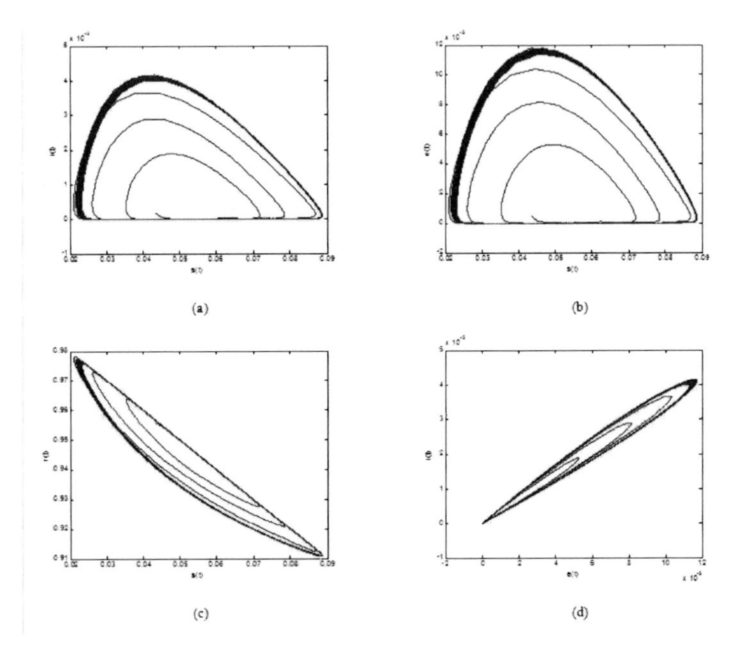

Fig. 14.6 The projection of hyperchaotic attractor of system (14.4) with systemic parameter $\beta_1 = 0.28$ on plane (a)s-i; (b)s-e; (c)s-r; (d)e-i.

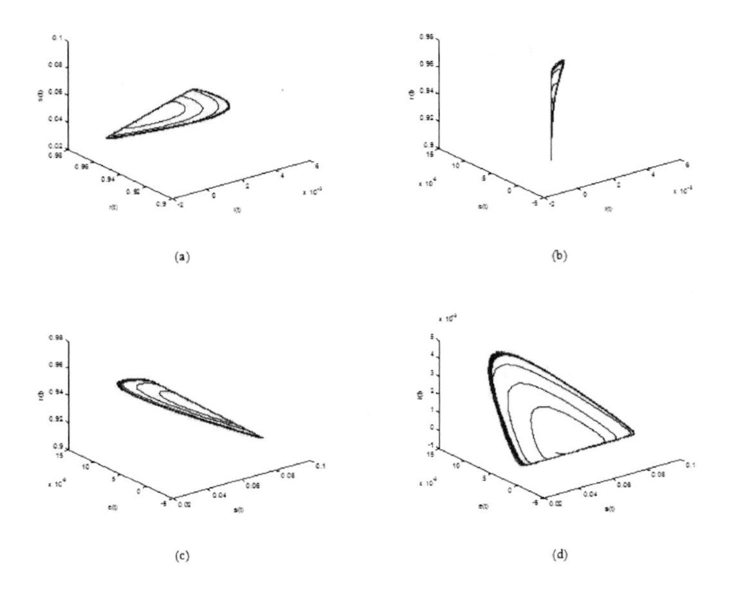

Fig. 14.7 Hyperchaotic attractor of system (14.4) with parameter $\beta_1 = 0.28$. (a) i-r-s; (b) i-e-r; (c) s-e-r; (d) s-e-i.

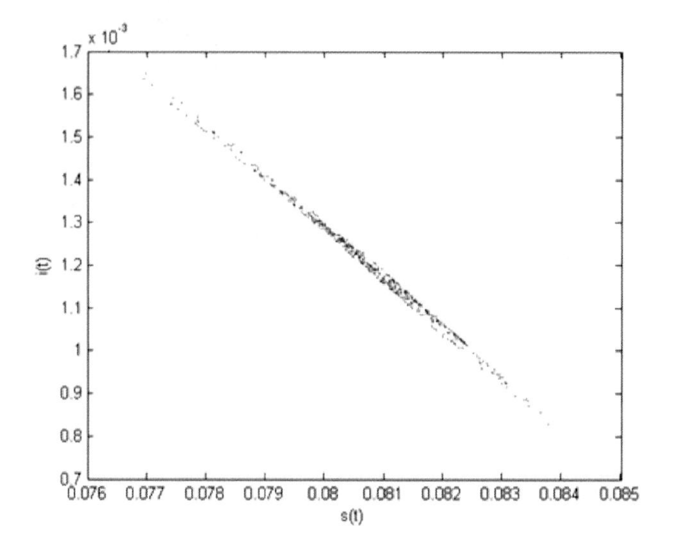

Fig. 14.8 Poincaré section of system (14.4) with $\beta_1 = 0.28$.

14.3.1.2 Two and Three Varying Parameters

It is well known that systemic parameters are varying in many practical problems. In this subsection, we consider the cases of two and three varying parameters. H. Broer et al. [1, 2, 3] introduced an algorithm on a Lyapunov diagram and the diagram is used to scan the parameter plan. To observe clearly the dynamical behaviors, the Lyapunov diagrams Fig. 14.9 and Fig. 14.10 are applied in this chapter. Lyapunov diagram is a plot of a two-parameter plane, where each color corresponds to one type of attractor, classified on the basis of Lyapunov exponents $\lambda_i > \lambda_{i+1}, i = 1,2,3,4$, according to the colour code in Table 14.2.

Table 14.2 Legend of the color coding for Fig. 14.9 and Fig. 14.10: The attractors are classified by means of Lyapunov exponents $(\lambda_i, i = 1,2,3,4,5)$.

Color	Lyapunov exponents	Attractor type
Red	$\lambda_1 \geq \lambda_2 > \lambda_3 = 0 > \lambda_4 \geq \lambda_5$	hyperchaotic attractor
Green	$\lambda_1 > \lambda_2 = 0 > \lambda_3 \geq \lambda_4 \geq \lambda_5$	chaotic attractor
Blue	$\lambda_1 = 0 > \lambda_2 \geq \lambda_3 \geq \lambda_4 \geq \lambda_5$	periodical attractor

For the case of two varying parameters, we discuss three sub-cases as follows:

Case 1. Fixing $b = 0.02, \beta_0 = 1800, \gamma = 100$, let parameter β_1 and α be varying parameters. Taking $\beta_1 \in [0,1]$ and $\alpha \in [34,36]$, Lyapunov diagram is given in Fig. 14.9(a).

Case 2. Fixing $b = 0.02, \beta_0 = 1800, \alpha = 35.84$, let parameter β_1 and γ be varying parameters. Taking $\beta_1 \in [0,1]$ and $\gamma \in [90,110]$, Lyapunov diagram is shown in Fig. 14.9(b).

Case 3. Fixing $b = 0.02, \beta_0 = 1800, \beta_1 = 0.28$, let parameter α and γ be varying parameters. Taking $\alpha \in [34,36]$ and $\gamma \in [90,110]$, Lyapunov diagram is given in Fig. 14.9(c).

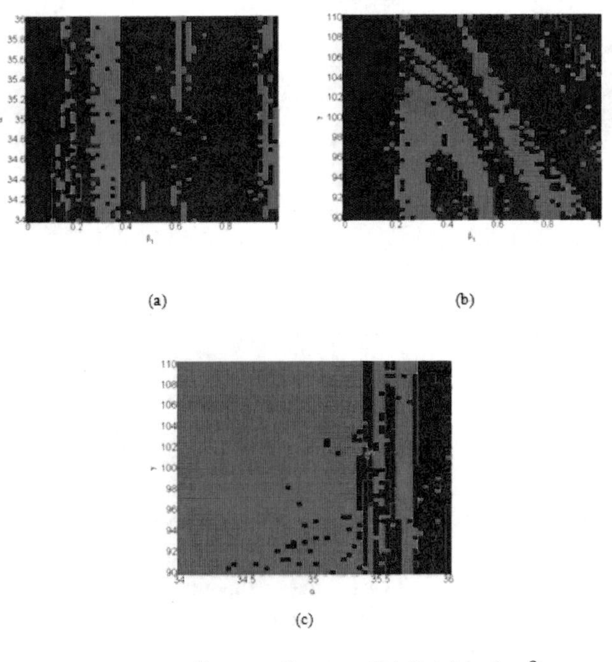

Fig. 14.9 Lyapunov exponents diagram of system (14.4) (a) in the $\beta_1 - \alpha$ parameter plane; (b) in the $\beta_1 - \gamma$ parameter plane; (c) in the $\alpha - \gamma$ parameter plane. For the color code, see Table 14.2.

For the case of three varying parameters, fixing the parameter $b = 0.02, \beta_0 = 1800$, and taking $\beta_1 \in [0,1]$, $\alpha \in [35.5,36]$, and $\gamma \in [96,104]$, the Lyapunov diagram is shown in Fig. 14.10.

According to the above-mentioned analysis, we know that system (14.4) has very complicated dynamical behaviors, such as period, chaos, and hyperchaos phenomena with some parameter values, respectively.

Hyperchaotic dynamical behavior is similar to chaotic dynamical behavior, multi-stabilities coexist in a system. The hyperchaotic attractor has the multi-direction adjacent orbit exponent divergent characteristic, as well as the complex characteristic of high tangle orbit. From Fig. 14.4 to Fig. 14.8, the hyperchaotic attractor has not only the general characteristic of low dimension chaotic attractor, but also the

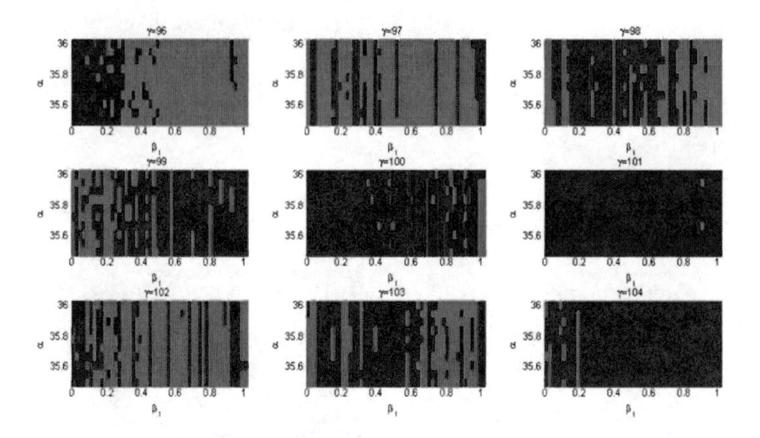

Fig. 14.10 Lyapunov exponents diagram of system (14.4) with the parameter $\gamma = 96:1:104$ in the $\beta_1 - \alpha$ parameter plane. For the color code, see Table 14.2.

following specialty: Hyperchaotic systems have shrinkable or radiation behavior at least on a plane or loop plane. Hereby, the projections of the hyperchaotic attractor on a phase plane are of more complicated fold and tensible trajectories. It is shown that the instability in a local region of hyperchaotic systems is stronger than low dimension chaotic systems. Hence, the control difficulty of hyperchaotic systems is increased.

Biologic signification of hyperchaos in epidemic models is that the epidemic disease will break out suddenly and spread gradually in a region at the period of the high incidence of the epidemic disease. This means that many people in the region would be infected by disease, and some of them would even lose their lives. Nevertheless, there exists uncertain prediction for the low period of the incidence of the epidemic disease. Therefore, it is important to control hyperchaos of the epidemic model.

14.3.2 Hyperchaos Control

In this subsection, we will control hyperchaos for system (14.4) and design a tracking controller $u(t)$ so that $i(t) \to 0$ when $t \to \infty$. That is, the disease gradually disappears and our aim is reached.

It is well known that there are three conditions for the epidemic transmission, i.e., sources of infection, route of transmission, and susceptible population. If we understand correctly the rule of the epidemic process of epidemic disease, take timely the valid measure, and prevent any one of the three conditions from being produced, the transmission of epidemic disease can be prevented. Therefore, we can reach the aim of controlling and eliminating epidemic disease. The susceptible, those with certain diseases who have or lack of immunity, can't resist the invasion of certain pathogens. The higher the percentage of the susceptible is, the larger the possibility of disease outbreak. Therefore, it is important to control the susceptible and it is easy to implement this measure. The new controlled system takes the following form:

$$\begin{cases} \dot{s}(t) = b(1-s(t)) - \beta_0(1+\beta_1\cos(2\pi t))s(t)i(t) + u(t), \\ \dot{e}(t) = \beta_0(1+\beta_1\cos(2\pi t))s(t)i(t) - (a+b)e(t), \\ \dot{i}(t) = \alpha e(t) - (\gamma+b)i(t), \\ 0 = s(t)+e(t)+i(t)+r(t)-1. \end{cases} \tag{14.6}$$

To simplify, we take the transformation $x(t) = 2\pi t$, and nonautonomy system (14.6) is equivalent to the following autonomy system:

$$\begin{cases} \dot{s}(t) = b(1-s(t)) - \beta_0(1+\beta_1\cos(x(t)))s(t)i(t) + u(t), \\ \dot{e}(t) = \beta_0(1+\beta_1\cos(x(t)))s(t)i(t) - (a+b)e(t), \\ \dot{i}(t) = \alpha e(t) - (\gamma+b)i(t), \\ \dot{x}(t) = 2\pi, \\ 0 = s(t)+e(t)+i(t)+r(t)-1. \end{cases} \tag{14.7}$$

System (14.7) could be written as the standard form of system (14.4). So let $X(t) = (s(t), e(t), i(t), x(t))^T, Y(t) = r(t)$,

$$f(X(t),Y(t)) = \begin{bmatrix} b(1-s(t)) - \beta_0(1+\beta_1\cos(x(t)))s(t)i(t) + u(t) \\ \beta_0(1+\beta_1\cos(x(t)))s(t)i(t) - (a+b)e(t) \\ \alpha e(t) - (\gamma+b)i(t) \\ 2\pi \end{bmatrix},$$

$$g(X(t),Y(t)) = \begin{bmatrix} 1 \\ 0 \\ 0 \\ 0 \end{bmatrix},$$

$$p(X(t),Y(t)) = s(t)+e(t)+i(t)+r(t)-1.$$

According to the definition of M derivative, take the output $h(X(t),Y(t)) = i(t)$ of system (14.7), and we obtain

$$M_g M_f^0 h(X(t),Y(t)) = M_g M_f^0 i(t) = 0,$$
$$M_g M_f h(X(t),Y(t)) = 0,$$
$$M_g M_f^2 h(X(t),Y(t)) = \alpha\beta_0(1+\beta_1\cos x(t))i(t) \neq 0.$$

It shows that the relevance degree is 3 (relevance degree is mentioned in (5.19) Chapter 5). Taking the following coordinate transformation,

$$\begin{cases} \xi_1(t) = i(t), \\ \xi_2(t) = \alpha e(t) - (\gamma + b)i(t), \\ \xi_3(t) = \alpha\beta_0(1 + \beta_1 \cos(x(t)))s(t)i(t) - \alpha(\alpha + 2b + \gamma)e(t) + (\gamma + b)^2 i(t), \\ x(t) = 2\pi t. \end{cases} \tag{14.8}$$

The following standard form can be obtained:

$$\begin{cases} \xi_1(t) = \xi_2(t), \\ \xi_2(t) = \xi_3(t), \\ \xi_3(t) = \bar{v}(t), \\ x(t) = 2\pi t, \\ 0 = \frac{\alpha+\gamma+b}{\alpha}\xi_1(t) + \frac{1}{\alpha}\xi_2(t) + \sigma(t) + r(t) - 1, \end{cases} \tag{14.9}$$

where

$$\sigma(t) = \frac{(\alpha + b)(\gamma + b)\xi_1(t) + (\alpha + 2b + \gamma)\xi_2(t) + \xi_3(t)}{\alpha\beta_0(1 + \beta_1 \cos(x(t))\xi_1(t))},$$

$$\begin{aligned} \bar{v}(t) = \ & [\frac{\xi_2(t)}{\xi_1(t)} + \frac{2\pi\beta_1 \sin(x(t))}{1 + \beta_1 \cos(x(t))} - \beta_0(1 + \beta_1 \cos(x(t))\xi_1(t))] \\ & \times [(\alpha + b)(\gamma + b)\xi_1(t) + (\alpha + 2b + \gamma)\xi_2(t) + \xi_3(t)] \\ & + b[\alpha\beta_0[1 + \beta_1 \cos(x(t)) - 2(\alpha + b)(\gamma + b)]\xi_1(t) \\ & + [\gamma + b - (\gamma + 2b)(\alpha + 2b + \gamma)]\xi_2(t) - (\alpha + 3b + \gamma)\xi_3(t) \\ & + \alpha\beta_0(1 + \beta_1 \cos(x(t))\xi_1(t)u(t)). \end{aligned}$$

Obviously, it shows that the differential equations of system (14.7) are divided into the linear subsystem of input-output behavior (that is from the first to third differential equations of system (14.9)), where the dimension is 3 and the other subsystem has dimension 1 (that is the fourth differential equation of system (14.9)), but this subsystem does not affect the output of system (14.7). In order to research output tracking of system (14.7), we consider only from the first to third differential equations of system (14.9) and the algebraic restrict equation. Our aim is that the output trajectory of system (14.7) tracks an ideal state $i_d(t) = 0$; this means that the disease gradually disappears.

Theorem 14.1. *The controller of controlled system (14.7) is*

$$u(t) = \frac{-Z(s(t), e(t), i(t)) + v(t)}{\alpha\beta_0(1 + \beta_1 \cos(2\pi t)i(t))}, \tag{14.10}$$

where

$$\begin{aligned} & Z(s(t), e(t), i(t)) \\ = \ & \alpha\beta_0(1 + \beta_1 \cos(2\pi t))i(t)[b - (\alpha + 4b + 2\gamma)s(t) - \beta_0(1 + \beta_1 \cos(2\pi t)s(t)i(t)] \\ & + \alpha e(t)[\alpha\beta_0(1 + \beta_1 \cos(2\pi t))s(t) + (\alpha + b)(\alpha + 2b + \gamma) + (\gamma + b)^2] \\ & - (\gamma + b)^3 i(t) - 2\pi\alpha\beta_0\beta_1 s(t)i(t)\sin(2\pi t), \end{aligned}$$

$$\begin{aligned}v(t) &= -c_0 i(t) - c_1[\alpha e(t) - (\gamma+b)i(t)] - c_2\alpha\beta_0(1+\beta_1\cos(2\pi t))s(t)i(t)\\ &\quad + c_2\alpha(\alpha+2b+\gamma)e(t) + (\gamma+b)^2 i(t),\end{aligned}$$

where constants c_0, c_1, c_2 satisfy that all roots of equation $p^3+c_2p^2+c_1p+c_0=0$ lie in the left half plane of p, the output of system (14.6) $h(X(t),Y(t)) = i(t) \to 0$ when $t \to 0$.

Proof. Let the error variable $\omega(t) = (\omega_1(t), \omega_2(t), \omega_3(t))^T \in R^3$, $\omega(t) = \xi(t) - \xi_d(t)$, where $\xi(t) = (\xi_1(t), \xi_2(t), \xi_3(t))^T, \xi_d(t) = (\xi_{1d}(t), \xi_{2d}(t), \xi_{3d}(t))^T$. We can get the following error system:

$$\begin{cases}\dot\omega_1(t) = \omega_2(t),\\ \dot\omega_2(t) = \omega_3(t),\\ \dot\omega_3(t) = \bar v(t),\\ \dot x(t) = 2\pi,\\ 0 = \frac{\alpha+\gamma+b}{\gamma}\omega_1(t) + \frac{\omega_2(t)}{\alpha} + \theta(t) + r(t) - 1,\end{cases} \tag{14.11}$$

where $\theta(t) = \dfrac{(\alpha+b)(\gamma+b)\omega_1(t) + (\alpha+2b+\gamma)\omega_2(t) + \omega_3(t)}{\alpha\beta_0(1+\beta_1\cos(x(t))\omega_1(t))}$.

By substituting (14.8) and (14.10) into (14.11), we can obtain the following subsystem:

$$\begin{cases}\dot\omega_1(t) = \omega_2(t),\\ \dot\omega_2(t) = \omega_3(t),\\ \dot\omega_3(t) = v(t),\end{cases} \tag{14.12}$$

where $v = -c_0\omega_1(t) - c_2\omega_2(t) - c_3\omega_3(t)$. According to the theory of [39], choose appropriate constants c_0, c_1, c_2 satisfying all roots of equation $p^3+c_2p^2+c_1p+c_0 = 0$ that lie in the left half plane of p; thus, subsystem (14.12) after feedback is an asymptotically stable system. That is $\omega(t) \to 0$ when $t \to \infty$, thus $i(t) \to 0$ when $t \to \infty$, which means the output of system tracking is the ideal trajectory.

Remark 14.3. There is important practical significance of the control for the susceptible in Theorem 14.1. First, we can vaccinate the susceptible and enhance immunity using exercise. Second, it is necessary to decrease contact with the infectious.

14.3.3 Case $\beta_1 = 0$

In this subsection, we discuss the stabilities of trivial equilibrium and nontrivial equilibrium for system (14.4) with $\beta_1 = 0$, respectively. We further study the bifurcation of the system and design an isolation control such that the disease is eliminated gradually.

System (14.4) with $\beta_1 = 0$ can be rewritten as follows:

$$\begin{cases} \dot{s}(t) = b - bs(t) - \beta_0 s(t)i(t), \\ \dot{e}(t) = \beta_0 s(t)i(t) - (\alpha + b)e(t), \\ \dot{i}(t) = \alpha e(t) - (\gamma + b)i(t), \\ 0 = s(t) + e(t) + i(t) + r(t) - 1. \end{cases} \tag{14.13}$$

In order to obtain equilibrium of system (14.13), let

$$\begin{bmatrix} b - bs(t) - \beta_0 s(t)i(t) = 0, \\ \beta_0 s(t)i(t) - (\alpha + b)e(t) = 0, \\ \alpha e(t) - (\gamma + b)i(t) = 0, \\ s(t) + e(t) + i(t) + r(t) - 1 = 0. \end{bmatrix}$$

It follows that disease-free equilibrium $P_0(1,0,0,0)$ and endemic equilibrium $P_1(s^*, e^*, i^*, r^*)$,
where $s^* = \dfrac{b}{b + \beta_0 i^*}, e^* = \dfrac{(\gamma + b)i^*}{\alpha}, i^* = \frac{\alpha b}{(\alpha+b)(\gamma+b)} - \dfrac{b}{\beta_0}, r^* = \dfrac{bi^*}{\gamma}$.

For simplicity, let $f(X_1, Y_1, \beta_0) = \begin{bmatrix} b - bs(t) - \beta_0 s(t)i(t) \\ \beta_0 s(t)i(t) - (\alpha + b)e(t) \\ \alpha e(t) - (\gamma + b)i(t) \end{bmatrix}, g(X_1, Y_1, \beta_0) =$ $s(t) + e(t) + i(t) + r(t) - 1$, where $X_1(t) = [s(t), e(t), i(t)]^T$, $Y_1(t) = r(t)$ and β_0 is a bifurcation parameter of system (14.13). Since $D_{Y_1} g = 1$, we can get

$$\Sigma = D_{X_1} f - D_{Y_1} f (D_{Y_1} g)^{-1} D_{X_1} g = \begin{bmatrix} -b - \beta_0 i & 0 & \beta_0 s \\ \beta_0 i & -(\alpha + b) & \beta_0 s \\ 0 & \alpha & -(\gamma + b) \end{bmatrix},$$

The following theorem shows the stability of disease-free equilibrium $P_0(1,0,0,0)$.

Theorem 14.2. *Disease-free equilibrium $P_0(1,0,0,0)$ of system (14.13) is globally asymptotically stable in Γ as $\beta_0 < \beta_0^*$; if it is unstable, $\beta_0 > \beta_0^*$ where* $\beta_0^* = \dfrac{(\alpha + b)(\gamma + b)}{\alpha}$.

Proof. The Jacobian matrix of system (14.13) at equilibrium P_0 is

$$\Sigma_{P_0} = \begin{bmatrix} -b & 0 & -\beta_0 \\ 0 & -(\alpha + b) & \beta_0 \\ 0 & \alpha & -(\gamma + b) \end{bmatrix},$$

and it leads the characteristic equation, which gives that

$$\det(\lambda \Upsilon - \Sigma_{P_0}) = (\lambda + b)[\lambda^2 + (\alpha + \gamma + 2b)\lambda + (\alpha + b)(\gamma + b) - \alpha\beta_0] = 0,$$

where Υ is a unit matrix.

We can see that one of the eigenvalues is $-b$ and the other two are the roots of

$$\lambda^2 + (\alpha+\gamma+2b)\lambda + (\alpha+b)(\gamma+b) - \alpha\beta_0 = 0.$$

If $\beta_0 < \beta_0^*$, all three eigenvalues have negative real parts and the equilibrium P_0 is locally asymptotically stable. If $\beta_0 > \beta_0^*$, the nontrivial equilibrium emerges and the trivial equilibrium P_0 becomes unstable.

Theorem 14.3. *If $\beta_0 > \beta_0^*$, equilibrium P_1 of system (14.13) is locally asymptotically stable; if $\beta_0 < \beta_0^*$, P_1 is unstable.*

Proof. The Jacobian matrix of system (14.13) at P_1 is as follows:

$$\Sigma_{P_1} = \begin{bmatrix} -b-\beta_0 i^* & 0 & -\beta_0 s^* \\ \beta_0 i^* & -(\alpha+b) & \beta_0 s^* \\ 0 & \alpha & -(\gamma+b) \end{bmatrix},$$

and it leads the characteristic equation, which gives that

$$\det(\lambda \Upsilon - \Sigma_{P_1}) = \lambda^3 + C_1\lambda^2 + C_2\lambda + C_3 = 0,$$

where $C_1 = \alpha+\gamma+2b+\dfrac{b\alpha\beta_0}{(\alpha+b)(\gamma+b)}$, $C_2 = \dfrac{\alpha b\beta_0(\alpha+\gamma+2b)}{(\alpha+b)(\gamma+b)}$, and $C_3 = b[\alpha\beta_0 - (\alpha+b)(\gamma+b)]$.

By using Routh-Hurwitz criterion [20], it is easy to show that equilibrium P_1 is locally asymptotically stable if $\beta_0 > \beta_0^*$. Equilibrium P_1 is unstable if $\beta_0 < \beta_0^*$.

Remark 14.4. According to Theorem 14.2 and Theorem 14.3, we note that if $\beta_0 < \beta_0^*$, system (14.13) is stable at equilibrium P_0, which corresponds to the disappearance of the disease. If $\beta_0 > \beta_0^*$, system (14.13) is stable at equilibrium P_1 where the endemic disease is formed. The stability of equilibrium P_0 produces the transformation at $\beta_0 = \beta_0^*$. It shows that a bifurcation may happen at $\beta_0 = \beta_0^*$.

Theorem 14.4. *System (14.13) undergoes transcritical bifurcation at the disease-free equilibrium P_0, when bifurcation parameter $\beta_0 = \beta_0^*$.*

Proof. When bifurcation parameter $\beta_0 = \beta_0^*$, it is easy to show that the matrix

$$\Sigma_{P_0} = \begin{bmatrix} -b & 0 & -\frac{(\alpha+b)(\gamma+b)}{\alpha} \\ 0 & -(\alpha+b) & \frac{(\alpha+b)(\gamma+b)}{\alpha} \\ 0 & \alpha & -(\gamma+b) \end{bmatrix}$$

has a geometrically simple zero eigenvalue with right eigenvector $\Phi = \left(-\dfrac{(\alpha+b)(\gamma+b)}{\alpha b} \;\; \dfrac{\gamma+b}{\alpha} \;\; 1 \right)^T$ and left eigenvector $\Psi = \left(0 \;\; \dfrac{\alpha}{\alpha+b} \;\; 1 \right)$.

There is no other eigenvalue on the imaginary axis and

$$\Psi(D_{\beta_0 X_1} f)\Phi = \frac{\alpha}{\alpha+b} \neq 0, \Psi((D_{X_1 X_1} f)(\Phi,\Phi)) = -\frac{2(\alpha+b)(\gamma+b)^2}{b\alpha} \neq 0.$$

According to the literature [12], system (14.13) undergoes transcritical bifurcation at the disease-free equilibrium P_0.

Remark 14.5. Note that when $\beta_0 = \beta_0^*$, the endemic equilibrium P_1 translates the disease-free equilibrium P_0. We must effectively control the transmission rate β_0, such that the value β_0 is less than β_0^*. It is important to effectively control the value β_0.

We take the isolation control method to reach our aim. Take isolation control, then the system (14.13) can be rewritten as follows:

$$\begin{cases} \dot{s}(t) = b - bs(t) - (\beta_0 - \eta)s(t)i(t), \\ \dot{e}(t) = (\beta_0 - \eta)s(t)i(t) - (\alpha+b)e(t), \\ \dot{i}(t) = \alpha e(t) - (\gamma+b)i(t), \\ 0 = s(t) + e(t) + i(t) + r(t) - 1, \end{cases} \tag{14.14}$$

where isolation rate $0 < \eta < \beta_0$.

According to Theorem 14.2, Theorem 14.3, and Theorem 14.4, we know that if $\beta_0 - \eta < \beta_0^*$, system (14.14) is stable at P_0; if $\beta_0 - \eta > \beta_0^*$, system (14.14) is stable at P_1. Obviously, when $\beta_0 - \beta_0^* < \eta < \beta_0$, the aim of elimination of the disease can be reached, and it is easy to implement in real life. Nevertheless, the investments in human, material, and financial resources are larger as the isolation rate increases, and it is hard to realize. Therefore, we take the isolation rate $\eta \to \beta_0^* - \beta_0^*$ to achieve our aim.

Remark 14.6. By enhancing the immunity of the susceptible, quarantining the infectious, and decreasing the contact between the infectious and the susceptible, we can obtain the isolation rate η.

14.3.4 Numerical Simulation

In this subsection, numerical examples are used to demonstrate the validity of the controller.

Case I. The parameters of system (14.4) are supposed as follows: $b = 0.02$, $\alpha = 35.84$, $\gamma = 100$, $\beta_0 = 1800$, and $\beta_1 = 0.28$. In this case, system (14.4) is hyperchaotic. According to Theorem 14.1, we design the controller of controlled system (14.6),

$$u(t) = \frac{-Z(s(t),e(t),i(t)) + v(t)}{35.84 \times 1800(1+0.28\cos(2\pi t))i(t)},$$

where

$$\begin{aligned} Z(s(t),e(t),i(t)) = {} & 64512(1+0.28\cos(2\pi t))i(t)[0.02-235.92s(t) \\ & -1800(1+0.28\cos(2\pi t))s(t)i(t)] \\ & +35.84e(t)[64512(1+0.28\cos(2\pi t))s(t)+14873.9396] \\ & -100.02^3 i(t)-36126.72\pi s(t)i(t)\sin(2\pi t), \end{aligned}$$

$$\begin{aligned} v(t) = {} & -6i(t)-11[35.84e(t)-100.02i(t)] \\ & -6[64512(1+0.28\cos(2\pi t))s(t)i(t)-4869.9392e(t)+100.02^2 i(t)], \end{aligned}$$

where $p^3+6p^2+11p+6=0$ lies in the left half plane of p; figures of $i(t)$ trajectory with uncontrolled system and controlled system are shown in Fig. 14.11 and Fig. 14.12.

From Fig. 14.12, we can easily see that the infectious trajectory of system (14.6) tracks an ideal state $i_d(t)=0$ via designing a tracking controller and it is shown that the disease would gradually disappear.

Case II. The parameters of system (14.13) are supposed as follows: $b=0.02$, $\alpha=35.84$, $\gamma=100$, $\beta_0=180$, and $\beta_0^*=100.0758$. Make the different isolation rate η, and the response of $i(t)$ is shown in Fig. 14.13.

From Fig. 14.14, we can see that the larger isolation rate η is, the better the effect of control is, and the smaller the infectious is. When $\eta=70<\beta_0-\beta_0^*$, the controlled system is stable at the endemic equilibrium. It shows that the endemic disease forms. When $\eta=80>\beta_0-\beta_0^*$, the number of the infectious is gradually zero with time, i.e., when $\beta_0-\eta\to\beta_0^*$, the disease is ultimately eliminated. To avoid forming endemic disease in a certain region, isolation control is an effective measure, which is also a common method.

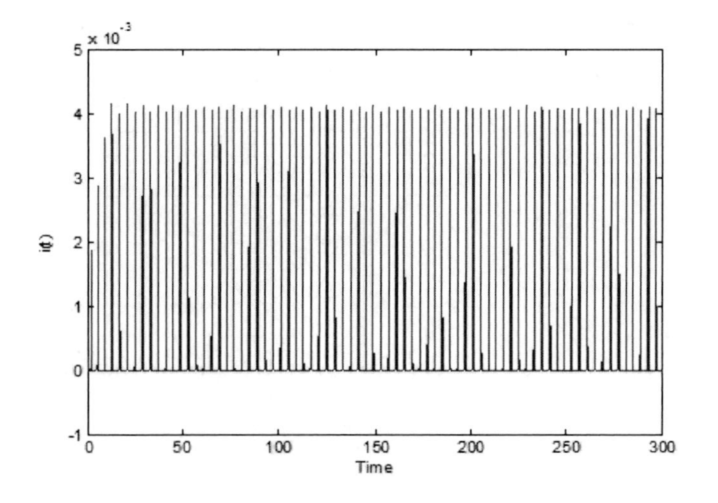

Fig. 14.11 The dynamic response of $i(t)$ trajectory under uncontrolled system.

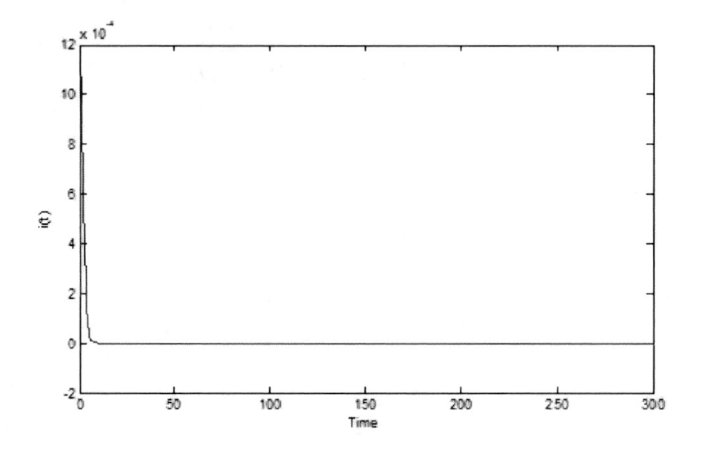

Fig. 14.12 The dynamic response of $i(t)$ trajectory under controlled system.

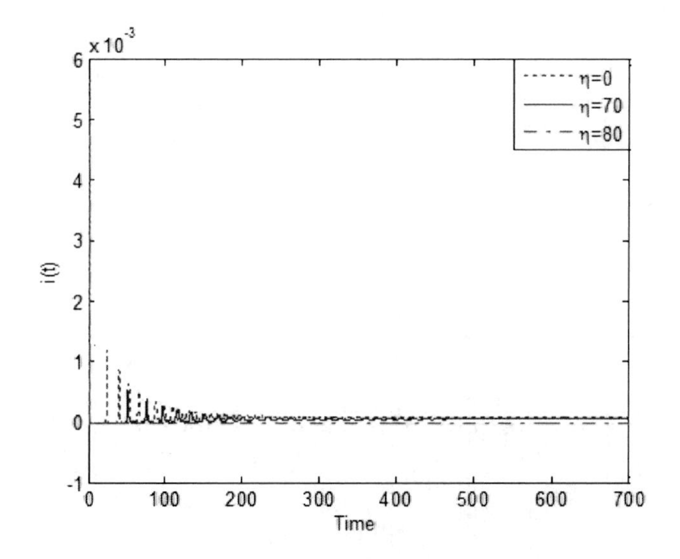

Fig. 14.13 The response of $i(t)$ of system (14.14) for different isolation rate η at initial value (0.579, 0.02, 0.001, 0.4).

14.4 Conclusion

Bifurcation or chaos dynamical behavior exists in many epidemic models. These dynamical behaviors are generally deleterious for biologic systems and often lead to disease spreading gradually or breaking out suddenly in a certain region. In other words, lots of people in the region would be infected by disease and some of them would even lose their lives. Therefore, it is important to effectively control bifurcation or chaos dynamical behavior of epidemic models. In this chapter, we study an SEIR epidemic model, which is a differential algebraic system with seasonal forcing in transmission rate.

We consider three cases: only one varying parameter, two varying parameters, and three varying parameters. For the case of only one varying parameter, we analyze the dynamics of the system by using the bifurcation diagrams, Lyapunov exponents spectrum diagram, and Poincaré section. For the cases of two and three varying parameters, a Lapunov diagram is applied in the analysis of dynamical behaviors. Furthermore, for the hyperchaotic dynamical behavior of the system, we design a tracking controller such that the disease gradually disappears. In particular, we discuss the stability and the transcritical bifurcation for the degree of seasonality $\beta_1 = 0$. The disease is eliminated by taking isolation control, which is an effective measure. Finally, numerical simulations are given to illuminate the proposed control methods.

References

1. Broer, H., Naudot, V., Roussarie, R., Saleh, K.: Dynamics of a predator-prey model with non- monotonic response function. Disc. Cont. Dyn. Syst. A 18, 221–251 (2007)
2. Broer, H., Simo, C., Vitolo, R.: Hopf saddle-node bifurcation for fixed points of 3D-diffeomorphisms: Analysis of a resonance bubble. Phy. D 237, 1773–1799 (2008)
3. Broer, H., Simo, C., Vitolo, R.: The Hopf-saddle-node bifurcation for fixed points of 3D- diffeomorphisms: The Arnol'd resonance web. Bul. Bel. Math. Soc. Ste. 15, 769–787 (2008)
4. Chen, L.S., Chen, J.: Nonlinear Biologic Dynamic Systems. Science Press, Beijing (1993)
5. Cooke, K.L., Driessche, P.V.: Analysis of an SEIRS epidemic model with two delays. J. Math. Bio. 35, 240–260 (1996)
6. Dai, L.: Singular Control Systems. Springer, Heidelberg (1998)
7. Fan, M., Michael, Y.L., Wang, K.: Global stability of an SEIS epidemic model with recruitment and a varying total population size. Math. Bio. 170, 199–208 (2001)
8. Ghoshal, G., Sander, L.M., Sokolov, I.M.: SIS epidemics with household structure: The self-consistent field method. Math. Bio. 190, 71–85 (2004)
9. Glendinning, P., Perry, L.P.: Melnikov analysis of chaos in a simple epidemiological model. J. Math. Biol. 35, 359–373 (1997)
10. Greenhalgh, D., Khan, Q.J.A., Lewis, F.I.: Hopf bifurcation in two SIRS density dependent epidemic models. Math. Comp. Model. 39, 1261–1283 (2004)

11. Greenhalgh, D.: Hopf bifurcation in epidemic models with a latent period and non-permanent immunity. Math. Compu. Model 25, 85–107 (1997)
12. Guckenheimer, J., Holmes, P.: Nonlinear Oscillations. Dynamical Systems and Bifurcations of Vector Fields. Springer, New York (1983)
13. Hilker, F.M., Michel, L., Petrovskii, S.V., Malchow, H.: A diffusive SI model with Allee effect and application to FIV. Math. Bio. 206, 61–80 (2007)
14. Isidori, A.: Nonlinear Control System. Springer, Berlin (1985)
15. Jia, Q.: Hyperchaos generated from the Lorenz chaotic system and its control. Phy. Lett. A 366, 217–222 (2007)
16. Jiang, M.J., Chen, C.L., Chen, C.K.: Sliding mode control of hyperchaos in Rossler systems. Chaos Soli. Frac. 14, 1465–1476 (2002)
17. Jyi, M., Chen, C.L., Chen, C.K.: Sliding mode control of hyperchaos in Rossler systems. Chaos Soli. Frac. 14, 1465–1476 (2002)
18. Kamo, M., Sasaki, A.: The effect of cross-immunity and seasonal forcing in a multi-strain epidemic model. Phys. D 165, 228–241 (2002)
19. Kermack, W.O., McKendrick, A.G.: A contribution to the mathematical theory of epidemics. Proc. Roy. Soc. London A 115, 700–721 (1927)
20. Kot, M.: Elements of Mathematical Biology. Cambridge University Press, Cambridge (2001)
21. Kuznetsov, Y.A., Piccardi, C.: Bifurcation analysis of periodic SEIR and SIR epidemic models. Math. Bio. 32, 109–121 (1994)
22. Li, X.Z., Gupur, G., Zhu, G.T.: Threshold and stability results for an age-structured SEIR epidemic model. Comp. Math. Appl. 42, 883–907 (2001)
23. Liu, W.M., Hethcote, H.W., Levin, S.A.: Dynamical behavior of epidemiological models with nonlinear incidence rates. J. Math. Bio. 25, 359–380 (1987)
24. Lu, Z.H., Liu, X.N., Chen, L.S.: Hopf bifurcation of nonlinear incidence rates SIR epidemiological models with stage structure. Comm. Nonl. Sci. Num. Sim. 6, 205–209 (2001)
25. May, R.M., Oster, G.F.: Bifurcation and dynamic complexity in simple ecological models. Amer. Nat. 110, 573–599 (1976)
26. Michael, Y.L., Graef, J.R., Wang, L.C., Karsai, J.: Global dynamics of an SEIR model with varying total population size. Math. Bio. 160, 191–213 (1999)
27. Olsen, L.F., Schaffer, W.M.: Chaos versus periodicity: Alternative hypotheses for childhood epidemics. Science 249, 499–504 (1990)
28. Rafikov, M., Balthazar, J.M.: On control and synchronization in chaotic and hyperchaotic systems via linear feedback control. Commu. Non. Sci. Num. Sim. 13, 1246–1255 (2008)
29. Rosehart, W.D., Canizares, C.A.: Bifurcation analysis of various power system models. Elec. Pow. Ene. Syst. 21, 171–182 (1999)
30. Sun, C.J., Lin, Y.P., Tang, S.P.: Global stability for a special SEIR epidemic model with nonlinear incidence rates. Chaos Soli. Frac. 33, 290–297 (2007)
31. Venkatasubramanian, V., Schattler, H., Zaborszky, J.: Analysis of local bifurcation mechanisms in large differential-algebraic systems such as the power system. In: Proc. 32nd Conf. Deci. Cont., vol. 4, pp. 3727–3733 (1993)
32. Xu, W.B., Liu, H.L., Yu, J.Y., Zhu, G.T.: Stability results for an age-structured SEIR epidemic model. J. Sys. Sci. Inf. 3, 635–642 (2005)
33. Wang, J., Chen, C.: Nonlinear control of differential algebraic model in power systems. Proc. CSEE 21, 15–18 (2001)
34. Yan, Z.Y., Yu, D.: Hyperchaos synchronization and control on a new hyperchaotic attractor. Chaos Soli. Frac. 35, 333–345 (2008)

35. Yau, H.T., Yan, J.J.: Robust controlling hyperchaos of the Rossler system subject to input nonlinearities by using sliding mode control. Chaos Soli. Frac. 33, 1767–1776 (2007)
36. Zeng, G.Z., Chen, L.S., Sun, L.H.: Complexity of an SIRS epidemic dynamics model with impulsive vaccination control. Chaos Soli. Frac. 26, 495–505 (2005)
37. Zhang, H., Ma, X.K., Li, M., Zou, J.K.: Controlling and tracking hyperchaotic Rossler system via active backstepping design. Chaos Soli. Frac. 26, 353–361 (2005)
38. Zhang, J.S.: Economy cybernetics of singular systems. Tsinghua University Press, Beijing (1990)
39. Zhang, Y., Zhang, Q.L., Zhao, L.C., Liu, P.Y.: Tracking control of chaos in singular biological economy systems. J. Nor. Uni. 28, 157–164 (2007)
40. Zhang, Y., Zhang, Q.L.: Chaotic control based on descriptor bioeconomic systems. Cont. Deci. 22, 445–452 (2007)
41. Zhou, X.B., Wu, Y., Li, Y., Xue, H.Q.: Adaptive control and synchronization of a novel hyperchaotic system with uncertain parameters. Appl. Math. Compu. 203, 80–85 (2008)

Chapter 15
Discussion

Singular system theory is applied to biological complex systems, which provides a theoretical basis for the quantitative study of complex systems, especially for biological complex systems.

Further discussions will be made about the applications of theoretical results obtained in this chapter into the field of harmful algal blooms with a biological controlling strategy.

It is well known that harmful algal blooms are a disaster to marine ecology. Severe depletion of dissolved oxygen (DO) and poisonous toxins resulting from the die-out or crash of algal blooms not only cause serious destruction to marine ecosystem, but also lead to massive mortality of tourism and the farmed fisheries [1]. Furthermore, some uncertain potential toxins released by algae may do harm to people's normal lives. Consequently, it is of foremost importance to provide efficient strategies to control and eliminate harmful phytoplankton blooms.

Generally speaking, it is a complex dynamic process for the occurrence of harmful algal blooms. The traditional ways to investigate the harmful algal blooms mainly focus on experimental analysis. With the rapid need of research, the shortage of traditional research ways become more and more evident. Since marine systems are usually nonlinear systems with high complexity, they can't be investigated thoroughly using traditional research methods. In recent years, some nonlinear dynamic models are proposed to simulate the dynamic process of harmful algal blooms and are utilized to investigate the occurrence reason that harmful algal blooms occur.

The research results obtained in this chapter can be applied to investigate the reason that harmful algal blooms occur and provide certain constructive controlling strategies.

$$\begin{cases} \dot{x}_1(t) = x_1(t)[\alpha_1 - \beta_1 x_1(t) - v_1 x_2(t) - \gamma x_1(t) x_2^2(t)], \\ \dot{x}_2(t) = x_2(t)[\alpha_2 - \beta_2 x_2(t) - v_2 x_1(t)], \end{cases} \tag{15.1}$$

where $x_1(t)$ and $x_2(t)$ are the densities of two competing phytoplankton species (measured as the number of cells per liter); $E(t)$ represents the harvest effort on economic phytoplankton species $x_2(t)$; α_1 and α_2 denote the cell proliferation rate per day for the first and second species, respectively; β_1 and β_2 represent the rate of

Q. Zhang et al.: Complex., Analy. and Cntrl. of Singular Biological Systems, LNCIS 421, pp. 227–229.
springerlink.com

intraspecific competition term for the first and second species, respectively; and v_1 and v_2 stand for the rate of interspecific competition for the first and second species, respectively. Therefore, the growth of either species follows the usual logistic law of growth with intrinsic growth rate α_1, α_2 and environmental carrying capacity α_1/β_1 and α_2/β_2, respectively. Interspecific competition for limited resources is modeled by the law of mass action. γ denotes the rate of toxic inhibition for the species $x_1(t)$ released by the second species $x_2(t)$.

By similar theoretical analysis with theoretical analysis obtained in Section 4.1 of this book, it follows that the model system has a singularity induced bifurcation when the economic interest of harvesting m increases through zero, and a stability switch occurs as m increases through 0.

As stated in Remark 4.4 of this book, the singularity induced bifurcation can result in an impulse phenomenon, which may lead to the collapse of the proposed model system. In the two competing phytoplankton ecosystems, the impulse phenomenon of the model system is usually connected with the rapid growth of the phytoplankton species. If this phenomenon lasts for a period of time, the density of the species in the marine ecosystem will be beyond the carrying capacity of the surrounding environment, which is disastrous for the ecological balance of the marine ecosystem. On the other hand, people always try to make a profit of harvesting, so they are usually interested in the case of positive economic interest. However, it follows from Theorem 4.1 of this book that the model system becomes unstable in the case of positive economic interest of harvesting and is disadvantageous to the persistent property of harvesting. With the purpose of dealing with such the dilemma, some related measures should be applied to eliminate the stability switch of the model system and maintain the economic interest of harvesting at an ideal level.

By similar theoretical analysis with theoretical analysis obtained in Section 4.2 of this book, it follows that the model system is stable around the interior equilibrium when the economic interest of harvesting m increases through zero. The proof is omitted.

The biological background of such a controlling strategy is also obvious. Over the past decades, many ways such as chemical methods [10, 11] and physical strategies [2, 8] have been used to control or mitigate harmful phytoplankton blooms. However, their potential damage to the marine ecosystem needs further identification. Recently, several mathematical models [7, 3] have been established to investigate the mechanism of algal blooms; these models are all described by the systems of differential equations, which provide some useful theoretical guidance for controlling and eliminating algal blooms from a view point of mathematics. With the rapid development of biological technology, the biological strategies of mitigating and controlling algal blooms received a great deal of attention [4, 5], and it has become an essential subject of current research. Varieties of organisms that are available and environment-friendly in the marine ecosystem are conceivably used to control and mitigate algal blooms. A case in point is that macroalgae plays an important role in affecting the marine phytoplankton sequence, either by secreting extracellular organic substances (allelopathy) into the environmental medium or by competing for the available nutrient supply. Consequently, the macroalgae are suggested to be potential agents for controlling and mitigating algal blooms (see reference [6]). Especially, some research results reveal that gracilearia lemaneiformis, the macroalgae distributed widely in coastal areas, has a negative effect on the growth of causative algal bloom species from a view of nutrient competition (see references [12, 9]).

According to the design of the state feedback controller, it is revealed that both abrupt fluctuation of species density can be avoided and the ideal economic interest of harvesting can also be maintained by adopting appropriate the harvest effort on the toxin liberating phytoplankton. With the introduction of a state feedback controller, the model system can be stabilized around the interior equilibrium and the singularity induced bifurcation is also eliminated, which relates to the restoration of ecological balance and sustainable development of species in the harvested marine ecosystem.

References

1. Anderson, D.M.: Turning back the harmful red tide. Nature 388, 513–544 (1997)
2. Cao, X.H., Song, X.X., Yu, Z.M.: Removal efficiency of red tide organisms by modified clay and its impacts on cultured organism. Envi. Sci. 25(5), 148–152 (2004)
3. Huppert, A., Blasius, B., Olinky, R., Stone, L.: A model for seasonal phytoplankton blooms. J. Theo. Bio. 236, 276–290 (2005)
4. Jeong, J.H., Jin, H.J., Sohn, C.H., Suh, K.H., Hong, Y.K.: Algicial activity of the seaweed Corallina pilulifera against red tide microalgae. J. Appl. Phy. 12, 37–43 (2000)
5. Jordi, S., Marta, E., Emilio, G.L.: Biological control of harmful algal blooms: A modeling study. J. Syst. 61, 165–179 (2006)
6. Nakai, S., Inoue, Y., Hosomi, M., Murakami, A.: Growth inhibition of blue-green algae by allelopathic effects of macrophytes. Wat. Sci. Tech. 39, 47–53 (1999)
7. Sarkar, R.R., Chattopadhayay, J.: Occurrence of planktonic blooms under environmental fluctuations and its possible control mechanism-mathematical models and experimental observations. J. Theo. Bio. 224, 501–516 (2003)
8. Sugawara, T., Taguchi, S., Hamasaki, K., Toda, T., Kikuchi, T.: Response of natural phytoplankton assemblages to solar ultraviolet radiation (UV-B) in the coastal water. Hydro. 493, 17–26 (2003)
9. Wang, Y., Yu, Z.M., Song, X.X., Zhang, S.D.: Effects of macroalgae on growth of two species of bloom microalgae and the interactions between the microalgae in laboratory culture. Envi. Sci. 27(2), 274–280 (2006)
10. Yu, Z.M., Zou, J.Z., Ma, X.N.: Application of clays to removal of red tide organisms. Chi. J. Ocea. Limi. 12, 193–200 (1994)
11. Zhang, H., Yang, W.D., Chao, J., Liu, J.S.: Inhibition and elimination of chlorine dioxide on phaeocystis globosa. Chi. J. Appl. Ecol. 14, 1173–1176 (2003)
12. Zhang, S.D., Yu, Z.M., Song, X.X., Wang, Y.: Competition about nutrients between gracilaria lemaneiformis and prorocentrum donghaiense. Acta Ecol. Sin. 25(10), 2676–2680 (2005)

Index

Lecture Notes in Control and Information Sciences

Edited by M. Thoma, F. Allgöwer, M. Morari

Further volumes of this series can be found on our homepage:
springer.com

Vol. 421: Zhang, Q.; Liu, C.; Zhang, X.:
Complexity, Analysis and Control of
Singular Biological Systems
234 p. 2012 [978-1-4471-2302-6]

Vol. 420: Trinh, H.; Fernando, T.:
Functional Observers for Dynamical Systems
218 p. 2012 [978-3-642-24063-8]

Vol. 419: Samy, I.; Gu, D.-W.:
Fault Detection and Flight Data Measurement
170 p. 2012 [978-3-642-24051-5]

Vol. 418: Alberer, D.; Hjalmarsson, H.; del Re, L.:
Identification for Automotive Systems
348 p. 2012 [978-1-4471-2220-3]

Vol. 417: Johansson, R.; Rantzer A.:
Distributed Decision Making and Control
412 p. 2012 [978-1-4471-2264-7]

Vol. 416: Varga, A.; Hansson, A.; Puyou, G.:
Optimization Based Clearance of Flight
Control Laws
451 p. 2012 [978-3-642-22626-7]

Vol. 415: Chesi, G.:
Domain of Attraction
283 p. 2011 [978-0-85729-958-1]

Vol. 414: Imine, H.; Fridman, L.; Shraim, H.;
Djemai, M.:
Sliding Mode Based Analysis and Identification
of Vehicle Dynamics
127 p. 2011 [978-3-642-22223-8]

Vol. 413: Eleftheriou, E.; Reza Moheimani, S.O.:
Control Technologies for Emerging Micro and
Nanoscale Systems
289 p. 2011 [978-3-642-22172-9]

Vol. 412: Fridman, L.; Moreno, J.; Iriarte, R.:
Sliding Modes after the first Decade of the
21st Century
XXX p. 2011 [978-3-642-22163-7]

Vol. 411: Kaczorek, T.;
Selected Problems of Fractional Systems Theory
344 p. 2011 [978-3-642-20501-9]

Vol. 410: Bourlès, H., Marinescu, B.:
Linear Time-Varying Systems
637 p. 2011 [978-3-642-19726-0]

Vol. 409: Xia, Y., Fu, M., Liu, G.-P.:
Analysis and Synthesis of
Networked Control Systems
198 p. 2011 [978-3-642-17924-2]

Vol. 408: Richter, J.H.;
Reconfigurable Control of
Nonlinear Dynamical Systems
291 p. 2011 [978-3-642-17627-2]

Vol. 407: Lévine, J., Müllhaupt, P.:
Advances in the Theory of Control,
Signals and Systems with
Physical Modeling
380 p. 2010 [978-3-642-16134-6]

Vol. 406: Bemporad, A., Heemels, M.,
Johansson, M.:
Networked Control Systems
appro. 371 p. 2010 [978-0-85729-032-8]

Vol. 405: Stefanovic, M., Safonov, M.G.:
Safe Adaptive Control
appro. 153 p. 2010 [978-1-84996-452-4]

Vol. 404: Giri, F.; Bai, E.-W. (Eds.):
Block-oriented Nonlinear System Identification
425 p. 2010 [978-1-84996-512-5]

Vol. 403: Tóth, R.;
Modeling and Identification of
Linear Parameter-Varying Systems
319 p. 2010 [978-3-642-13811-9]

Vol. 402: del Re, L.; Allgöwer, F.;
Glielmo, L.; Guardiola, C.;
Kolmanovsky, I. (Eds.):
Automotive Model Predictive Control
284 p. 2010 [978-1-84996-070-0]

Vol. 401: Chesi, G.; Hashimoto, K. (Eds.):
Visual Servoing via Advanced
Numerical Methods
393 p. 2010 [978-1-84996-088-5]

Vol. 400: Tomás-Rodríguez, M.;
Banks, S.P.:
Linear, Time-varying Approximations
to Nonlinear Dynamical Systems
298 p. 2010 [978-1-84996-100-4]

Vol. 399: Edwards, C.; Lombaerts, T.; Smaili, H. (Eds.):
Fault Tolerant Flight Control
appro. 350 p. 2010 [978-3-642-11689-6]

Vol. 398: Hara, S.; Ohta, Y.; Willems, J.C.; Hisaya, F. (Eds.):
Perspectives in Mathematical System Theory, Control, and Signal Processing
appro. 370 p. 2010 [978-3-540-93917-7]

Vol. 397: Yang, H.; Jiang, B.; Cocquempot, V.:
Fault Tolerant Control Design for Hybrid Systems
191 p. 2010 [978-3-642-10680-4]

Vol. 396: Kozlowski, K. (Ed.):
Robot Motion and Control 2009
475 p. 2009 [978-1-84882-984-8]

Vol. 395: Talebi, H.A.; Abdollahi, F.; Patel, R.V.; Khorasani, K.:
Neural Network-Based State Estimation of Nonlinear Systems
appro. 175 p. 2010 [978-1-4419-1437-8]

Vol. 394: Pipeleers, G.; Demeulenaere, B.; Swevers, J.:
Optimal Linear Controller Design for Periodic Inputs
177 p. 2009 [978-1-84882-974-9]

Vol. 393: Ghosh, B.K.; Martin, C.F.; Zhou, Y.:
Emergent Problems in Nonlinear Systems and Control
285 p. 2009 [978-3-642-03626-2]

Vol. 392: Bandyopadhyay, B.; Deepak, F.; Kim, K.-S.:
Sliding Mode Control Using Novel Sliding Surfaces
137 p. 2009 [978-3-642-03447-3]

Vol. 391: Khaki-Sedigh, A.; Moaveni, B.:
Control Configuration Selection for Multivariable Plants
232 p. 2009 [978-3-642-03192-2]

Vol. 390: Chesi, G.; Garulli, A.; Tesi, A.; Vicino, A.:
Homogeneous Polynomial Forms for Robustness Analysis of Uncertain Systems
197 p. 2009 [978-1-84882-780-6]

Vol. 389: Bru, R.; Romero-Vivó, S. (Eds.):
Positive Systems
398 p. 2009 [978-3-642-02893-9]

Vol. 388: Jacques Loiseau, J.; Michiels, W.; Niculescu, S-I.; Sipahi, R. (Eds.):
Topics in Time Delay Systems
418 p. 2009 [978-3-642-02896-0]

Vol. 387: Xia, Y.; Fu, M.; Shi, P.:
Analysis and Synthesis of Dynamical Systems with Time-Delays
283 p. 2009 [978-3-642-02695-9]

Vol. 386: Huang, D.; Nguang, S.K.:
Robust Control for Uncertain Networked Control Systems with Random Delays
159 p. 2009 [978-1-84882-677-9]

Vol. 385: Jungers, R.:
The Joint Spectral Radius
144 p. 2009 [978-3-540-95979-3]

Vol. 384: Magni, L.; Raimondo, D.M.; Allgöwer, F. (Eds.):
Nonlinear Model Predictive Control
572 p. 2009 [978-3-642-01093-4]

Vol. 383: Sobhani-Tehrani E.; Khorasani K.;
Fault Diagnosis of Nonlinear Systems Using a Hybrid Approach
360 p. 2009 [978-0-387-92906-4]

Vol. 382: Bartoszewicz A.; Nowacka-Leverton A.;
Time-Varying Sliding Modes for Second and Third Order Systems
192 p. 2009 [978-3-540-92216-2]

Vol. 381: Hirsch M.J.; Commander C.W.; Pardalos P.M.; Murphey R. (Eds.)
Optimization and Cooperative Control Strategies: Proceedings of the 8th International Conference on Cooperative Control and Optimization
459 p. 2009 [978-3-540-88062-2]

Vol. 380: Basin M.
New Trends in Optimal Filtering and Control for Polynomial and Time-Delay Systems
206 p. 2008 [978-3-540-70802-5]

Vol. 379: Mellodge P.; Kachroo P.;
Model Abstraction in Dynamical Systems: Application to Mobile Robot Control
116 p. 2008 [978-3-540-70792-9]

Vol. 378: Femat R.; Solis-Perales G.;
Robust Synchronization of Chaotic Systems Via Feedback
199 p. 2008 [978-3-540-69306-2]

Vol. 377: Patan K.
Artificial Neural Networks for the Modelling and Fault Diagnosis of Technical Processes
206 p. 2008 [978-3-540-79871-2]

Vol. 376: Hasegawa Y.
Approximate and Noisy Realization of Discrete-Time Dynamical Systems
245 p. 2008 [978-3-540-79433-2]

DESIGNING COMPLEX SYSTEMS

Foundations of Design in the Functional Domain

Erik W. Aslaksen

CRC Press is an imprint of the
Taylor & Francis Group, an **informa** business
AN AUERBACH BOOK

Complex and Enterprise Systems Engineering Series

Auerbach Publications
Taylor & Francis Group
6000 Broken Sound Parkway NW, Suite 300
Boca Raton, FL 33487-2742

Printed in the United States of America on acid-free paper
10 9 8 7 6 5 4 3 2 1

International Standard Book Number-13: 978-1-4200-8753-6 (Hardcover)

Library of Congress Cataloging-in-Publication Data

Aslaksen, E. (Eric)
Designing complex systems : foundations of design in the functional domain / Erik W. Aslaksen.
p. cm. -- (Complex and enterprise systems engineering)
Includes bibliographical references and index.
ISBN-13: 978-1-4200-8753-6 (alk. paper)
ISBN-10: 1-4200-8753-3 (alk. paper)
1. Systems engineering. I. Title. II. Series.

TA168.A73 1991
620.001'171--dc22 2008022872

Visit the Taylor & Francis Web site at
http://www.taylorandfrancis.com

and the Auerbach Web site at
http://www.auerbach-publications.com

Contents

Preface

This book attempts to develop a rigorous basis for carrying out that early part of the design process that converts a set of requirements on the service to be provided by a system into requirements on a set of interacting functional elements, which then form the point of departure for the classical part of the design process — the conversion of functional requirements into a physical entity that, through its operation, will satisfy those requirements.

The reason for including this book in a series dedicated to Complex Systems, that is, a class of systems where the elements are predominantly independent agents behaving and interacting in a dynamic fashion, is that, while the systems to which the design methodology applies are not generally in this class, the environment in which the design takes place exhibits these characteristics. In the not too distant past, engineers would exclude consideration of much of this environment from their scope of work, but this narrow focus and limitation of responsibility are no longer accepted, neither by our clients nor by society. On the contrary, the increasing intrusiveness of such engineered objects as highways, railways, ports, dams, mines, and factories, just to name a few, into our daily lives and environment, coupled with a much greater awareness of the wider consequences of engineering works, has led to the demand of both clients and society for a holistic approach to the design of these objects and to a greatly increased legal regime. As a result, the complexity of the environment in which design takes place is reflected in the design process itself, and it is this complexity, rather than any complexity of the objects themselves, that systems engineering needs to address.

The immediate purpose of the book is to introduce students and practitioners in the field of system design to the basic issues raised by this complexity and to a methodology that addresses those issues in a rigorous and consistent, top-down fashion. A much more indirect purpose, and one with regard to which the book can, at best, initiate a discussion within the engineering profession, is to reassess the characteristics of engineering and its place within the field of intellectual activity, in particular, to examine the creative aspects of design, as reflected in the difference between an engineer and a technician.

A central theme is the necessity for developing standardized functional elements, the building blocks of design in the functional domain and the counterpart of standardized construction elements in the physical domain. The argument for this is simple: Without standardized construction elements, such as nuts, bolts, bearings, beams, resistors, capacitors, etc., the design of physical equipment would be hopelessly inefficient, and engineers would be forever bogged down with redesigning these elements over and over again. Instead, a large part of design is now the application of these standard elements according to fixed rules, and can be carried out by technicians and design drafters, freeing the engineers for the unique and creative part of each project. Why should not the same be true in the functional domain, that is, in the domain of ideas and performance requirements, which must precede any physical realization? Only through such standardization will the significant increase in efficiency and quality, which systems engineering promises, be realized. The attitude that mental activity is spontaneous and somehow "effortless" only because it does not expend material resources must by now surely be well and truly outmoded, and all the effort that has gone into improving the efficiency of physical work, through time and motion studies and the like, now needs to be replicated for mental work. And just as machine tools increased the productivity of factory workers and CAD systems increased the productivity of drafters, new software tools can be developed to increase the productivity of system designers (as distinct from software for managing the design process).

The work reported on in this book was undertaken in the years between 1998 and 2001 and subsequently reviewed by Terje Fossnes from the Royal Norwegian Navy. The work was sporadically updated and modified as a result of other research undertaken by the author in the years since; however, the present version owes its genesis to the initiative and support of a group of engineers at MITRE, led by Brian White (as editor of the Taylor & Francis series *Complex and Enterprise Systems Engineering*), and, in particular, to the thought-provoking review carried out by Duane Hybertson. The support of the publisher, Rich O'Hanley, is also gratefully acknowledged.

The book is divided into three parts, each with a somewhat different view of the complexity of the design process. Part A considers the purpose and basic features of design, and how the concept of value can provide a quantitative measure of that wider interaction of the engineered object with its environment. Part B develops the domain in which functional design takes place, and explores how the system concept, as the key to handling complexity, can be embedded in that domain. Finally, Part C proposes a number of functional design elements and develops them in considerable detail, and outlines how they could be applied as part of a coherent functional design framework, supported by a software tool.

Finally, my sincere thanks go to my wife, Elfi, without whose unstinting support my systems engineering work would not be possible.

Chapter 1

Introduction

1.1 How the Subject Matter Is Approached

The purpose of this book is to introduce the reader to a particular design methodology and to develop a rigorous theoretical basis for that methodology. Somewhat similar methodologies have been proposed and developed to varying degrees of completeness in the past, often under the general heading of "Systems Engineering,"[1] but because they have lacked a solid foundation in engineering they have not been able to provide a basis for an ongoing development to which a wide cross section of the engineering community could contribute in a coherent manner, and have therefore failed to make a significant impact in many areas of engineering. Also, some of them have created the impression that a systems approach to anything necessarily means a fuzzy, verbose, and hand-waving approach, far from the intellectual rigor that characterizes classical engineering. Hopefully that impression will be dispelled by the approach taken here; a careful, detailed, and, to the extent that it is reasonable and useful, rigorous, step-by-step development of the methodology. However, as the ideas and concepts may not be familiar to some readers, a great deal of explanations, examples, and analogies have been included.

The use of models in engineering and, in particular, in systems engineering, is very well established. With its roots in software engineering, there is the Unified Modeling Language (UML) and its extension to systems engineering, SysML, both managed by the Unified Modeling Group (UMG).[2] The International Council on Systems Engineering (INCOSE) has a working group dedicated to modeling and associated tools,[3] and there are numerous models promoted by individuals or small groups.[4] It is probably fair to say that most of these modeling approaches result in descriptions of aspects of the behavior of defined physical objects, even if most often based on the requirements of yet-to-be-designed objects, and they are therefore basically different to the approach proposed in this book, which links engineering with business through the concept of the service provided by the system. On the other hand, the modeling of economic performance is also well established,[5] but lacks the direct link to engineering and design.

The first part of the book introduces the basic ideas and concepts, in particular the concept of the functional domain. This deceptively simple concept lies at the heart of the methodology, and it is discussed from a number of points of view so that the reader can get a thorough understanding of it. We then look at the purpose of the design process, which leads, when pursued to its logical conclusion, to a particular view of the design process as the core activity within engineering. In this view, design becomes a value-neutral activity, guided only by the desire to provide the most cost-effective solutions to predetermined requirements, but with the value of achieving those requirements prescribed by the stakeholders.

By stakeholders we will understand the totality of persons who are directly affected by a project, such as owners, operators, maintainers, and local residents, and of organizations that become involved as a result of their purpose, such as statutory bodies, environmental conservation organizations, and unions, just to mention a few. From the point of view of complex systems, predetermined requirements may appear to be an anachronism and the distinction between stakeholder and designer artificial; the two together form the complex system, with requirements evolving through the dynamic interaction between them. However, it is a fact that engineering (and design) takes place within a contractual framework, which by its nature distinguishes parties to the contract and defines interfaces between them in the form of requirements, and any design methodology that does not recognize this is unlikely to be successful.

Based on this understanding, the design methodology is outlined and its main characteristics discussed in some detail, and a number of rules for how to apply the methodology are developed.

The second part contains the detailed development of the components of the design methodology. Functional elements are defined and their properties investigated before considering the nature of interactions between functional elements. This leads to the concept of systems in the functional domain, and it is shown how functional systems can be developed to the point where they provide all the features required by the design methodology.

The third part develops some of the major applications of the methodology in considerably more detail, resulting in specific models and procedures for carrying out system design and optimizing system performance in these cases. Again, while optimality may be an elusive concept in the context of complex systems, it must be understood here to relate to the contractual framework in which the design is undertaken.

When introducing a new methodology, we are always faced with the choice of either first explaining the process without having defined the elements on which the process operates, and then developing the elements later, or first developing the elements without any justification for why they are needed, and then introducing the process afterward. The approach taken here is a compromise, an iterative approach that alternates between elements and process. It starts out by introducing what at first appear to be four relatively unrelated topics, and then goes on to show how they come together like the pieces of a puzzle to form an initial view of the subject matter, a view that is then expanded and refined in subsequent chapters. The topics are introduced in the next four sections, and are brought together in the last section of this chapter.

1.2 Engineering

Engineering is an activity, it is a profession, it means different things to people associated with different aspects of the activity, and to most other people it is only a vague concept. For our purpose it will be necessary to have a clear definition of engineering in order to precisely delineate the domain to which our methodology is to apply, even though this precision may result in a more limited definition than that which would be accepted by most of the professional engineering bodies:

> Engineering consists of two largely distinct activities, developing technology and applying technology, where technology is the ability to achieve a desired effect in a repeatable and controlled manner through the application of natural science.

Within this definition of engineering, our focus will be entirely on the application of technology. This application of technology is a process, which we shall call *the process of engineering*, and it may be made more precise by the following definition:

> The process of engineering converts an expressed need into an entity which, through its operating lifetime, meets the need in the most cost-effective manner.

This definition contains the seeds of many of the concepts that will be crucial in developing our design methodology. First, the conversion process that results in the entity (e.g., equipment or a system) contains all activities from concept design to production, supported by management and logistics. Second, the "need" places the focus on something outside engineering as the point of departure and the measure against which the success of the process is measured. Third, the "operating lifetime" emphasizes that the benefit of the entity comes only through its operation over time, necessitating a life-cycle approach to the whole conversion process and leading to a recognition of the importance of maintenance. Fourth, the words, "meets the need," highlight the importance of what the entity does for the stakeholders, rather than any physical property of the entity itself. And finally, "cost-effective" sets engineering off from both science and art and encapsulates the need for engineering to be *useful*.

Some of these concepts will be further detailed and defined more precisely in later chapters, but we can already at this point formulate the following definitions:

> The need is defined in the stakeholder requirements (often in the form of a requirements definition document or RDD).

> The activities related to meeting the need, and the timeframe in which they take place define the project. The process of engineering is a part of the project.

Within the process of engineering, we shall focus even more narrowly on the activity of *design*, a subprocess that converts the need into a complete description of the entity that is to meet the need, complete in the sense that it contains all the information required to produce the entity. As an activity, design has a long and well-documented history, with early testimony in the form of civil structures ranging from the pyramids through the Roman aqueducts to Brunelleschi's dome in Florence; more recently in the form of machines such as steam engines, textile machines, cars, and airplanes; and most recently in the form of electrical and electronic devices. In every case, the design process started with the designers being able to visualize the final physical object in their minds, documenting this vision in some form, and then utilizing their knowledge of the natural sciences to dimension the various elements of the design correctly. But this process did not start from scratch each time; over the centuries a vast collection of *standard construction elements* emerged, and engineering students were taught the properties of these elements and developed their visions of new objects in terms of them.[6] Today we have literally millions of standard elements at our disposal, and the efficiency of the design of physical objects is only possible because of them. Imagine if we had to design every bolt and nut from scratch each time we needed one!

For simple objects, or objects similar to ones that have been previously produced, this design process is a straightforward *bottom-up* process; it synthesizes new objects out of a set of existing construction elements, and the proof of correctness occurs only when the new object is created, through a test procedure that establishes that the performance of the object does indeed meet the original requirements. For more complex objects, the design process, that is, the synthesis step, requires considerable skill and experience on the part of the designer, and it is likely that the outcome of the test is that the object does not quite meet the requirements. A revision of the design is required before a repeated test establishes that the requirements have been met. As the complexity of the object increases, more and more iterations are required to reach a satisfactory result, and the design process becomes increasingly *inefficient*.

An obvious solution to this problem of inefficiency is to somehow subdivide the process into a number of subprocesses, each of which results in a simpler object, but such that when all these simple objects are brought together and allowed to interact, they form an object that satisfies the original, complex requirements. That is, we precede the bottom-up process by a *top-down* process, in which the functional requirements are analyzed and partitioned into interacting subsets of functional requirements, or functional elements, each of which is simple enough to be efficiently realized through a bottom-up process. The two processes have two main differences, as illustrated in Figure 1.1, where the (identical) requirements are shown as given in both cases.[7] (The upper part of the figure should be thought of as a requirements space and the lower as an element space, with the design process making the transition between them.) First, in the case of the bottom-up design, the physical realizability in terms of known elements is guaranteed (manufacturability is something else), whereas, at the end of the top-down design, the system elements are still functional elements and their physical realizability is not guaranteed. Second, in the top-down approach, the requirements are (ideally) always, as an intrinsic part of the methodology, satisfied through every step of the design process, whereas, in the bottom-up approach, the methodology provides no assurance of this.

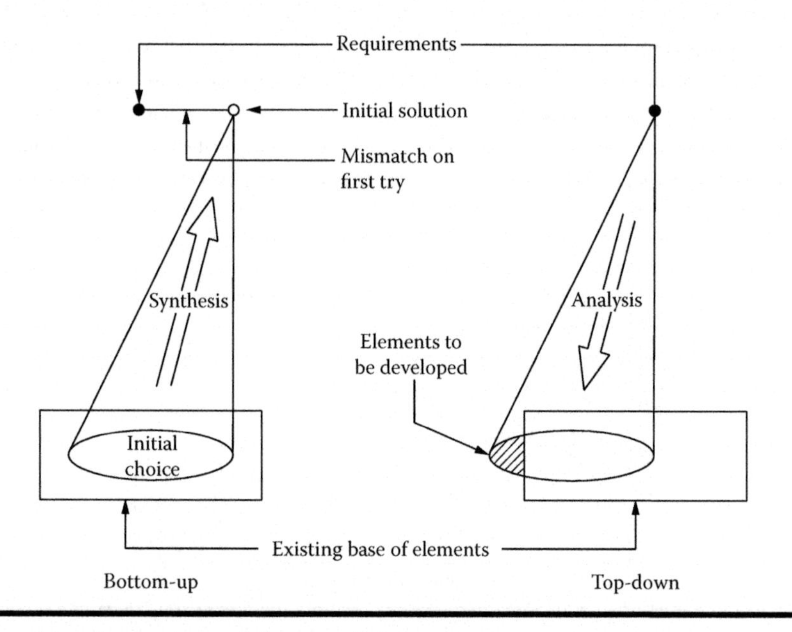

Figure 1.1 Comparison of the two design processes.

This composite approach is, essentially, the approach taken in creating complex systems today, and there have been many notable successes (e.g., the public telephone network and the Apollo project), but also many less successful projects, and some disasters. The reason for this great variability in the outcome of the process lies in the fact that there is no established methodology or theoretical basis for the partitioning into interacting, physical objects; it is very much a case of using previous experience and the skill of the system designer. So we are again faced with a problem that is very similar to the one encountered in the synthesis process, except that the scale is increased by orders of magnitude, which makes iterations very expensive.

The extent to which the top-down process is applied as a complexity-reducing pre-process depends on the complexity of the product to be designed. As it gets more and more complex, that is, consisting of more and more interacting construction elements and characterized by more and more parameters, the probability of picking a combination of elements that will result in a performance anywhere near the requirements becomes less and less, as does the probability of picking a new combination that will result in an improvement. Consequently, it becomes advantageous to spend more effort on the complexity-reducing pre-process, which is the core of systems engineering. This is illustrated in Figure 1.2.

To conclude this section, there is one further point that needs to be brought out, which is that, in reality, particular engineering projects start at very different stages in the process of engineering. At the one end of the scale there is the project that starts at the pre-concept stage, as an idea for a possible investment, in which case the RDD contains almost only functional requirements. At the other end of the scale are projects where the design has been completed and the construction phase is about to begin, in which case the RDD contains only physical requirements. As our methodology is concerned only with the very early part of the design phase, we shall from now on understand that the stakeholder requirements are all functional requirements, unless particular reference is made to any physical requirements.

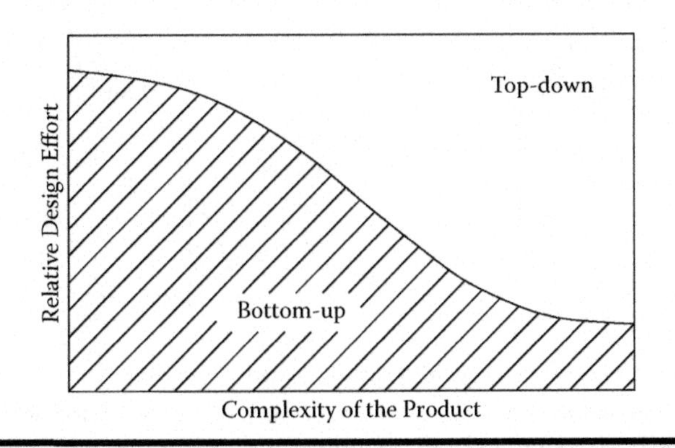

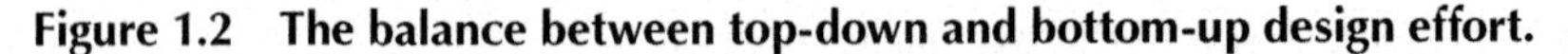

Figure 1.2 The balance between top-down and bottom-up design effort.

1.3 Epistemology and Functionality

Epistemology is that part of philosophy concerned with the theory of knowledge, and in the present case the question is: What can we know about an object? In particular, we are concerned with objects resulting from the process of engineering. Our knowledge of such objects falls into two groups; knowledge about what the object *is*, and knowledge about what the object *does*. In the first group, the *physical description*, the physical characteristics of objects are defined in terms of their physical parameters, such as size, shape, weight, color, and material, and the values of these parameters. In the second group, the effect of objects on their environments, called variously performance, purpose, function, etc., but which we shall call their functionality, is defined in terms of functional parameters and their values, that is, the *functional description*.

These two types of descriptions may be considered complementary descriptions of a given object, but they relate to the object in quite different ways. The physical description, if it is complete (see also chapter 4), will allow the object to be reproduced or manufactured without any reference to what its purpose is, and while the same physical object may have many different uses, the physical description is unique to that object. We would say that two objects that have the same physical description are identical. But to a functional description there corresponds a whole set, perhaps with infinitely many members, of different physical objects that all have the same functionality. If we call the totality of all functional descriptions of objects the *functional domain*, and the totality of physical descriptions the *physical domain*, then we see that neither the mapping from the physical to the functional domain nor its inverse is single-valued.

This may be best illustrated by considering an example, the functional element that effects the extraction of the cork from a corked bottle. You will be familiar with numerous different physical objects that satisfy this purpose, including many variants of the ordinary cork screw, a two-pronged device that is inserted between the cork and the bottle, and a gas-operated device that forces the cork out under pressure. But what is the functional element? Is it the collection of physical objects that have this functionality? No, it is not a physical object nor a collection of such objects; it is the *description* or definition of the functionality, that is, the sentence "An object has the functionality of a decorker if, by applying it to a corked bottle, it allows the cork to be extracted."

Throughout this book, the term "functional" plays a major role; it appears in such central concepts as functional element, functionality, functional system, and functional parameter, and it is therefore essential to have an absolutely clear and unambiguous understanding of its meaning in the present context. It is arguably the case that many of the problems that arise in the development of large and/or complex systems have their roots in an inadequate understanding of this term and, in particular, in the attitude that it is not important to be precise about anything but the physical characteristics of systems. We shall therefore first develop a basic understanding of the term, and then refine and strengthen this understanding in numerous instances as we progress and get into the details of the design methodology.

In the case of a physical description, we, as humans, can relate to it by *visualizing* it; even though the object has not been produced yet, we can imagine what it will look like, see it with our "inner eye." This is possible even if the object has never existed, but because we have demanded that the physical descriptions must be complete in the sense of containing all the information necessary to (re)produce the object (at least theoretically, if not practically), there is a one-to-one mapping (isomorphism) between points in the physical domain and physical objects, and it is therefore often not necessary to differentiate between the two. (That is, we shall often, somewhat loosely, refer to either set as the physical domain.)

But how can we relate to a functional element? There is no one physical object that corresponds to it; if it were to be visualized at all, it would have to be as an *action*. It can be difficult to visualize an action without at the same time visualizing an object that performs the action, particularly in such often-observed cases as that of removing a cork from a bottle, but it is not difficult to visualize (or imagine) actions that cannot possibly be performed by any physical object, such as transporting something over a finite distance in an arbitrarily short time. Being able to separate an action from any particular object is the starting point of lateral thinking; being able to recognize which actions are physically realizable differentiates thinking from dreaming. At this point it is simply important to realize that we find it considerably more difficult to think about functional elements than physical objects, and that to do so effectively may require some practice.

Sometimes physical objects may be used to perform functions for which they were not designed, such as the many uses found for surplus military equipment. However, as we are concerned with design within engineering, such unintentional functionality will be mostly disregarded.

A physical object does not have to have only one function; a Swiss army knife has many other functions besides that of a decorker. This example leads us into another aspect of physical objects and functional elements — what we naturally think of as objects and elements. We naturally think of the Swiss army knife as a single object, and would not generally think of its individual blades and other implements as individual objects unless we were, for example, looking at the process of manufacturing such knives. But the functionality of the corkscrew on the knife is an obvious element, whereas the functionality of the whole knife is not easy to formulate, except as the sum of individual elements. The underlying issue here is *complexity*, the topic of the next section.

Before leaving this introduction to the concept of a functional element, we recognize that a functional element does not have to represent the complete functionality of a physical object. It can describe some part of that functionality (such as the decorker on the Swiss army knife), so that the complete functionality may be, on the one hand, represented by a single (large, complex) functional element or, on the other hand, by a set of functional elements, thus introducing a form of *hierarchical ordering* in the space of all functional elements. But, more importantly, even where the object appears to have only one function, in the common sense of the word, the functionality, that is, what the object does, can have many *aspects,* each one representing the view of a person with a different relationship to the object. The views of the manufacturer (return on investment), the retailer (profit, customer satisfaction), the user (utility), and the repairer (source of work) may all be represented by functional elements. Throughout this book, we shall use *service* to designate the more restricted meaning of functionality, the part concerned with the physical processes of the object.

1.4 Complexity

The third topic that plays a central role in understanding the methodology is *complexity.* We encountered this concept already in section 1.2, but what do we mean when we say that something is complex? That it has many sides or aspects to it, needs many variables or parameters to describe it, or consists of many parts? Or that it is hard to understand, needs many words to explain, or is difficult to predict? Usually we mean an unspecified combination of some or all of these and similar definitions, with the emphasis depending on the particular case, but in one way or the other, complexity is related to the number of parameters required to describe behavior. In particular, complexity increases enormously as soon as humans are involved.[8] Or any living organism, for that matter, as illustrated by the fact that we can predict the position of the planets five years from now, but not the position of a dog five minutes from now.

Until recently, human behavior was considered to be outside the scope of engineering, an attitude that has resulted in engineering losing some of its importance in the business world. However, in the last twenty years or so there has been a renewed emphasis on the fact that if engineering is about providing solutions for people, a thorough understanding of the users and their requirements has to provide the point of departure for any engineering project. For example, in the case of the decorker, its esthetic appeal and ornamental features are significant aspects of its functionality, but not easy to measure in "engineering terms." In addition, the rise of information technology has provided a close coupling between humans and technology, which results in design requirements that go far beyond the simple ergonomics of mechanical equipment.

But if taking human nature and behavior into account increases the complexity of a design project, it is equally true that complexity itself is a thoroughly human concept. Something is considered complex because it is difficult for us, as humans, to come to grips with and to work with; it has to do with the capabilities of our brain. It makes no sense to say that something is complex in itself, without putting it in the context of whatever entity is going to operate on it; what is complex to a human may be very simple for a computer, and vice versa. The difficulty we have in conceiving of something as a single entity once it has more than about ten parameters is a characteristic of the brain, and indeed, the success of our whole design methodology will depend on how well it exploits the strengths and avoids the weaknesses of our brains.

In equating complexity to the number of parameters required to describe something, it is important to recognize that there are two essentially different cases. In the first case, all (or almost all) of the parameters relate to different aspects of the entity being described, as would typically be true in a user requirements specification. In the second case, the parameters fall into subsets of the same types of parameters, such as the case of a volume of gas, where, if each molecule is treated as a mass point, the state of each molecule is described by six parameters (or variables), but the number of each type of parameter might be in the order of 10E24, equal to the number of molecules.

In the former case, one approach to handling the complexity is to combine sets of parameters into single "composite" parameters; one example of such a parameter is quality; another is cost. Another approach is to divide the parameters into classes of diminishing importance, and then treat the parameters in the first class exactly, and those in the following classes as perturbations of increasingly higher order.

In the latter case, the complexity is handled by taking a statistical approach, exemplified by statistical mechanics, and while this exact case does not occur often in engineering, a similar case does. Consider that the molecules in the gas volume all have slightly different mass; then it is possible to revert to the previous case by defining an "average" molecule (this may not be the arithmetic mean), and this approach is used frequently in engineering applications where system behavior is dependent on the behavior of a large number of similar elements, such as integrated circuits (numerous individual transistors), telecommunications systems (numerous subscriber loops), and defense systems (numerous units, down to individual soldiers).

In a well-known paper,[9] Weaver considered these two cases and called them *organized complexity* and *disorganized complexity*, respectively. He pointed out that in the case of, for example, a gas, we understand and can give an exact quantitative account of what happens when two molecules collide, whereas in the case of chemical substances we do not understand why one molecule is a poison while another, made up of exactly the same atoms but assembled in a mirror-image pattern, is completely harmless. In the former case, the complexity arises solely as a result of the large number of molecules in a useful volume of gas, whereas in the latter the complexity arises from the necessity of simultaneously treating a sizable number of factors that are interrelated into an organic whole. However, from the point of view of engineering (not science) there is no sharp distinction between these two cases; it is a matter of degree. Even if the details of the molecular interaction between the poison and the human body were understood, the macroscopic effect of the poison would still have a random component, for example, in the form of the probability of death as a function of the amount of poison ingested.

So, for the purpose of design, in either case the complexity requires a statistical or probabilistic approach, and introduces *uncertainty.* This inherent uncertainty leads to *risk* in various forms, and this risk must be handled within the design methodology in the same manner as we handle acquisition cost and operating cost. In the design of complex systems, risk becomes itself a major design parameter, and the uncertainty is handled by letting the functional parameters become stochastic variables, characterized by probability density functions. The various measures of risk become expressions involving integrals of these function.

Of course, uncertainty also leads to *opportunity*; risk and opportunity are the two sides of the same coin. And in some areas of engineering, particularly in the development of technology, this is an accepted and exploited part of the process. Another area where risk and opportunity go hand in hand is in the prospecting and exploration for minerals and gas and oil. But in engineering design the contractual framework has traditionally been heavily skewed toward the avoidance of risk through such measures as liquidated damages and other penalties; only in the last ten years or so have more innovative contracting methods, in the form of *alliances*, become accepted, in which the parties share gain and pain in a "best for project" culture.[10]

Figure 1.3 is a simplified illustration of the prevailing situation in "hard dollar" contracts. The dash-dot line is the benefit, as defined by the contract, increasing linearly up to the required performance (which in this case has the value 1.0), but

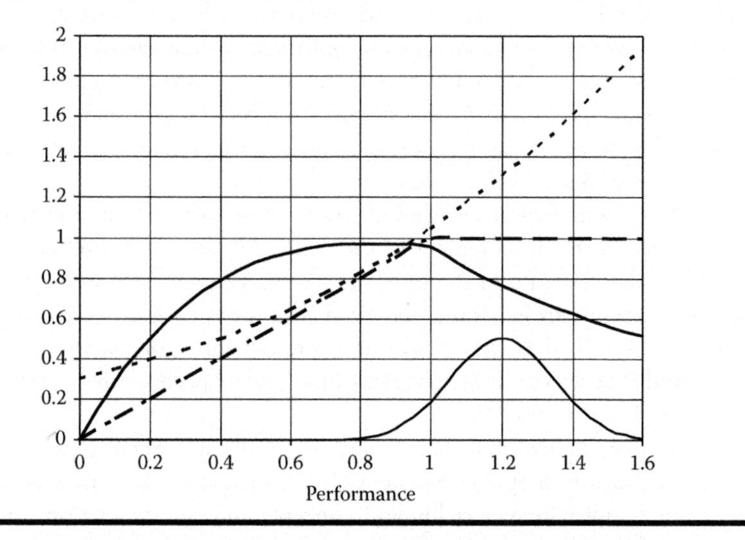

Figure 1.3 A simplified illustration of the major parameters involved in a "hard dollar" contract. (The y-axis is the relative value of each parameter.) The thin, full line is the performance probability distribution, as estimated by the designer; the dash-dot line is the benefit, as stipulated by the contract; the dotted line is the estimated cost at completion; and the heavy, full line is the benefit/cost ratio.

then remaining constant above that. In other words, the contract acknowledges no benefit for achieving a greater performance than specified, but would enforce penalties to make up for any deficit in the performance. Consequently, the designer has centered the performance distribution, represented by the thin, full line, around a performance value of 1.2 in order to leave only a very small probability of incurring a penalty. In choosing this performance value, he would have balanced the probability of winning the contract against the expectation value of the penalty. The result is that the cost, represented by the dotted line, is about 25% higher than it would be for a design centered around the required performance value, and the benefit/cost ratio, represented by the heavy, full line, is correspondingly lower.

1.5 Systems

The discussion on the nature of complexity in engineering leads quite naturally to the fourth topic — systems — because systems can be viewed as a means of handling complexity. The idea of partitioning an entity into a set of interacting elements for the purpose of getting a better mental grasp of it, and the recognition that the entity can have features that are not present in any of the elements, is not new. It lies at the center of what is usually known as general systems theory,[11] and has been applied with varying degrees of rigor to most areas of our experience. Indeed, the word "system" has become so common that it is used without any precise understanding of its meaning or significance; it vaguely signifies something that consists of many parts, or has many sides to it — a few examples are telephone system, monetary system, and solar system. Dictionaries give such general definitions as "complex whole, set of connected things or parts, and organized body of things;" we need to discuss how we want to use the concept.

First of all, even though we shall restrict our attention to engineered systems when it comes to the development of the methodology, which excludes such systems as the solar system and the immune system, engineered systems are not different from systems in general as far as their system nature is concerned. They are only different in that they reflect an expressed intention on the part of their designer, and that we can therefore talk about such aspects as their cost-effectiveness in achieving that intent.

Second, we recognize that there is nothing in the system concept that restricts it to physical systems; it applies equally well to concepts or ideas, in which case the elements are simpler (or better known) concepts and the interactions are formal relationships between these elements (i.e., inherent in their definitions). Indeed, a major feature of our methodology will be the formulation of a functional description as a system of interacting, smaller descriptions, called *functional elements*. However, because physical systems are much easier to visualize, it is often easier to use well-known physical system to illustrate some important features of systems in general that therefore apply also to abstract or functional systems.

Third, because of the interactions between the elements, a system can be thought of as introducing an ordering in a previously unordered set of elements, or of *structuring* the set. An example of this is when a set of non-interacting water molecules (as in a dilute gas of water molecules) are brought close enough to interact and form water, and finally brought even closer (by lowering the temperature) to form ice. The set of non-interacting molecules do not form a system, the molecules in the ice crystal do.

This example also shows that the same set of elements can have quite different properties, depending on the interactions of the elements, and this can be further illustrated by a related example — a set of 12 hydrogen and 5 carbon atoms. When brought together they form pentane, but the pentane molecule comes in three varieties, the so-called isomers, each with a different structure and with different physical properties, such as boiling point. The concept of structure will play an important part in the theoretical developments in later chapters.

Fourth, the elements in a system may themselves be treated as systems (or, perhaps, subsystems); in the case of pentane, in a system of pentane molecules (say, a small volume of penthane liquid) the elements are molecules, but the molecules may be treated as systems of atoms.

Fifth, some of the elements in the systems we shall consider will consist of, or include, people in the form of operators and maintainers. We shall return to this most important aspect of our systems in various places throughout this book; here we just note that it is the inclusion of people that allows our systems to have a steady state through the capability of people to generate negative entropy (i.e., create or restore order) from an interaction with the environment, and also that we now have two groups of people, those inside the system and those outside. The latter group is the *stakeholder group*, and our design methodology will treat the two groups quite differently.

Finally, we note that while, strictly speaking, a system is a *mode of description* of an object in terms of a set of interacting elements, common language uses the word "system" to identify the object itself irrespective of how it is described. We shall use both meanings of the word; which one is meant in a particular case should be obvious from the context and not cause any confusion.

1.6 Bringing It All Together

In section 1.2, we argued that design is the core activity of engineering; it is the creative activity that distinguishes the engineer from the technician. It is a mental activity that takes place within the brains of individual engineers and consists of synthesizing a solution from a set of known construction elements. But it is limited by the capability of the brain to handle large or complex tasks and becomes inefficient as the complexity increases. What is required is a process of reducing the complexity *before* making the transition into the physical domain and the start of the classical design activity, a process we may describe as doing *design in the func-*

tional domain. The discussion in section 1.3 indicates two reasons why functional elements provide this means. First, because the nature of functional elements, representing classes of physical objects, allows us to abstract from the details of a specific physical object and therefore in itself constitutes a reduction of complexity. Second, because the functional domain offers the possibility of proceeding from the stakeholder requirements in a top-down fashion, arriving at a set of functional elements of such limited complexity that each can be realized as physical objects without the iterations that make the current design methodology inefficient.

But the elements in the set are not independent; it is the fact that they interact that allows relatively simple elements to represent complex behavior. The discussion in section 1.5 shows that the general theory of systems provides a framework within which functional elements can interact and thereby produce a behavior that is not evident in any one of the elements and one that also displays the correct hierarchical ordering of the elements.

Finally, as handling *complexity* lies at the heart of the design activity, the discussion of complexity in section 1.4 shows that our design methodology will have significant statistical aspects. In some ways, this design methodology is to engineering what statistical mechanics is to mechanics.

What emerges from this is a design methodology in which the design elements are functional elements and the end result is a system of functional elements that provides a service that satisfies the stakeholder requirements, and in which the design rules ensure that the physical system that results from converting the functional elements into physical objects will represent the most cost-effective solution to meeting the stakeholder requirements.

Before closing this introduction, it is appropriate to make a comment about the relationship between this proposed design methodology and the methodologies used to develop software, in particular structured and object-oriented programming. It will no doubt have occurred to most readers that the approach outlined above is very similar to that used in modern software development, and that functional elements are somehow related to the software representation of objects. To see what that relationship is, we recall that the objects of object-oriented programming are real, physical objects, such as persons, assets, accounts, etc., with real, physical attributes, or *properties*, such as name, address, quantity, date, etc., and the software describes interactions between objects in terms of operations on these attributes, called *methods*. The software is a *language* to describe what takes place in the physical world; it has no substance in itself. Functional elements, on the other hand, are a priori to any physical realization, and have an existence and meaning (and therefore "substance," even though not physical) in their own right. Functional elements, describing what physical objects are capable of doing, are therefore related to methods rather than to objects; they are a type of generalized methods. They can be described by software, and if an object-oriented approach is taken, they become a new set of abstract "objects."

Notes

1. A few representative references in the development of systems engineering, with particular emphasis on design rather than on management, are Blanchard, B.S. and Fabrycky, W.J., *Systems engineering and analysis,* 4th ed., New York: Prentice Hall, 2005; Bode, H., The systems approach, in *Applied science — Technological progress,* report to Committee on Science and Astronautics, US House of Representatives, 1967; Chestnut, H., *Systems engineering methods,* John Wiley & Sons, New York, 1969; Johnson, R.A., Kast, F.W. and Rosenzweig, J.E., *The theory and management of systems,* McGraw-Hill, New York, 1963; Kalman, R.E. et al., *Topics in mathematical system theory,* McGraw-Hill, New York, 1969; Mahelanahis, A., *Introductory systems engineering,* John Wiley & Sons, New York, 1982; Miles, R.F. (ed.), *System concepts,* John Wiley & Sons, New York, 1973; Sage, A.R., *Methodology for large-scale systems,* McGraw-Hill, New York, 1977; Warfield, J., *A science of generic design: Managing complexity through systems design*, Iowa State University Press, Ames, IA, 1994; Wymore, A.W., *A mathematical theory of systems engineering — The elements,* John Wiley & Sons, New York, 1967; *Systems engineering methodology for interdisciplinary teams,* John Wiley & Sons, New York, 1976; *Model-based systems engineering,* CRC Press, Boca Raton, FL, 1993.
2. The roots of UML lie in object-oriented software development, as set out, for example, in Jakobson, I., *Object-oriented software engineering; a use case driven approach,* rev. print. ed., Addison-Wesley, Reading, MA, 1993. The UMG Web site is http://www.omg.org, and an introduction and numerous links can be found in Wikipedia, at http://en.wikipedia.org/wiki/Unified_Modeling_Language.
3. INCOSE, Web site of the Model Based Systems Engineering working group, http://www.incose.org/practice/techactivities/modelingtools/mdsdwg.aspx. See also Wymore, A.W., *A mathematical theory of systems engineering — The elements,* John Wiley & Sons, New York, 1967.
4. A listing of modeling tools can be found at http://www.SysMLforum.com.
5. A journal dedicated to economic modeling is *Economic Modelling*, Hall, S. and Pauly P., eds., Elsevier, New York; otherwise, a good starting point is the Web site http://zia.hss.cmu.edu/econ/.
6. Being a well-established subject, engineering design has a vast literature associated with it. A few references that document well the classical design process and the issues arising within it, albeit with a grounding mainly in mechanical/structural engineering, are Dhillon, B.S., *Advanced design concepts for engineers*, Technomic Publishing, Lancaster, PA, 1998; Ertas, E. and Jones, J.C. *The engineering design process*, John Wiley & Sons, New York, 1993; Lewis, W.P. and Samuel, A. E. *Fundamentals of engineering design*, Prentice Hall, New York, 1989.
7. This figure and the following one first appeared in E.W. Aslaksen, Going up? Coming down!, *Engineering World*, 1, 4, 1991.

8. The issue was discussed at some length in the book by Stephen Kline, *Foundations of multidisciplinary thinking,* Stanford University Press, 1995; some other, more recent references to the complex systems literature are Braha, D., Minai, A., and Bar-Yam, Y., eds., *Complex engineered systems: Science meets technology,* Springer-Verlag, Heidelberg, 2006; Axelrod, R. and Cohen, M., *Harnessing complexity: Organizational implications of a scientific frontier,* Simon and Schuster, New York, 1999; INCOSE, Web site of the Systems Science group, at http://incose.org/practice/techactivities/wg/sseg/.
9. Weaver, W., Science and complexity, *American Scientist,* 36, 536, 1948, available online at www.ceptualinstitute.com/genre/weaver/weaver-1947b.htm.
10. Some background to the development of this contracting method is contained in *A resource and research bibliography,* available online at www.mcmullan.net/eclj/Alliance_Contracting.htm. Legal aspects are discussed in *Commonwealth procurement guidelines,* available online at www.ags.com.au/publications/agspubs/legalpubs/commercialnotes/comnote04.htm.
11. Two of the most significant references in general systems theory are Laszlo, E., ed., *The relevance of general systems theory,* George Braziller, New York, 1972, and Von Bertalanffy, L., *General Systems Theory,* George Braziller, New York, 1941.

Chapter 2

The Purpose of Design

2.1 The Design Process and Measures of Success

If we are to develop a methodology for designing systems, we first have to answer the question: What constitutes a good design methodology? In a qualitative sense, the requirements of the methodology must include a high probability of the object resulting from the design meeting the user requirements (i.e., a successful outcome), low cost of performing the design, and applicability to as wide a range of potential design problems as possible. In addition, we would prefer a methodology that is logical and easy to learn and remember, and, perhaps particularly important, it should not be a hindrance to creative thinking, but on the contrary support it by providing reusable models that link our creative ideas to reality, without which they are just dreaming or wishful thinking.

In this regard, it is useful to contrast what we are trying to achieve with other design methodologies, exemplified in particular by TRIZ, the inventive problem solving technique developed originally by Altshuller.[1] Briefly, the technique consists of four steps, as illustrated in Figure 2.1.

The starting point is to view design as a problem-solving exercise, so the first step is to understand (i.e., classify) exactly what the problem is. The second step is then to search through past experience to find where analogous problems arose (they form a class), and in the third step select the most appropriate solution from those used to solve the problems in the class. The fourth step is then the adaptation of this solution to the specifics of the current problem.

The technique is based on the observation that the vast majority of design problems are solved by modifying (to a greater or lesser extent) solutions to previous

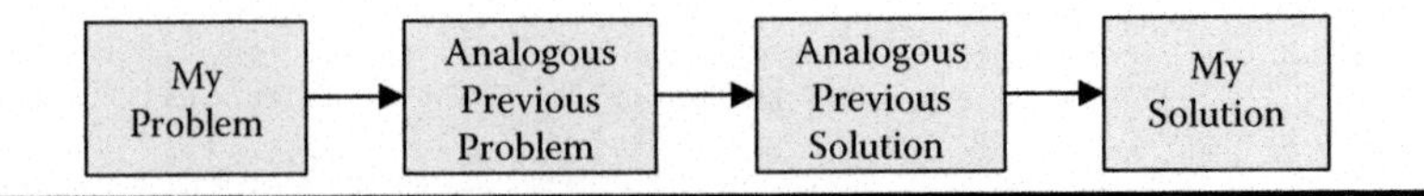

Figure 2.1 Four-step representation of the TRIZ process.

problems. In effect, this creates a second level of reusable (standard) elements above the level of construction elements, and using this picture, what we are proposing is to create a third level, where reusable models (functional elements) are used to rationalize and support the creative or project-specific part of the design process, as illustrated in Figure 2.2.

These qualitative requirements will be useful for guiding us through the development of the methodology, but it would be good to have some quantitative measure to tell us how successful we are as we go along, to somehow reassure us that we are on the right track. The difficulty with this is that the proof of success can only lie in having successful outcomes when the methodology is applied, which requires the methodology to be, first, more or less fully developed and, second, accepted by at least some part of the engineering community. Thus, the proof of success could only arrive some considerable time after the methodology has been developed. However, the situation is not that we want to develop a completely new methodology; many of the general ideas and concepts we shall be using have been used successfully before, and the basic approach of treating complex objects as systems is very old.

The situation is that the design of systems runs into a number of problems well known to system engineers, the primary one being low efficiency (or high cost), which often makes the design process not cost-effective, and which leads to the adoption of existing solutions in new situations where they may not be

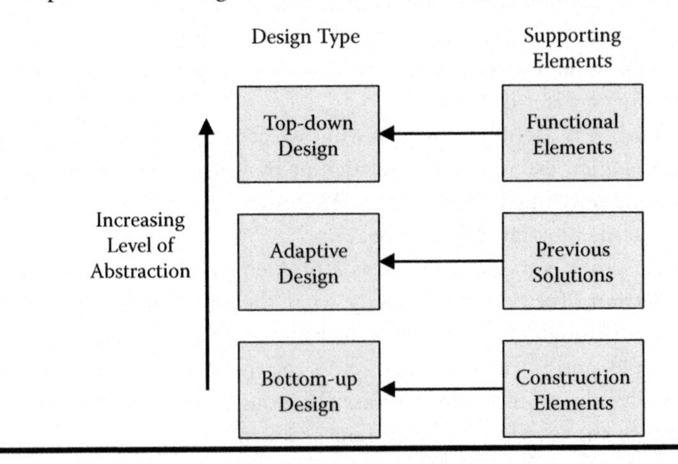

Figure 2.2 A view of design as consisting of three types of processes at differing levels of abstraction, with each type relying on supporting elements for efficiency.

entirely appropriate. It is the underlying assertion of the present work that most of this inefficiency comes from a lack of a complete and rigorous foundation for the design process and, as a result, a lack of the standardization that is a prerequisite for efficiency. Therefore, if we are able to develop a basis for system design that is sufficiently rigorous to demonstrably support a program of standardization of functional elements, we shall deem our undertaking to have been successful. Realizing a standardization program will take many years.

In answering the question about what constitutes a good design methodology, a significant component was, obviously, a high probability of achieving a successful outcome. But what is "a successful outcome"? What is good design? What are we trying to achieve by the design process? If we look upon the object resulting from the design purely as a physical object, that is, for what it *is*, we could try to come up with criteria of good design along the lines of "most appropriate materials," "no waste of material," "easy to manufacture," "low energy consumption," and so on, and these are all among the criteria traditionally used to judge a design. They are all characterized by being criteria that can be applied to the object of the design as such, without knowing the application of the object, and where the judgment can be made by the designer's peers. They are most significant when applied to objects at the bottom of the complexity scale, that is, to the basic building blocks, to the components that represent our technology, but as we move upward on the scale, through modules, equipment, and up to systems, this group of criteria is displaced (or at least augmented) by a group concerned with what the object *does*, that is, its functionality. Functionality can only be judged when the object is actually operating, and the judges are the users, not designers. In the case of a semiconductor, the judges of good design are bound to be engineers, but the judges of good design of a telephone system are mainly the subscribers, and they do not care about the design of the semiconductors that are used in the system. This shift in focus from technology to functionality and from the judgment of the designers to that of the users is a major source of the complexity of system design, and must therefore have a great influence on our design methodology.

It follows from the above that the design of a system cannot be judged by considering the system by itself; a system design can only be judged relative to its intended *stakeholder group*. Conversely, the stakeholder group must be defined and fully characterized, in the form of the relationships of its constituents to the project, prior to the start of the design process. The *stakeholder requirements* should be defined prior to the start of design; this is commonly accepted and generally the case. And it is also accepted that there should be a defined process for handling the inevitable changes to these requirements, although the implementation of this process often leaves something to be desired. However, the clear and detailed definition of the stakeholder group is much less common, and a unique relationship between the group and the requirements, in the sense that a change to the stakeholder group will result in a change to the requirements, is even more uncommon. A result of this is that dissatisfaction with the performance of a new system will just as often be due to an inadequate definition and understanding of the stakeholder group and its dynamics as to any shortcoming in the subsequent design process itself, and herein lies a main cause of the complexity of systems engineering.

The term "stakeholder group" should not be taken to imply that it is a group in the mathematical sense. It is actually just a set, and "subgroups" are subsets. The members of the group may include not only the users in the narrow sense, such as "the users of public transport" being the people who travel on the system, but in principle anyone who has a stake in the behavior of the system, such as the equity and debt providers. The members are not individuals; they represent a particular relationship between a system and a person, such as "investor," and the same person can be represented by more than one member, such as "investor" and "traveler" in the case of the public transport system. This can also be expressed by saying that each member is interested in only particular *aspects* of the functionality. The members form a set by virtue of placing requirements (which may be very different) on the same system (or subsystem). In the following, we shall, for simplicity, refer to members of the stakeholder group as stakeholders, because the people within the system, although stakeholders in the sense defined in section 1.1, play no role in the functional domain.

In chapter 1 we had noted that there are two groups of people related to a project, those within the system and those outside the system, that is, the group. In view of the foregoing paragraph, we should now strictly reformulate this in terms of roles rather than people, as the same person can be inside the system, for example, as an operator on the transport system, during working hours and outside, for example, as a traveler, outside working hours. However, for simplicity we shall not make the distinction between person and role, so that when we say "operator," we mean the person occupying that role.

The issue here is not the existence of a stakeholder group or not, because eventually every system is, of course, exposed to its stakeholder group. The issue is how to include the stakeholder group in the design methodology in such a manner as to increase the probability of a successful outcome of the design process, and the first step in this direction is to develop an operational definition of "stakeholder group," in the sense that the concept can be manipulated as an object in the methodology. The definition we shall adopt, and develop in more detail shortly, is that the stakeholder group is the set of persons, not included in the system, who determine the success of a project, where "project" denotes the whole life cycle of a system, in accordance with our definition in chapter 1. At present this is a circular definition; the stakeholder group is defined in terms of the concept of success, but success is defined by the stakeholder group, so it all comes back to the basic problem of finding an objective, universally applicable *measure* of the success of an engineering project. The *value* that this measure takes on in a particular case can then be determined by the stakeholder group.

For any given project, the measure of success will generally be very complex, as it must somehow be related to the fulfillment of the requirements that define the purpose of the project, and the documents defining these requirements can run into hundreds of pages even for a medium-sized project. We can try to visualize this by thinking of the requirements as points in a "requirements space" and simplify this space to a plane; a project is then characterized by a certain area of this space. Similar projects will have areas that overlap to a large degree, dissimilar projects will have areas that overlap to only a small degree, as shown in Figure 2.3; but can two projects have completely disjoint areas? Or is there some purpose that is common to all projects, and which might therefore be the universal measure we are seeking?

Before attempting to answer this question, let us explicitly recognize that we have now introduced two closely related, but still completely distinct concepts. One is that of *functionality*, which arose as a means of expressing what a physical object *does*, and as such is something that the designer can manipulate and about which he can make decisions. The other is that of *purpose*, which is an expression of what the stakeholders want. It is embodied in the stakeholder requirements, and it is the fulfillment of this purpose that has a *value*. Value, purpose, and stakeholder requirements are all related to the stakeholders and are things over which the designer has no influence; they are given at the outset of the design process, and will evolve in a dynamic fashion during the design. The two are closely related; first, because the designer's task is to find that functional element (or system of functional elements) that has as its output a functionality that meets the stakeholder requirements and thereby fulfills the stakeholders' purpose, and, second, because the parameters describing the service must be related to the parameters of the stakeholder requirements in that the purpose must be able to be expressed as a function of the functional parameters. If this latter condition is not met, there is no way of connecting to the value and therefore no way of optimizing the design. But it is a serious mistake to simply equate functionality (or, more precisely, the service) with purpose (as was, unfortunately, implied in *The Changing Nature of Engineering*[2]), and a first illustration of that is given in the next section. As a very rough analogy, equating functionality with purpose is like equating the symptoms with

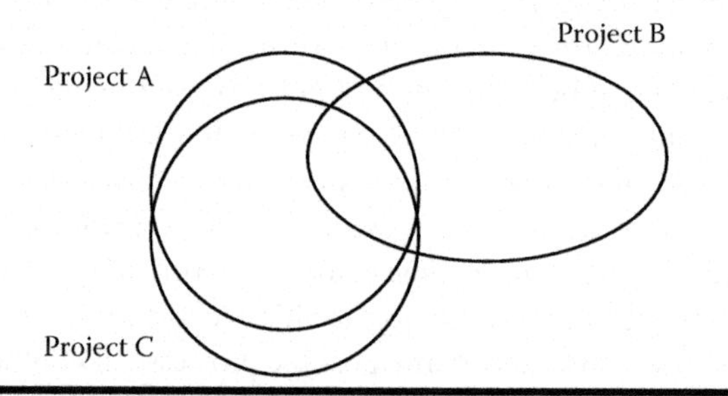

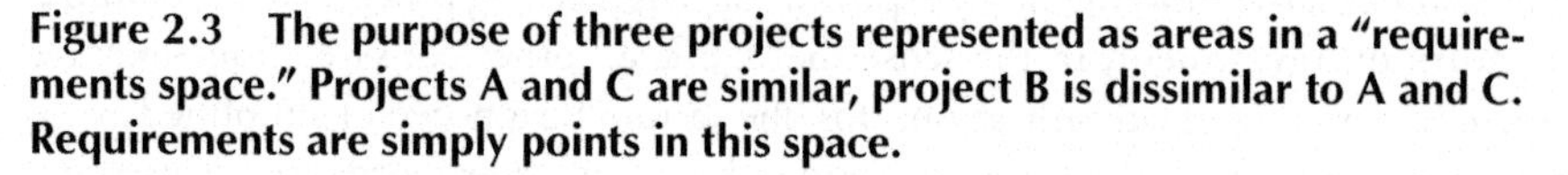

Figure 2.3 The purpose of three projects represented as areas in a "requirements space." Projects A and C are similar, project B is dissimilar to A and C. Requirements are simply points in this space.

the disease. By having a complete set of symptoms we can, presumably, identify any disease, but this is not the same as knowing the disease, in the sense of what causes it, which becomes apparent when we look for a cure. Similarly, having a set of functional parameters that would be adequate for describing the functionality is not the same as knowing the purpose, which becomes apparent when we try to optimize our design.

In summary, the point of departure for our design methodology is a set of requirements for the to-be-designed object, the stakeholder requirements. These requirements fall into two main groups: physical parameters describing what the object should *be*, and functional parameters describing what the object should *do*. But in addition, the stakeholder requirements must describe what the object is supposed to *achieve* from the point of view of the stakeholders (which is its purpose), and this is done by associating a *value* with each of the functional parameters.

2.2 Return on Investment

If we recall the idea, introduced in section 1.6, of describing the functionality of a complex physical system in terms of a system of functional elements, the question posed at the end of the last section can be reformulated as "Is there a functional element that is common to all systems?" This would be the functional element that fulfills the purpose common to all projects, if there is such a basic purpose. The argument for the existence of such an element, and the definition of its parameters, is a crucial step in the development of the design methodology; because of the hierarchical ordering of functional elements alluded to in section 1.3 and discussed in more detail in chapter 4, every functional element will be related to this common element, and the development of all functional elements must start from this one element.

The creation and operation of any system must involve an expenditure of resources in some form, such as labor, energy, information, and materials, and even forms that are not normally or easily measured in monetary terms. But would anyone incur a cost without any prospect of a return? The return might not be directly in terms of money; it can be in the form of personal well-being, absence of illness, peace of mind, a sense of achievement, and so on, but indirectly, as with the cost, this can, in principle, always be measured in monetary terms.

The return can only occur once the system has been created and put into operation, so that the expenditure must come before the return and is therefore an *investment*. All engineering projects are subject to this cycle of investment, creation, and return, and as there are infinitely many possible projects, all competing for a finite set of resources, the fundamental purpose of the engineering process is to maximize the return on investment, in the sense that while all other considerations may be ignored in a process of simplification, this one cannot be ignored or simplified away without making the design process irrational. Or, in other words, the constraint of competition for limited resources is the essence of the process of engineering, that which makes it fundamentally different from a science. Engineering is not about truth, but about cost-effectiveness, with the effectiveness being judged by the stakeholders. Consequently, the functional element representing this fundamental purpose, return on investment (ROI), can be said to be an *irreducible* element, and it is not dependent on the particular system, but is, because it has its genesis in the process that creates all systems, universal to all systems.

In order to produce a return on the investment, any system must put its functionality into operation, and that operation is described by the values taken on by the functional parameters. The design identifies the value each of the parameters needs to take on in order to meet the requirements of the stakeholders; they are the *nominal design values* for the particular system. The degree to which the system actually meets these requirements can, in principle, be characterized by a single variable, the system *quality of service*, or QoS, which is a function (often the weighted average) of the degree to which the individual parameters meet their nominal design values. This single performance parameter is denoted by S, and while it is called "quality," it may in fact often be mostly, or even wholly, a measure of the quantity of whatever the system produces. It just depends on what is the most important feature of the system's functionality, what Hitchins calls "prime directive."[3]

As already stated, the creation and operation of any system must involve an expenditure of resources. This expenditure may take many forms, but in the context of engineering, they must all be converted to monetary terms, for example, dollars, and together they shall be termed the *cost* of the system, denoted by C.

No system can continue to operate, producing a service and incurring costs (or expending resources), without receiving something in return that sustains the operation. This shall be called the *revenue*, and no matter in what form it is received, it must, within the context of the process of engineering, be expressed in monetary terms. As with the costs, this may not always be a simple matter, as is exemplified by such a return as "quality of life" in, for example, a foreign aid project; this is part of the vast increase in complexity that occurs once engineering goes beyond the traditional boundaries of objects and outcomes directly describable in terms of physics. The revenue will be denoted by R.

All of the above takes place within a time period; no system can perform any service in an instant of time, nor can it be created instantly. The time period over which all activities associated with a system take place is its *life cycle*; in the terms of a living organism, this is the time period between conception and death. The duration of the life cycle, which may just be called the *life* of the system, will be denoted by L, and measured in units of time. Both the cost and the revenue are referenced to the beginning of the life cycle, that is, they are the sums of the present values of costs and revenues, respectively, incurred throughout the life cycle (see section 7.3).

These two parameters L and S, and the two variables R and C, are necessary and sufficient for expressing the concept of ROI, which in the following will be denoted by Q,

$$ROI = Q = 100\left[\frac{R(S,L)}{C(S,L)} - 1\right]\%,$$

and constitute what we shall call *the basic set.*

The two parameters S and L appear only implicitly, but that does not make them any less necessary. If the cost and the revenue did not depend on both the quality and the duration of the operation, well, then, why not ask for perfection and that it last forever?

We can now define a very special functional element, which we shall call the *irreducible element*, as that element which is common to all engineered objects and which has the unique property of unifying functionality and purpose. It can be represented graphically as shown in Figure 2.4, with the functional parameters (i.e., those parameters that enter into the definition of the functionality) on the right-hand side, and the functional variables (i.e., those parameters describing the interaction with the outside world required in order for the element to provide its functionality) on the left-hand side.

This first example of a small functional element (albeit a very special one) already gives an idea of why we must differentiate between functionality and purpose. The functionality of the irreducible element is to deliver a service, characterized by the two parameters S and L, by using resources, parameterized by the cost, C, and attracting a revenue, R. This is the most basic description of the functionality of any object in that it is an abstraction away from any specifics of the functionality; it can be expanded by expanding each of the four variables, that is, the cost can be subdivided into cost types and the revenue into revenue types, the service can be characterized by more parameters, and the lifetime can be subdivided into different parts.

The ROI is the most basic expression of the purpose, and the four variables of the irreducible element are sufficient for expressing the ROI as a function of them. A given object, such as a complex system, may be intended to fulfill a multitude of purposes, usually represented by different members of the stakeholder group, and in order to fulfill them, we need a very complex functional element, or a system of less complex elements, with many variables. However, and this is the crux of the matter, even in the case of a very detailed (i.e., complex) description of functionality, ROI (for example) may

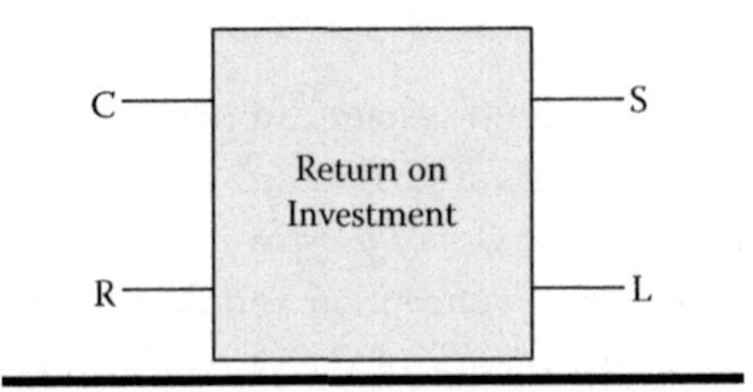

Figure 2.4 The irreducible element return on investment.

be the only purpose. That is, *the functionality must be complex enough to allow us to fulfill the purpose, but the purpose of an object with complex functionality may be simple.*

2.3 Philosophical Motivation

The approach to the development of some of the basic concepts put forward above was inspired by the approach taken by Kant in his *Critique of Pure Reason*,[4] in particular, by the section called "Transcendental Analysis." While Kant's purpose was to identify the basic characteristics and capabilities of the mind and so determine the limitations on mental activities, in particular as pertains to those that do not rely on any sensible input, the purpose of this work is very much more modest and of an altogether more limited and practical nature — to investigate the characteristics of the concepts used to think about engineering problems. Above all, there is no transcendental aspect involved in this investigation; engineering is firmly rooted in reality as experienced through our senses. However, despite these fundamental differences, there are strong parallels between the two approaches.

In the *Critique of Pure Reason*, Kant argued that a person's thoughts appear to that person as a coherent whole (the unity of consciousness), within which thoughts can be related to one another, because they are generated by the mind using a fixed set of rules, or *concepts*. We interact with the external world through our senses, and through the representation of such inputs, or sensations, within a space-time framework, the mind creates intuitions. That is, we (i.e., our minds) become aware of the external world through intuitions, but knowledge is only generated through the processing of intuitions by the faculty of understanding.

The analogy between this and engineering can then be expressed by the following correspondences:

Kant	*Engineering*
Creating knowledge through the faculty of understanding	Converting requirements into solutions through the process of engineering
Unity of consciousness	All systems have a functional element in common (the irreducible element)
Categories	Basic set
Concepts	Functional elements
Intuitions	User requirements

In Kant's theory of the mind, the concepts generated by the understanding and used by it to give synthetic unity to a manifold of intuitions are themselves formed according to a set of rules; these are the *categories*, and they are the manifestation of pure reason. Correspondingly, in engineering, the functional elements arise from the irreducible element by the application of certain rules, and the starting point for formulating these rules is the *basic set* of parameters associated with the irreducible element.

But the most important point of this analogy is the correspondence between the unity of consciousness and the irreducible functional element. The former is illustrated by the well-known example of the difference between six men each reading one word of a six-word sentence and one man reading the whole sentence; the meaning of the sentence emerges only when the six words are perceived by the same brain. The synthesis that takes place when the representations of the words are brought together is the "I think" of Descartes, and the fact that they can be brought together, that is, that completely disparate representations have something in common, is the unity of consciousness. It is what allows me to perceive the external world as a connected whole and that constitutes "me."

Similarly, what allows functional elements to be brought together and form complex functional systems is the fact that they are all derived from a common element, the irreducible element. The purpose expressed by the irreducible element, which we have termed ROI, is what binds the elements of the system together; they all participate in achieving that purpose. And also, without this unity of purpose, individual functional systems remain just that; they have nothing in common that identifies them as elements of the same process, the design process. Conversely, the purpose of ROI can, in this sense, be considered to be the *essence* of engineering; when we extract from all other differences between engineering tasks, this purpose remains, common to them all.

2.4 The Concept of Value

The return arises through the *value* of the service provided by the system; the revenue is determined by (in the simplest case equal to) what the users perceive to be the value of the service *to them*, that is, what they are (or would be) willing to pay for the service. Clearly, the value of a given service is not an absolute measure; it exists only in relation to a particular user group. Indeed, this dependence of the value on the user group provides an alternative definition of the user group — the group of all persons who have a say in determining the value of the service provided by the system.

The concept of value is essential to the process of engineering; without it, design ceases to be a rational process.[2] In the author's experience, a greater proportion of poor designs is due to erroneous or inadequate value definitions than to any other cause, and the reasons why it is difficult to define value adequately are easily discernible. They include such practical difficulties as achieving consensus within a diverse user group, making users understand the significance of the parameters used to characterize the service, and getting users to articulate and quantify what are often feelings, prejudices, and impressions rather than rational judgments. But they also include a lack of understanding and definition of the process of engineering itself. On the one hand, there is the view, supported by a significant segment within the engineering community, that engineering is a science, albeit an applied science, and must not only remain securely anchored in the natural sciences, but must not allow itself to become tainted or diluted by "soft" issues that belong more properly to such fields as psychology, linguistics, philosophy, economics, etc. This view essentially makes engineering an incestuous occupation; "good engineering" becomes whatever advances our stock of technology, without any further measure of the usefulness of this technology. Of course, in the past, this was not a problem, because there was such a backlog of unsatisfied needs that almost any new technology was eagerly picked up by the community and put to use in satisfying those needs through the production of capital goods. This situation is one of the hallmarks of the industrial society, but that is starting to change as the industrial era comes to an end, as elegantly argued by Danielmeyer.[5]

On the other hand, there is a view, among both engineers and in the community in general, that there is something morally, or even religiously, wrong with putting a (monetary) value on all services. In the extreme, this argument says that as all services are there to meet human needs, and humans, made in God's image, are infinitely valuable, satisfying their needs is also infinitely valuable. Nonsensical as this argument is by any practical standards, it is astonishing to see how shades of it creep into many of the debates in the community about the level and priority of services.

Both of the above views have the effect of limiting the relevance of engineering to society. Properly understood, engineering is a potent methodology for meeting many of the needs of society, but this must include an understanding of the value concept. It is the value concept that provides the coupling between the system and the stakeholder group, and it is only in terms of such two-component entities that engineering becomes relevant to society. Designing systems in isolation is an exercise in self-gratification. But the coupling of system and stakeholder group allows the design *methodology* to become value neutral; while the definition of a value function may be developed with the help of the designer, assigning numerical values to the parameters of the function is the responsibility of the stakeholder group. Questions relating to how value is to be allocated, what is the proper value in a given case, cannot be answered by the engineering methodology, and trying to make it do so only destroys its effectiveness. However, this does not mean that engineers, as members of the community, and as particularly well-informed members of the community in many cases, should not contribute to finding the answers; quite on the contrary.

Defining the value of a service is perhaps the most difficult part of the process of converting a need into a set of stakeholder requirements, the stakeholder requirements definition process. And again, it is possible to draw a parallel with Kant's view of the mind; the requirements definition process corresponds to the

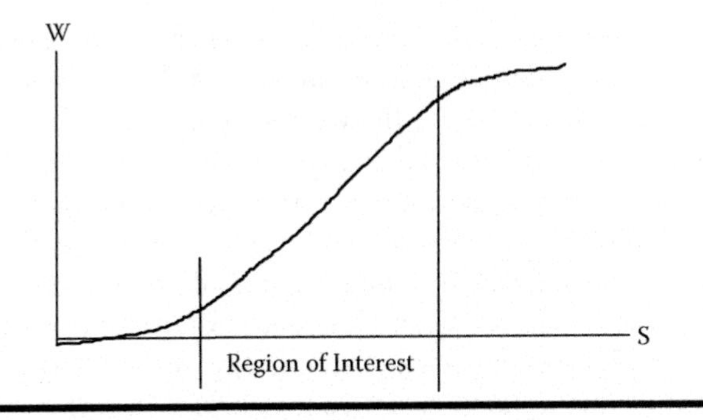

Figure 2.5 Typical form of the value function W(S).

sensibility, that part of the faculty of representation that converts a sensation into an intuition.

The value of a particular service will, in general, depend on the values of the parameters describing the service; if a parameter can be varied without affecting the value, the parameter is superfluous. Therefore, as the description of a service is developed in a top-down fashion, with an increasing number of parameters to express the increasing level of detail, the expression for the value will need to be developed in this top-down manner as well. At the top level, the service is always represented by the irreducible element, ROI, and the value W is a function of the single parameter S, the quality.

The function W(S) is often a highly non-linear function; for a number of reasons there is a relatively narrow range of S in which W(S) increases rapidly with increasing S.[6] Below this range, the quality of the service is so poor that it is basically useless. Above this range, an increase in S brings hardly any further increase in the value; it is a region of saturation or "overkill," as shown in Figure 2.5.

2.5 The Central Role of Money as a Measure

The development in this chapter has led to a view of engineering design as "creativeness under the constraint of cost-effectiveness." At first glance this seems very reasonable, even maybe a bit of a "motherhood statement." Designers are always told to develop "cost-effective solutions," and mostly believe this is what they are doing. However, once we start to look more closely at providing a precise definition of "cost-effective," such as the concept of ROI, we find that what most solutions represent is "an acceptable solution at minimum cost." This focus on the cost side of the equation does not arise because of an inherent parsimoniousness on the part of engineers, but because it is so much easier to determine the cost of a service than its value (which is the monetary expression of effectiveness). As anyone who has participated in value management or value engineering workshops knows, after a bit of lip-service to "value for money," the workshop often quickly turns into a cost-cutting exercise, and relatively rarely is a monetary figure put on the value of a feature deleted as a cost-cutting measure.

This one-sided approach does not allow any proper optimization, in the sense of finding the optimum balance between the opposing factors of value and cost. If the creativeness in design is considered to consist of recognizing a range of options as possible solutions to a requirement, then the lack of a well-defined measure of value, and thereby the lack of a criterion for deciding which one is the better option, is bound to result in a preference for options that have already been successfully applied. The uncertainty in the values of the untried options translates into a risk that often removes them from any further consideration.

This restriction on creativeness is a manifestation of one of the central issues in the design of complex systems. It is not that we cannot, in principle, see how to find an optimal solution; it is that the cost of doing so, in any one particular case, is so high that it appears unlikely to have a reasonable payback. Any design methodology must therefore not only demonstrate that it will lead to optimized solutions, but that it will do so at a cost that is clearly considerably less than the benefit expected from optimization.

Before delving any further into the issue of assigning monetary value to a service, we note that there exists a very significant industry group where this is relatively straightforward and not in general controversial. That is the resource industry, where the functionality is to produce a given product (ore, coal, oil, gas), but with relatively little responsibility for the features of that product, so that the value is proportional to the quantity produced. A typical case is the design of a mine. The service is to deliver ore to the processing plant, and optimizing the design means minimizing the total (i.e., on a life cycle basis) cost per ton of ore, and also taking into account the cost of inadequate delivery rate (i.e., the processing plant being idle, penalties on delivery contracts, etc.). In this case it is well-known how to develop, in a top-down fashion, a cost model of the mine and its operations that allows any design option or feature to be evaluated in monetary terms (but, as previously noted, it is time consuming). All that is required to arrive at a truly optimized design is then the "creativeness" part of the design process, a systematic search for and development of options for every facet of the operations. An example of where this approach, even in a rudimentary form, led to a good result is the Northparkes E26 underground copper mine in New South Wales, a joint venture between North Ltd. and Sumitomo, where the extraction costs are at or below world's best practice levels.[7] Because the top-down part of the design of this mine was limited and easy to explain, it will be used as an example in chapter 7, and will be referred to simply as "the Mine."

In advocating that money be used as a unit of measure, it needs to be emphasized that within the process of engineering there is no difference in principle between the use of money as a measure and the use of any physical measure, such as mass, and the choice of currency, for example, dollar or mark, is equivalent to the choice of mass unit, for example, kilogram or pound. However, in practice there are, of course, significant differences arising out of both issues relating to a currency as a unit of measure, such as a lack of time invariance (inflation) and lack of a fixed ratio between currencies, both of which are easily handled, and the subjective character of money as a measure, which is quite difficult to handle. We shall return to the practical difficulties associated with assigning a cost to a functional requirement in chapter 3.

Why, then, is it so important to assign a monetary value to a service? Are not other parameters, such as human dignity and quality of life, more appropriate in certain circumstances? There is no argument about the importance of these and other parameters of a service; they have emerged as essential concepts in the discussion and assessment of broad aspects of our existence. But, besides the fact that some of these parameters are qualitative in nature, they are all limited in their applicability to a specific context. That is, not only the value of the parameter will vary from case to case, but the very meaning of the measure by which a value is assigned to the parameter will depend on the particular context in which the parameter is applied. If we want to develop a generally valid design methodology, the quantities with which the methodology operates must be given in terms of a context-free measure. Any theory of physical objects operates on quantities with measures defined in terms of the three basic measures for time, distance, and mass (i.e., the second, the meter, and the kilogram); in our present case, the composition of the basic set shows that we need a measure for cost and value. The only generally accepted context-free measure of these two quantities is money.

2.6 The Dynamics of the Design Process

The preceding discussion of the design process and the importance of the stakeholders and their requirements at the start of the process may have given the impression that the process is a linear one, progressing in an immutable sequence of steps based on fixed requirements. If so, that needs to be emphatically corrected at this point; indeed, if it were the case, the methodology being developed would be of limited value. The complexity the methodology is intended to handle arises mainly because of the changes to what might be termed the boundary conditions of the project over time, both during the design and throughout the lifetime of the project.

The two main sources of change are the stakeholder requirements on the one hand, and, on the other hand, the developments taking place in the technology available to meet those requirements. The latter are, in general, beyond any control of the engineer in the types of project considered here (i.e., mainly infrastructure in the widest sense), whereas the stability of the stakeholder requirements can be influenced by a component of project management called stakeholder management (sometimes also called "managing stakeholder expectations," although this is a somewhat more restricted and, to some extent, negative view of the matter). Stakeholder management has received a great deal of attention in the last decade,[8] often in conjunction with customer relationship management (CRM) and/or quality management. Much of the literature is focused on corporations rather than on projects, but the basics and the principles are the same in both cases. It is not our purpose to go into this subject matter as such, what is important for us is to recognize that design is strongly influenced by the changes to the stakeholder requirements and that the manner in which the design process handles these changes is a significant factor in the success (or otherwise) of the project.

Combining the available technology and the stakeholder requirements under the concept of boundary conditions, there are several aspects to the issue of changes to these conditions. First, there is the *dynamics* of the interaction between boundary conditions and design, and in view of what was said earlier in this chapter, we can conceptualize this in terms of the dynamics of the change in the ROI, Q, in the following manner: Starting out with an optimized design (but at a very high level) with an ROI value of Q^*_0, then, if no account is taken of the changing boundary conditions as the design progresses into more detail, two changes take place. On the one hand, the value of Q will decrease due to the fact that the original requirements are not seen as so important by the stakeholders any more. On the other hand, the value of the optimal value of the ROI, Q^*, will increase due to the possibility of taking advantage of new technology and the possibility of achieving additional value by meeting the new requirements. The result is a widening gap, $\Delta Q(t)$, between the ROI of the design and the optimal ROI value, as illustrated in Figure 2.6.

This figure shows two critical times. The first, t_1, is the time at which ΔQ has become so large that a decision is taken to abandon the present design and start over again; the second, t_2, is the time at which the ROI has fallen below the financially viable value, and the project is either abandoned or restarted in a different form. Which one of these comes first depends on the particular project and, of course, they may never occur, because what is missing from Figure 2.6 is the time scale in terms of the project stages. If the engineered object has been constructed and gone

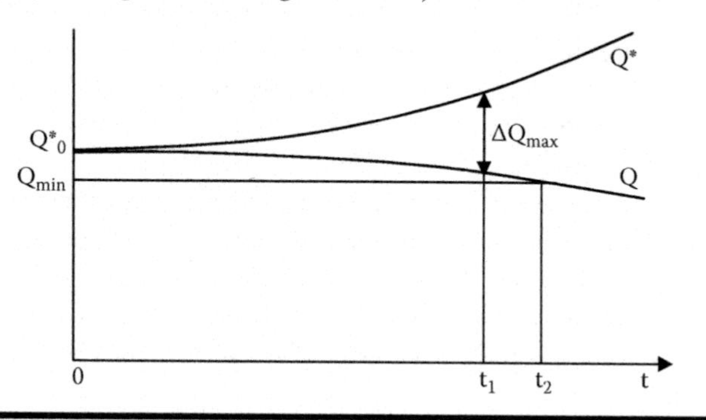

Figure 2.6 The actual, Q, and possible optimal, Q*, values of ROI, as functions of time for a project in which the design takes no account of changing boundary conditions.

into operation before these points occur, then t_1 becomes irrelevant and t_2 the point at which a decision has to be taken whether to refurbish or shut down operations.

However, any professionally led design process would not let the situation depicted above develop without taking any action; it would both respond to changing boundary conditions during design (and, to the extent possible, during construction) and endeavor to predict and take into account future changes. The extent to which this is done varies widely, and one reason for a less-than-adequate treatment in many cases is the cost and time of doing so. That is, the effort required to do this by starting from scratch on a project-by-project basis is seen as detracting more from the "core" design effort and its completion time than it is worth. Another reason is that, as also illustrated in Figure 2.6, the changes to the boundary conditions usually change gradually, in many small steps, so that it is easy to dismiss each one with the argument that its effect on the project is too small to bother with, and by the time the accumulated effect becomes apparent, it is judged to be too late and disruptive to do anything about it except perhaps a quick, knee-jerk reaction, which often only makes matters worse. That is why the proposed design methodology emphasizes the importance of building a model of the project, in the form of a system of functional elements, from the very beginning, and then continually expanding and refining this as the design progresses, so that we are at all times able to make an accurate assessment of any change to the boundary conditions.

Notes

1. No publication by the inventor of TRIZ, Genrich S. Altshuller, is known to the author, but a translation, by L. Shulyak, of a book by his son, Henry Altshuller, is available as *The art of inventing (and suddenly the inventor appeared)*, Technical Innovation Center, Worcester, MA, 1994. The main promoter of TRIZ is Ideation International (www.ideationtriz.com), and a good introduction is available online at www.mazur.net/triz.
2. Aslaksen, E.W., *The changing nature of engineering*, McGraw-Hill Australia, 1996, ch. 7.
3. Hitchins, D.R., *Systems engineering: A 21st century systems methodology*, John Wiley & Sons, New York, 2007.
4. Kant, I., *Kritik der reinen Vernunft*, 2nd ed, 1787. A later complete reprint is that published by Th. Knaur Nachf., Berlin, and the standard English translation is that by N. Kemp-Smith, Macmillan Press Ltd, which can also be found on line at http://humanum.arts.cuhk.edu.hk/Philosophy/Kant/cpr/. A very good discussion of Kant's ideas is given by R.P. Wolff, *Kant's theory of mental activity*, Peter Smith, Gloucester, MA, 1970.
5. Danielmeyer, H.G., *The industrial society*, reported at the Portland Int'l Conf. on the Management of Engineering and Technology, Portland, WA, USA, July 1997; published in *European Review*, October 1997.
6. Aslaksen, E.W., *The changing nature of engineering*, McGraw-Hill Australia, 1996, ch. 13.

7. The design is described in Tota, E.W., and Aslaksen, E.W. *Implementing a high-technology mining strategy at Northparkes Mines*, Conference on Robotics & Automation in Mining, Sydney, 4–5 September 1995 (AIC Conferences), and the successful operation of the mine is described in Aslaksen, E.W., Award-winning Northparkes Mine, *What's New in Process Engineering*, February 1999.
8. Two recent books on stakeholder management (albeit from a corporate standpoint) are Post, J.E., and Preston, L.E., *Redefining the corporation*, Stanford University Press, 2002, and Huber, M., and Pallas, M., *Customizing stakeholder management strategies*, Springer-Verlag, Heidelberg, 2006. This latter book discusses the close connection between stakeholder management and such quality management processes as Six Sigma.
   An interesting publication, which emphasizes the fact that different stakeholders view success and failure quite differently, is Hart, D., and Warne, L., *Roles apart or bedfellows? Reconceptualising information systems success and failure,* Information Systems Foundations Workshop, 2006, available online at: http://epress.anu.edu.au/infosystems02/mobile_devices/ch08.html.
   A view from the defense industry is presented in Bullard, S.G., *A qualitative assessment and analysis of stakeholder expectations*, Master's Thesis, Naval Postgraduate School, 2003, available online at: www.stormingmedia.us/35/3558/A355814.html

Chapter 3

The Design Methodology

3.1 Outline

Design in the functional domain starts with a set of stakeholder requirements and ends with a set of interacting functional elements, such that

1. the collective behavior of the elements satisfies the requirements,
2. the choice of elements allows an efficient transition into the physical domain (physical realizability), and
3. the most cost-effective (physical) solution is among the set of solutions corresponding to the set of elements.

From this rough description we can discern a number of the characteristics of the design process and some important issues that must arise in developing a methodology:

Intellectual content of the process. The development of the functional requirements into a set of interacting functional elements is not a simple process of dividing up the requirements in the sense of dividing up a cake into pieces such that no matter what size or shape the pieces are, they always add up to the whole. To each possible partitioning, that is, choice of interacting elements, there is a corresponding set of physical systems that is a subset of the set of systems corresponding to the stakeholder requirements, and there is a considerable intellectual effort involved in finding that choice which results in the smallest subset still containing the most cost-effective solution. This effect of choosing a partitioning is illustrated in Figure 3.1.

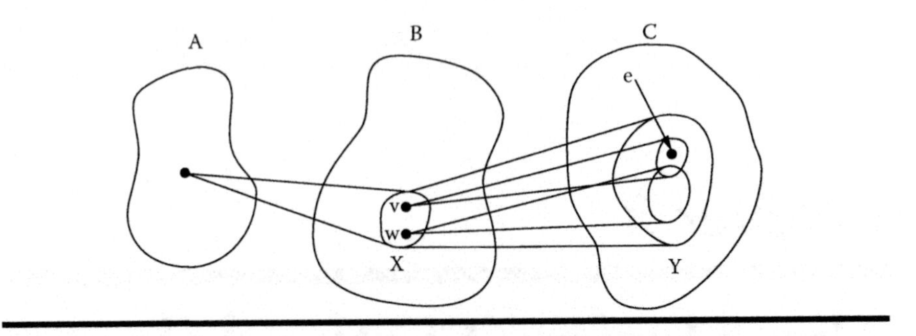

Figure 3.1 The effect of choosing a particular partitioning, illustrated by a good (V) and a poor (W) choice. On the left is the space of all sets of requirements, A, in the middle is the functional domain, B, and on the right is the physical domain, C. X is the set of all functional systems that satisfy the set of requirements, and Y is the set of all physical systems that satisfy the requirements. The most cost-effective solution is denoted by e.

The use of functional elements. The design process is a manipulation of functional elements; they are the conceptual entities that our minds can conveniently grasp as forming single functional units, and they are the design elements in the same sense that capacitors, inductors, and resistors are the design elements of passive circuits and in the same sense that numbers are the elements of arithmetic. Consequently, the methodology must contain both rules for developing (or choosing) functional elements in a particular case, and the rules and techniques for manipulating functional elements.

Top-down process. The purpose of the functional design process is to reduce the complexity of the initial set of requirements, which may consist of hundreds of individual functional parameters, by representing it in terms of a set of interacting functional elements. In order for this process to stay within the complexity of single functional elements, it must start out by defining a single functional element that represents the set of requirements, but to a very high level of abstraction (or, conversely, to a very low level of detail), and then, in a step-wise, top-down process, partition each element on one level into a small number of elements on the next level down that describe the same functionality in more detail, until all the detail in the set of requirements is represented. This immediately identifies an essential feature of the methodology; it must contain the means of determining the correct interactions between the elements, such that the functionality of the original element is preserved.

The ability to handle cost. It is not sufficient to find a possible solution; out of all the possible solutions the methodology should identify the most cost-effective. This means that the methodology, even though it operates in the functional domain, must be able to include cost as a design parameter.

Efficiency. After the end of the functional design process comes the transition into the physical domain and the classical design, which then allows the system to be produced. Again, even though the methodology operates in the functional domain, it must ensure that the subsequent bottom-up design is as simple as possible.

The top-down approach is an essential feature; each design project must start with the irreducible element and the basic set of variables, and then develop that element into a system of elements in a stepwise process that limits the size of each step such that both the identification of the possible options for further partitioning as well as the verification of the chosen option can be performed efficiently. From this, then, emerges an outline of a design methodology consisting of the following steps:

1. Determine a description of the irreducible element that is appropriate to the project. In particular, this means deciding on an expression for the quality of service, S, in terms of parameters relevant to the functionality of the object to be designed.
2. Determine cost and revenue (value) as functions of S.
3. Carry out an initial estimate of a set of parameter values, called an *operating point*, and an acceptable range for each parameter. (There may, of course, be no operating point that meets all the requirements, in which case one needs to go back to the stakeholders to see if the requirements can be modified.)
4. Investigate the extent to which these parameter values determine system features (architecture, technology, etc., as appropriate to the level of detail), and use this to determine how to expand (subdivide) the parameters to increase the level of detail.
5. Regroup the new parameters into new functional elements.
6. Determine functions between the parameters within each element and the interactions between the elements such that the original functions are preserved and the ROI remains optimized.
7. Repeat (4), (5), (6) until all the requirements are covered by parameters.

The repetitive subprocess consisting of steps (4), (5), (6) is called the *basic design process (BDP)*, which we shall examine in more detail at the end of this chapter. And we remind ourselves again that, as was pointed out at the end of the last chapter, although one could get the impression that this subprocess is applied in a linear, ever-advancing fashion, the dynamic nature of what we called the project boundary conditions will require us to go back, reassess previous decisions, and possibly set out in a new direction several times during a project.

Buried in this outline are a number of difficulties that will have to be overcome in order to make the methodology practical. For example, "decide on an expression for the quality of service" is easily said, but how should we go about actually doing it? Is there a unique solution? If not, how do we measure what is the "best" solution? And similar difficulties arise in carrying out the other steps in the above outline of the methodology. Each step will be addressed in the following sections of this chapter, but before doing so, it is worthwhile emphasizing an issue that arises again and again when discussing the concept of design in the functional domain. Just as in the physical domain, design remains a creative activity, and no matter how much we support this activity and improve its efficiency through a rigorous foundation and a program of standardization for the elements, the process retains a significant demand for human input in the form of what we term intuition, flair, insight, imagination, and so on. In particular, the choices involved in partitioning a set of requirements into a set of interacting functional elements will depend on the individual designer, and some designers will consistently come up with better designs than others. (And we have already agreed on what "better" means here.) It is perhaps tempting to imagine that we could try out all possible choices, much in the same way that a chess program tries out all possible moves, but while in chess the possible moves are strictly limited by the rules of the game, in partitioning a set of requirements there is no limitation on the number of possibilities, because the interactions between the elements can be arbitrarily complex.

3.2 Defining Quality of Service

The quality of service (QoS) expresses both the purpose and the performance of a system in a single parameter; the definition of the QoS in terms of the functional parameters describes the purpose, and the value assigned to the QoS is a measure of the level of performance of the system. But as a requirements definition document for a complex system may contain hundreds of functional parameters, deciding on an initial definition of the QoS in terms of a few parameters may appear a daunting task. In practice, this turns out not to be the case, and there are a couple of reasons for this. First of all, in the current context of ROI, the stakeholder requirements fall into two broad classes:

a. Those that are effectively characterized by a single binary parameter (present or not present, true or false), in that they need to be satisfied in order for the stakeholders to consider a system an acceptable solution. They may arise directly from the stakeholders or from standards, statutory requirements (e.g., safety-related parameters), and common practice. A typical example would be the requirement "The state (open or closed) of all doors shall be indicated to the operator."
b. Those requirements where the designer has a real choice, where there is a possible trade-off between value and cost of this requirement, that is, where a value function has been identified. This may also include requirements on functionality introduced by the designer as part of the requirements analysis process (see chapter 4).

Only parameters in the last class are of interest for the purpose of defining a QoS.

In connection with the above classification, we need to realize that the word "value" has two different meanings and that this can easily lead to confusion. On the one hand, "the parameter value" or "the value of the parameter" can mean its numerical value measured in its own units, such as meters for a length, or per unit time for a failure rate. On the other hand, this same wording can mean the value, measured in dollars, attached to a parameter in the form of a value function. We also note that it is a feature of the dynamics of the project boundary conditions that functional parameters may move between (a) and (b) above; this is part of the complexity that the design process should be able to handle.

The second reason is that most engineered objects have a main purpose, as was already mentioned in chapter 2, and if the parameters in class (b) above are ordered according to their influence on this purpose, it is generally true that a few stand out as being very significant, whereas the rest are of decidedly lesser significance. For example, if the purpose is to provide education or training of some form, the two significant parameters might be the number of students taking the course and the average grade achieved. In the case of the Mine, the purpose is to deliver a certain yearly tonnage of ore, and the two significant parameters are the nominal production rate and the availability.

However, the role of experience in choosing a useful and significant definition of the QoS should not be underestimated—not only the engineers' experience, but also that of the stakeholders, and this again shows the importance of the close interaction between engineers and stakeholders and the role this plays in the dynamic evolution of the stakeholder requirements. But important as this stakeholder management is, it is one of the many activities in an engineering project that sit under the general umbrella of "project management." Our concern in this book is with another activity, design, and in that activity, the stakeholder requirements are seen as part of the environment in which design takes place; in particular, that part of the environment we have called the project boundary conditions, consisting of the stakeholder requirements and the available technology. There are many other components of this environment, such as the skill level and morale of the design team, just to mention two, that have to be the concern of the engineers, but they are not considered in the present context. The sole purpose of the methodology put forward in this book is to improve the efficiency and effectiveness with which the complexity introduced by the greatly increased number of requirements, their interactions, and dynamic nature in any of today's engineering projects is handled by the design process.

Having chosen the relevant functional parameters, we are still faced with combining them to form a single parameter, the QoS. Again, with some knowledge of the industry in which the project is embedded, this is not usually a problem, and a few simple rules provide guidance. First, the QoS should represent the performance in an intuitive fashion, so that an increase in the value of the QoS is an increase in performance, and the magnitude of the increase in QoS should bear a simple relation to the magnitude of the increase in performance (the relationship does not necessarily have to be linear). Second, the QoS should be a single-valued function of its defining parameters. And third, even though the parameter values might in practice only vary over a limited range, the QoS should show a reasonable behavior for limiting values of the parameters. For example, in the case of the Mine, if either the nominal capacity or the availability go to zero, the QoS should go to zero, so that the combination is that of a product rather than a sum of the two parameters.

We recognize that expressing the performance of a possibly very complex system by a single parameter is a great simplification, and that the purpose of doing so is simply as part of creating a starting point for the top-down design process by making the irreducible element specific to a particular project. As the design progresses, the QoS will be expressed in terms of more and more parameters, and in the end by all the parameters in the class (b) above.

Two further examples illustrate this approach. The first is the well-known public switched telephone network (PSTN, also known as POTS, plain old telephone system). At the highest level, that is, at the beginning of the design process, the purpose of a telephone system, which is to connect subscribers, can be characterized by a single parameter, which is therefore also called the QoS, and defined as the probability of making a connection on the first try, assuming the called subscriber is not busy. The value of this parameter is the one over which the system designer has the greatest influence and which is most significant in the early design decisions (network topology, trunk capacities, etc.). There are a number of other parameters that are very important to the users, such as intelligibility, to take just one, but international design standards ensure that the value of these parameters stay within relatively narrow bounds. Then there are a number of features that contribute to the overall quality of the service, such as pulse or tone dialing, the availability of call waiting, caller identification, etc. And, finally, there are the parameters that are essentially given and that the designer has no influence over, such as the number of subscribers and their geographic distribution.

The second example is an air defense system.[1] At the highest level, the service provided by such a system could be defined as "ensuring that all enemy aircraft entering the country's airspace are destroyed before they reach the country," and the QoS could then be defined as the probability of meeting this requirement. This definition would be adequate for the high-level optimization implied by the irreducible element, but in order to make the concept precise and measurable, it needs to be defined in terms of more detailed parameters. For example, the QoS could be expressed as the product of the probability of detecting enemy aircraft and the probability of destroying a detected aircraft; the probability of detecting enemy aircraft would depend on detecting any aircraft and determining whether an aircraft is a friend or foe; and this decomposition can go on to any desired level of detail. Then there would be a number of parameters defining the meteorological conditions under which the system is to operate, parameters defining the characteristics of the aircraft to be detected (i.e., the scattering cross section), and so on.

This second example can be used to illustrate an important issue in system design — the degree of detail in the user requirements versus the definiteness or completeness of the user requirements. A requirement can be completely unambiguous and definite without being detailed; it just means that the user does not care about the details. The above definition of the service is not quite definite; the word "all" is open to interpretation. Does it mean "all existing aircraft," or "all aircraft known to the designer," or "all existing and future aircraft" (in which case, how far into the future?)? If we decide on "all existing aircraft," then this makes the requirement quite definite, but not very practical, as it puts the onus on the designers to find out about all existing aircraft, which would require them to have a sizable intelligence organization at their disposal. To make the requirement more useful to the designers, it needs to be more detailed.

3.3 Determining the Value of a Service

From the brief discussion in chapter 2, the concept of the value of a service is obviously very closely coupled to that of the QoS. Why have both? The QoS, defined above, and its refinement into a set of functional parameters, is characteristic of a *class* or *type* of system. It is a characteristic of the service provided by the entity that will satisfy the stakeholder requirements; it is a characteristic of the solution to the design problem. Value, and its refinement into components (in the same way as cost is refined into components), is a characteristic of the stakeholder group and its requirements, that is, of the environment, or market, in which the system operates. And while value and QoS are linked, it is precisely the task of determining the effect of that linkage that is an essential part of the design process.

If the value associated with the service is not defined explicitly in the requirements definition document in terms of the functional parameters belonging to the class (b) in the previous section, we could, in principle, find out what the stakeholder group is willing to pay for a service, and define this as its value, but in practice there are a number of well-known reasons why this is not a straightforward and unambiguous procedure. First, price is determined not only by the stakeholders, but also by the other suppliers, that is, by the competition. Second, in the case where the service has a social component, the price may be determined by legislation rather than by the stakeholders. Third, the perceived value of a service can be influenced by advertising. Fourth, the price obtainable for a service may be highly time-dependent, so that the value determined prior to design is quite different from the price actually obtained when the service is available. Fifth, stakeholders may not be willing to disclose the true value they place on a service (or feature of a service) for fear of alerting the competition. And finally, even disregarding all these difficulties, determining what value stakeholders put on a particular feature of a service may be very difficult, as is illustrated by the case of trying to determine what value travelers put on air safety. How much more would they be willing to pay for their tickets if the probability of a fatal accident could be reduced by a factor of two?

However, while these issues may make it more difficult to specify the value of a service, they do not make it any less imperative to do so. Every requirement in the user requirements must have been put there by at least one of the stakeholders, so this stakeholder (or group of stakeholders) should be able to give an indication of why this requirement had to be included and what the value of fulfilling it is, and there are many different, well-documented approaches to determining what value a user group assigns to a service.[2]

As far as our current concern with defining the irreducible element, we shall assume an expression for value of the form

$$W = W_0 \prod_{1}^{m} \left[1 + c_i \left(\frac{x_i}{x_i^0} - 1 \right) \right] \tag{3.1}$$

Here x_i is one of the m parameters used to define the QoS, x_i^0 is its nominal value, W_0 is the nominal value of the service, and c_i is a measure of the importance the stakeholder group assigns to the i-th service parameter. The reason for making this (simplifying) assumption is to be found in the S-shape of the function expressing the dependence of value on any one single parameter, as shown in Figure 2.3. The significant variation in value takes place over a limited range of variation of the parameter around its nominal value, and within this range of interest a linear approximation is usually a reasonable one. Consequently, that range has to be specified for each of the parameters in equation (3.1).

Again, as with the QoS, as the design progresses, the definition of W will depend on more and more of the parameters in the class (b) in the previous section, and the functional dependence may be different from the simple one assumed above.

3.4 Assigning Cost to a Functionality

By now, hopefully it is clear how necessary it is to be able to measure cost and value in monetary terms, and we have discussed some of the issues that arise when we attempt to put a monetary value on a service. That costs should be measured in monetary terms is not controversial, but how can we ascribe a cost to a functionality? Costs are normally associated with particular physical objects and depend on physical characteristics such as material, surface finish, and dimensional tolerance. And to complicate the issue even more, they depend on exchange rates, interest rates, salary and award rates, and so on. Is it sensible to talk about the cost of a concept?

Our approach to this issue must be the same as we employ in the physical domain; the costs are always *cost estimates* based on previous experience, and the accuracy of the estimates increases with the level of detail of the design. As the accuracy at the start of the physical design process, that is, conceptual design, is typically ±30%, the accuracy at the end of the functional design process should approach this value. However, the accuracy of the absolute values of costs is not always so important; often the accuracy of the relative values of cost components is more important when it comes to optimizing a design, trading off costs in one area against savings in another, and this will be illustrated in the case of the Mine in chapter 7.

The cost estimates associated with a functional element in any particular design project will be a matter for the individual designer, and will depend on the designer's experience, cost data available within the designer's organization, etc. But once there exists a body of standard functional elements, the parameterization of cost information will be uniform across projects and organizations, and the access to and exchange of cost data will be much improved. The situation is in one way similar to the current situation in the physical domain, where we have catalogs of standard (commercial-off-the-shelf, or COTS) components from which physical objects can be synthesized, and where price lists are available for many of these components. There must be millions of such components, organized by the areas of engineering to which they apply, and the same sort of database would have to be built up for functional elements. But there are two major differences.

First, as we have already remarked, in the functional domain the design process starts from the most general element and proceeds in a top-down fashion to a more detailed description using less general elements, which is the reverse of the process in the physical domain. Consequently, at the start of the process we need the estimate for a single cost parameter, the C in the basic set, but for a vast number of different applications. That is, in the physical domain we have a vast array of components, and each component, such as a metric hex nut, can be used in a vast number of different applications, but the cost is tied to the component, not to the application, whereas in the functional domain, we have a single element that can be used in a vast number of applications (in all, actually), but the cost is tied to the application. (The same goes for the other three variables in the basic set, too.) In this context, we note that the cost is dependent on not only the functional parameters in class (b) in section 3.2, but also on those in class (a). However, the latter cost does not (at least, not to a first approximation) *change* as the functional design progresses.

Second, even when the functional design process is completed and we have arrived at a set of interacting functional elements, there corresponds to each element a whole set of functionally equivalent physical objects, each of which will have a (somewhat) different cost. The final choice of the most cost-effective solution can only be made in the physical domain, usually determined by non-functional user requirements, as will be discussed later (see Figure 4.2).

However, if we keep these two differences in mind, there is no reason why a database of parameter values (including cost) could not be developed, once the parameters themselves are defined in terms of functional elements. The data, which simply reflect past experience, are, of course, present today; they are just not easily available in a usable form because the applications are parameterized (i.e., described) in so many different ways. The use of standardized functional elements would provide a framework into which all the data could be fitted and thereby made much more accessible.

3.5 Some Basic Rules for Developing Functional Elements

Because functional elements are the objects on which the mind has to operate if we want to do design in the functional domain, we need to ensure that there is compatibility between the objects and the capability of the mind. Just as production managers utilize the capabilities of their machine tools in order to obtain the greatest productivity, and would not try to produce a shaft using a milling machine, engineering managers must ensure that the objects with which they let their engineers grapple are defined in a manner that makes them suitable for the mind to handle. For example, engineering managers must know what they expect the mind to process and what they expect to be processed by computers, and they must realize that the capabilities of the two are very different. In the following, we formulate three rules that are useful in developing functional elements that are convenient for the brain to work with.

The first rule arises from the realization that all functional elements are related. Consider a particular function, such as the function "transport," that is, to move something from one location in space to another. This can be subdivided into a number of elements, for example, according to what is being transported, such as material goods, electrical power, or information. The latter element can be further subdivided according to whether it is point-to-point or broadcast, and so on. It simply illustrates that functional elements are related in a hierarchical structure, with an element on a particular level of the structure representing a *class* of elements on a lower level, as illustrated in Figure 3.2.

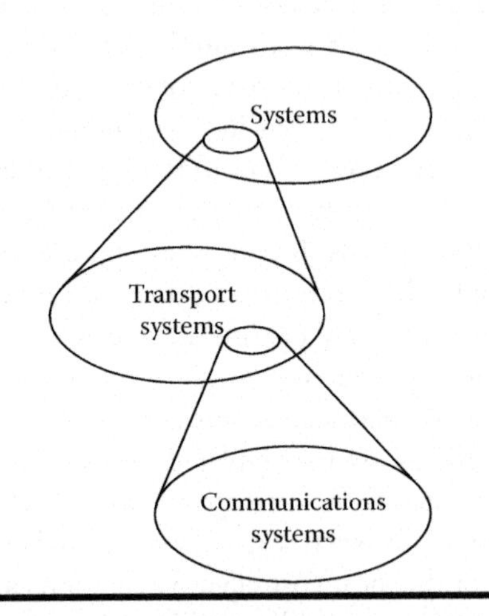

Figure 3.2 An illustration of the hierarchy of the classes of systems. (From Aslaksen, *The changing nature of engineering*, McGraw-Hill, New York, 1996, section 10.5.)

In particular, following this argument to its logical conclusion, it must be true that all elements in a class inherit the characteristics of the element from which the class was derived (the parent element), in the sense that the functionality of an element in a class is an instant of the functionality of the parent element, and that the class must contain all instances of the parent element. As an example of this, consider the class of elements that have the irreducible element as its parent element. All elements in this class must be instances of how one can obtain a return on investment and, more importantly, there can be no element outside this class that has this functionality. If we want to limit the number of elements in a class to a small number, then we must demonstrate that any system that has the functionality of the parent element can be represented by a set of interacting elements, each one of which belongs to the class. The rule that follows from this is therefore that *functional elements must be developed in a top-down fashion*, that is, from the general to the more detailed. Every element, except the irreducible element, must belong to a class, that is, must have a parent element, otherwise it is not possible to determine how it contributes to the ultimate purpose of producing a return on investment.

Three comments may be made with regard to this rule. First, when we say "developed," we understand a parent-to-child partitioning. However, the choice of the "children" may very well be made on the basis of common practice or previous experience, which most likely arose in a bottom-up fashion, because what we create are the individual items, and only later do we perceive the similarities and differences that lead to a structure.

Second, if we compare the functional domain with the physical domain, we see that in the former the higher up in the hierarchy the elements are, the greater applicability they have (but also the less powerful they are), whereas in the latter, elements lower down have the greatest applicability (e.g., an M6 hex nut has tremendously wide applicability). This is one illustration of how the two domains may be considered to be *conjugate.*

Third, a very significant characteristic of the mind is its ability to form associations; it is much easier to work with an object (model, concept) that relates to other objects already in the mind, that fits into our mental framework, so to speak. The above rule ensures that we develop such a framework, rather than a scattering of unrelated models.

The second rule is closely related to the first; it says that *the concepts (variables or functional parameters) used to describe functional elements must show a continuous development from higher to lower levels of the hierarchy*. The meaning of this rule is best illustrated by taking a particular concept, such as cost. Any project that aims to provide a service will have a cost associated with it, so that the concept of cost can be defined for the functional element that represents all projects. At the next level down, the element is partitioned into two sub-elements, one representing the creation of the object that is to provide the service, and one representing the operation of the object to actually provide the service. With the first we can associate a non-recurring cost, with the second a recurring cost. At the next level down, the concept of recurring cost can be developed in more detail by defining operating and maintenance costs, and these can again be developed in greater detail by defining further cost elements, and so on. At no level does a cost parameter suddenly appear that has no relationship to any cost parameter defined on a higher level.

A corollary to this second rule is that whatever concepts are used to describe the most general functional element (i.e., the element at the top of the hierarchy) must be adequate to encompass all possible functional parameters. Or, in other words, the set of all possible functional parameters can be partitioned into disjoint subsets such that all the parameters in one subset are related to one of the concepts used to describe the most general element, ultimately the irreducible element.

The third rule is again tied to the way our brain works — it arises from our desire as engineering managers to design tasks to be ergonomically correct, only in this case we are concerned with the "ergonomics" of the mind rather than of the body. In order for us to be able to think of something as an entity, as a single object, and to manipulate it in our minds, it must not be too complex in the sense of needing a lot of variables to describe it. We find that once something needs more than about 5–10 variables to describe it, it becomes too difficult to think of as a single entity,[3] and we either neglect some of the variables, or split the object up into two entities that can be thought about one at a time. The third rule can therefore be formulated by saying that *functional elements should not be more complex than that they can be described by at most ten variables.*

These three rules are important guidelines for developing functional elements, but they are not detailed or specific enough to uniquely define a consistent set of elements. There are innumerable ways is which different sets could be developed, and the choice of a particular set can only be justified by its *usefulness*. There may be many different sets that can be useful in different circumstances, but in any case the usefulness comes to a large extent from being able to utilize a set of *standardized* elements — elements that have already been developed and that fit together to form complete classes, and that are immediately recognized and understood by other people. If you were the only person in the world using metric thread, your metric hex nut would not be very useful, and similarly, the whole idea of design in the functional domain rests on our ability to develop standardized sets of functional elements.

3.6 Applying Functional Elements in Top-Down Design

In the outline of the design process given in section 3.1, the wording was in terms of a development process in order to demonstrate how the design progresses. From the above, we now know that the process has to be much more a case of selecting functional elements from an existing set rather than developing them for each application. So, assuming that we have a large collection of functional elements at our disposal, how do we go about picking the most appropriate ones? Well, the situation is not so different from two well-known ones from other fields of engineering. The first is to select a suitable complete orthonormal set of functions in which to expand the functions involved in a particular problem. We pick a set of coordinates that suits the geometry of the situation, and then the orthonormal set of functions on these coordinates that best fits the boundary conditions. The second is the choice of components for an electronic circuit. We first look at the environmental conditions, such as temperature and acceleration, to see if we need MIL-spec components, then we look at cost and design life to select the most appropriate encapsulation of components, power consumption limitations to select the technology, and finally, within the appropriate family of components, those with the right functionality and performance. Only if no suitable ones could be found would we consider designing our own, and then only for large production quantities.

Selecting the most appropriate set of functional elements is essentially the same as choosing the best partitioning of the functional user requirements, and the approach advocated here (to be developed somewhat further in chapter 7) is based on the assertion that any functionality (or more correctly, service, as will be explained in chapter 4) can be expressed in terms of three types of functional elements:

1. transport elements
2. storage elements
3. transformation elements

The first step in the selection process is therefore to determine which of these three types of functionality are involved in meeting the user requirements, and to group the user requirements into groups relating to each type of functionality. For certain parameters, such as reliability, this may entail subdividing the original requirement. If only one type is involved, we look within the subdivision of this type to see which main subtypes are involved. Having now partitioned the user requirements into a small number of groups of requirements, say, three to ten, each group can be considered as expressing the requirements of the performance of a physical object, for which an irreducible element can be found. However, these irreducible elements are not independent; by virtue of their derivation from a common element, there are relations between some or all of their parameters; they form a system of functional elements.

A word of caution is appropriate regarding the use of the word "irreducible" for the elements representing these system components. By definition, every physical object will have an irreducible element associated with it; indeed, there is only one such element that is common to all physical objects — the element representing ROI. But in order to *evaluate* the ROI for a particular object, we need the value function, and there is in general no simple or explicit relationship between the value of the service provided by the components in a system and the value provided by the system. The whole value of the service provided by the system may relate to an emergent property of the system, one that is not present in any of the components. Therefore, in the case of the components, the irreducible element is not of interest in order to evaluate the ROI, but in order to provide a point of departure for expanding the set of parameters describing the functionality of the component.

However, in addition to the user requirements that relate directly to what the system is supposed to do, there are a number of requirements that describe how well it should do it (e.g., how reliably), what it should cost to do it, and so on, and the functional elements describing these *aspects* of the functionality must also be subdivided in the partitioning process. But rather than a decomposition into different types of elements, they are decomposed into elements of the same type, but to a greater degree of detail. It can therefore be helpful to consider the partitioning process as two-dimensional; one dimension is the type of functionality, the other is the level of detail, and the process alternates between the two. Starting with the irreducible element for the whole, we increase the level of detail by including more and more parameters. Then we split this more complicated element into a small number of elements, each with fewer parameters. Then we expand the level of detail of each of these elements, then split each into a set of smaller elements, and so on. The theoretical foundations of this process, and the nature and definition of the two types of elements involved, are presented in chapter 4, but it is clearly an iterative process, with each iteration consisting of an expansion in the level of detail followed by a splitting, but with interactions between the new elements such that the overall functionality of the original element is preserved. This is the *basic design process* introduced earlier and discussed in the next section.

3.7 The Basic Design Process

The basic design process (BDP) was described in detail in *The Changing Nature of Engineering*,[4] and shall therefore just be summarized here. The process starts out with a given functional element, that is, a description of either the functionality of some physical object or some aspect of such functionality; in either case the description involves a number of variables and the relationships between them. Of these variables, a number are involved in describing *what* the element does (as opposed to *how well* it does it), what we might call the purpose of the element, and the first step in the BDP is to express this purpose as being achieved through a combination of two or more functions by a process of *analysis*. For example, in the air defense system introduced in section 3.2, the purpose of *detection* can be subdivided into the production of radiation of a suitable kind, the transport of this radiation to the target, the transformation process that takes place at the target (i.e., the scattering process), the transport of the scattered radiation back to a receiver, and the conversion of the received radiation into an electric signal. Each of these functions can be looked upon as the purpose of a separate functional element, and the description of this element can then be completed by adding further variables to describe how and how well this function is achieved, and functions between these variables to express the behavior of the element. However, these new elements are not independent; they *interact* by virtue of the fact that there are relationships between some of their variables. For example, each element will be characterized by reliability, but the sum of the failure rates of the elements must add up to the failure rate of the original detection element.

The second step in the BDP is to optimize the values of the variables, which generally means an *allocation* of a total value of a variable in the original element (e.g., failure rate) to the same variable in the individual elements or a *trade-off* between two different sets of variables, such as performance and cost. In either case we have an expression for a quantity that is to be optimized, such as cost-effectiveness, and we are faced with the familiar problem of determining the maximum or minimum of a function of several variables.

The third and final step in the BDP is to verify that the overall performance of the system of elements is identical to the performance of the element we started out with. We might expect that the optimization process in the previous step would guarantee this, and in simple cases that will be true, but in general the behavior of a set of elements (and of the original element) is much more complex than what can be expressed by fulfilling a single optimization criterion. Above all, the optimization usually relates only to the *static* functionality of the elements, whereas the representation of an element in terms of a system is often required to include both explicit time dependence (e.g., as in decay or wear) and the response to time-dependent external influences. Therefore, we need to *synthesize* the original element from the set of elements by prescribing interactions between the elements, and then *testing* that this synthesized element behaves identically to the original one.

The BDP is illustrated in Figure 3.3, and while the main features of this three-step process are relatively easy to understand, we also recognize that the above description has glossed over some of the details and problems, both conceptual and practical. But before we can make any further progress in detailing the methodology, we need to define and understand the properties of the functional domain in

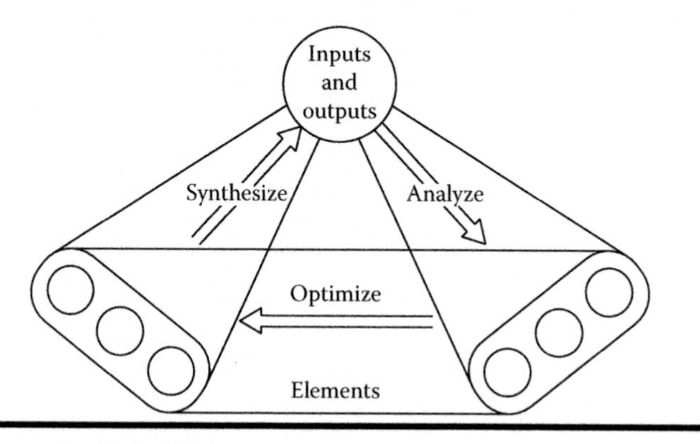

Figure 3.3 The basic design process.

much more detail, and that is the purpose of the next three chapters. At the end of chapter 6 we shall then return to the BDP and address much of the detail missing above.

Notes

1. Aslaksen, E.W., and Belcher, W.R. *Systems engineering,* Prentice Hall, Englewood Hills, NJ, 1992.
2. Aslaksen, E.W., *The changing nature of engineering,* McGraw-Hill, New York, 1996, section 10.5.
3. The seminal work in this area is a paper by G.A. Miller, The magical number seven, plus or minus two: Some limits on our capacity for processing information, *The Psychological Review,* 63, 81-97, 1956, available online at www.well.com/user/smalin/miller.html. References to subsequent papers can be found at http://citeseer.nj.nec.com/context. A very simple model of short-term memory, demonstrating why the brain might find it more efficient to work with smaller (but interacting) chunks, is contained in Aslaksen, E.W., *The changing nature of engineering,* section 3.2.
4. Aslaksen, E.W., *The changing nature of engineering,* section 2.4.

Chapter 4

Functional Elements and the Functional Domain

4.1 Functional Elements

As we discussed in section 1.3, to each physical object there correspond two basically different descriptions — the description of what the object *is* (its physical substance), and the description of what the object *does* (its functionality). The former consists of such data as size, shape, material specification, surface finish, etc., presented in the form of drawings, schedules, etc., and is what is needed by someone who wishes to reproduce (manufacture) the object. The latter consists of a set of performance specifications, in the form of mathematical functions, graphics, or tables, and is, for example, what is needed by stakeholders to determine if the object will meet their requirements. Both descriptions refer to the same entity — the physical object, and while we shall develop a theoretical framework in which it is possible to operate without any knowledge of this object, the development uses the fact that there is an underlying physical reality in several places. Consequently, our first step must be to define what we mean by a physical object.

Definition 4.1: A physical object is an entity consisting of one or more inseparable material parts, all connected together by definite physical relationships.

The definite physical relationship may be that part A is bolted on to part B, or that part A is screwed onto part B, or that part A rotates within part B, or even that part A just sits in a particular position on top of part B, and so on.

It follows that if two or more objects are connected together by definite relationships, they form a new object, but if there are no relationships between them, they form simply a collection of objects.

Definition 4.2: The *physical domain* is the set of all physical descriptions of physical objects.

In the case of physical descriptions, there is no limit to how many parameters we can use to describe a particular object; it just depends on the level of detail we want to go to. In principle we could give the type, position, and state of every atom in the object, but in practice just a few parameters describing shape and material are often adequate. The *completeness* of a physical description is therefore not an absolute characteristic of the object alone; it is defined only relative to a particular application.

Definition 4.3: A physical description is *complete* if it contains all the information required to reproduce the physical object in every detail required by its application.

Before leaving the physical description of an object, we note that the very important aspect of *safety* of the object, in the sense of *safety in design*, is essentially a condition on its physical design, not on its functionality. The analysis required, such as Hazop, requires the functionality as an input, but the requirements resulting from the analysis are always requirements of the physical design, such as handrails on platforms, guards over moving parts, lifting devices, showers, etc.

A physical object (or system) meets the stakeholder requirements by interacting with its *operating environment*, where the qualifier "operating" has been included only to distinguish it from the more popular meaning of "environment" as the natural environment of a species, in particular humans. The operating environment is, in principle, the complement of the object in the universe, that is, "everything else"; in practice it is, of course, much more restricted and determined (both explicitly and implicitly) by the stakeholder requirements. We now introduce a formal definition of functionality, and then discuss some implications of this definition:

Definition 4.4: The *functionality* of a physical object is its intended capability for interacting with its operating environment.

The word "intended" has been included in order to exclude such incidental interactions as a bull becoming enraged by a red sports plane making an emergency landing in its pasture; this is not part of the plane's functionality. On a more serious note, the word "intended" expresses a very significant difference between physical and functional descriptions; every statement in a physical description can be verified by an examination of the physical object, whereas the functionality of an object depends on the *intention* of the designer, which again is determined by the requirements of the stakeholder group. The functionality cannot, in general, be deduced from looking at or performing measurements on an object, and there is not necessarily any functionality inherent in a physical object, that is, disconnected from the intention of its designer. Deducing the functionality of a physical object is, of course, what we call reverse engineering, and its accuracy will depend on what additional information is available.

This definition of functionality is significant and not universally accepted; another definition is that the functionality of a physical object (or system) is what it does. In the author's opinion, that simply begs the question: For whom? An engineered object provides a service *to* somebody, so unless one wants to always explicitly link the definition of functionality with a definition of that body of people, the only body that is implicitly related to the object is the stakeholder group, that is, the group that formulated the original requirements, against which the object was designed. That the object can be used by other bodies for quite different purposes than originally intended is not a property of the object, but rather of the ingenuity of those bodies.

The word "capability" has been included for two reasons. First, the interactions between the object and its operating environment are described in terms of a set of *parameters*, and to each of these parameters belong two variables, one defining the range or *rated value* of the parameter, and the other being its *actual value*. Only the former is a property of the object; the latter is a consequence of the operating environment. An example would be an electrical power plant; its interaction with its operating environment includes supplying electrical power, and one parameter characterizing that supply is power output. There are two variables associated with this parameter; the capacity to supply power (its rating), and the actual power flow (generally given as a function of time in the form of a duty cycle). Second, "capability" signifies that the parameters involved in the interaction are related in a particular manner; the interaction takes place in a particular manner. This is the behavior of the object in response to changes in the operating environment.

Also, it should be noted already at this point (although it will be discussed comprehensively later) that to a given set of stakeholder requirements there may correspond a number of different functional elements, that is, a particular set of stakeholder requirements may be satisfied in a number of different ways.

It follows, then, that the descriptions of functionality are, as far as engineered objects are concerned (and we are, as previously noted, limiting our considerations to these), the primary entities, while physical objects and their descriptions are secondary, arising in response to a demand for a particular functionality. But it is obvious that to a description of functionality there corresponds a number of (generally infinitely many) physical descriptions; consider only the different materials and/or surface finishes that can be used to produce the same functionality. Thus, the description of the functionality of a particular physical object defines a whole set of *functionally equivalent* physical objects in the physical domain. This was already implied in Figure 3.1. And the greater the level of detail in the description of functionality, the smaller the set of functionally equivalent objects.

A description of functionality will generally require a number of variables and a number of functions defining the relations between these variables in order to fulfill the many clauses of a typical requirements definition document. These variables and functions can often be grouped according to distinct subsets of the stakeholder requirements, and it is therefore possible to regard a description of functionality as made up of a number of individual parts, each one describing some *aspect* of the functionality. Examples of aspects are the capability of generating earnings (cost effectiveness), the capability of providing continuity of service (availability), the capability of surviving in a given environment (reliability), and the capacity for producing its service (size or rating).

Definition 4.5: A *functional element* is a description of one or more aspects of the functionality of a physical object, and consists of a set of variables and a set of functions between them, as well as any values required of the elements of these two sets.

Definition 4.6: The *functional domain* is the set of all functional elements.

With regard to definition 4.5 and our use of the functional element concept in the following, the fact that the variables and functions take on numerical values and that some or all of these form part of the definition of a particular functional element will be understood and not mentioned explicitly every time. This is no different from what we do in the physical domain; for example, a resistor may be specified by two parameters, resistance and rated power dissipation, and it is understood that a particular resistor will have specific values attached to these two parameters.

The main reason for introducing the concept of "aspect" is clarity of language, as it would be easy (and quite natural) to think of "functionality" as only the immediate, physical function of an engineered object, such as generate electricity, provide education, etc., which are, perhaps, closest to the interests of most engineers, and to neglect the interests of the wider stakeholder group in what the object does, such as, for example, provide an opportunity to build expertise (technology transfer), support political stability, etc. This is clearly illustrated in terms of the irreducible element. This element was labeled "return on investment" because that was the purpose it was able to express; now we can say that it is labeled "return on investment" because that is the aspect it describes, without necessarily tying it to a particular physical function.

Two special types of functional elements, both of which play important roles in the development of the theory, can now be defined. The first of these arises if we imagine a collection of all those persons who could have any relation whatsoever to this object; this would be the largest stakeholder group possible. The complete description of functionality referred to the corresponding stakeholder requirements will be called a *maximal* functional element, and the following definition is an equivalent expression of the same concept:

Definition 4.7: A description of the functionality of a physical object is a *maximal* functional element if it describes all possible interactions of the object with the rest of the world.

Of all the possible interactions, only a small fraction will have a significant probability of actually occurring in the life-time of the object. Those interactions that the designer intended to occur (in response to the stakeholder requirements) would normally have a probability close to one, but there may also be many other, unintended interactions that have a very small, but non-negligible probability of occurring. It was pointed out by D. Hybertson in reviewing the manuscript for this book that this situation is somewhat analogous to the description of the state of a volume of gas by means of statistical mechanics. Phase space encompasses all possible states of the gas, but the probability of the system trajectory passing through a region corresponding to a macroscopic deviation from equilibrium (e.g., all the gas molecules in one half of the volume) is vanishingly small. However, it is a current trend for the designer to be made responsible for considering a widening range of such unintended interactions, with a decreasing lower limit on their probability, in the design, and for taking appropriate action to ensure that a duty of care is maintained.

Clearly, a maximal element is not an element we would ever use in carrying out a design activity; it is a limiting case we can approach more or less closely. The special feature of a maximal element is that it removes all reference to any particular stakeholder group and to any intention of a designer. Quite the opposite is the case with the second type of element — a *complete* element, the counterpart in the functional domain of a complete description in the physical domain. However, in order to define completeness for functional elements, we first need to look more closely at the definition of a functional element. As defined above, a functional element consists of a set of variables and a set of functions between them. But the set of variables can be divided into subsets that have quite marked differences, and it is recognizing and understanding these differences that allow the concept of a functional element to be further developed and made more precise.

To this end, we first have to introduce the concept of the *service*:

Definition 4.8: The *service* provided by a physical object is the immediate purpose of its operation, its *output* to the subset of the stakeholder group usually called the *users*.

Definition 4.9: *Functional parameters* are the parameters describing a service.

Referring to the definition of "aspect" and the example of power generation above, the service provided by a power station is the supply of electrical power. Its rating is a functional parameter, whereas the actual value of the power flow at any time is a variable characterizing the external demand.

We note that while a service is always an "output" as far as a functional element is concerned (in the sense of presenting the capability to the user), the functional parameter may actually describe an input to the physical object. As an example, consider a garbage disposal plant (e.g., incinerator). Its functionality is to dispose of garbage; this is the service it provides *to* the users (the community). It is a functional output, and its capability for disposing of garbage is the associated functional parameter, even though the physical flow of garbage is *into* the plant. Another example is a user command; the signal flow is into the equipment, but the equipment has the capability of sensing the user's command; it *provides* this capability *to* the user.

The second subset is the one consisting of the additional variables describing interactions required in order to provide the service, such as power, materials, manpower, waste heat to the environment, etc. As all these other variables used to describe the functionality are only there because they are required by the functional parameters to be there, no subdivision into parameters and variables is required, and they may all be called *dependencies*. They are all related to the functional parameters through the set of functions that forms part of the definition of a functional element and describes the behavior of the element.

Together with the functions between them, these two sets completely describe the functionality of the element, that is, the intended interactions with the rest of the world, and constitute what we might call the functional variables (or simply variables, when we are considering functionality). However, there is a third subset, consisting of those variables that, while they do not describe intended interactions, describe necessary interactions with the rest of the world. These are normal (i.e., non-functional) variables, usually describing those characteristics of the environment in which the object operates that affect the values of the functional parameters, such as interest rate, state of technological development, availability of trained manpower, and political stability, just to mention a few that are different from those that spring first to an engineer's mind, such as temperature and humidity. Their values are given by the environment and cannot be changed by the functional element. (This restriction is in reality only satisfied as an approximation, albeit most often a very good one, as any interaction implies the involvement of both parties to the interaction.) The variables in this third subset could be called *influences,* and a symbolic representation of a functional element could therefore be as shown in Figure 4.1. This symbol is consistent with the one shown for the irreducible element in Figure 2.2.

We can now define the concept of completeness for functional elements; it is similar to the one defined in relation to the description of physical objects, but with one difference:

Definition 4.10: A functional element is *complete* if the set of functional parameters is adequate for expressing (or defining) all the functional requirements in a set of stakeholder requirements.

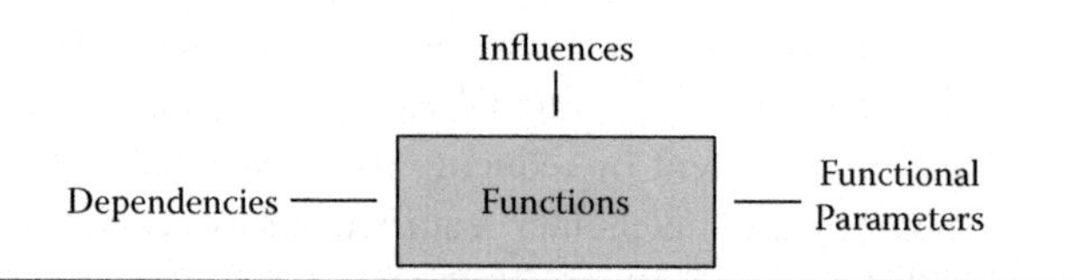

Figure 4.1 Suggested symbolic representation of a functional element.

The difference is, of course, that while the definitions of a physical object and its completeness were both in terms of properties of the object itself, the completeness of a functional element refers to something that exists prior to and independently of any object. This reflects our desire, in developing our design methodology, to consider stakeholder requirements as the primary entities; the design process starts with a set of stakeholder requirements and without reference to any specific physical object at all. It is a manifestation of the *abstraction* away from the physical domain that is an essential part of the theoretical foundations of the methodology; instead of defining functionality in terms of what any particular physical object does, we want to consider functional elements as descriptions of doing *per se* and therefore reusable as elements in models of the performance of various systems.

The two "worlds" of physical interactions and stakeholder requirements come together in the concept of maximal element:

Theorem 4.1: All maximal elements are complete.

Proof: A maximal element describes all interactions with the rest of the world, therefore also with all possible stakeholders, of which any set of stakeholders is a subset.

The converse of this theorem is, of course, not true. The irreducible element is complete with respect to the single stakeholder requirement of producing a return on investment; it is the requirement of a stakeholder group consisting of an investor only.

This immediately raises the issue of the *physical realizability* of a functional element. First, as functional elements are products of the mind, one can *think* of elements that cannot possibly be realized, because they contradict the laws of nature, for example, involving traveling at a speed exceeding that of light. Second, there are the elements that are currently not realizable, but for which there is no fundamental reason why they could not be realized in the future. Clearly, in order for our design methodology to be useful, we would want the elements that result from the functional design to be at least physically realizable, possibly even restricted to a smaller subset of existing objects (e.g., commercially off-the-shelf components).

With each maximal element there is associated a set of elements that describe all the aspects of its functionality and all the possible combinations of these; it will be called the *maximal set*.

Definition 4.11: The *maximal* set associated with a maximal element consists of all the complete functional elements that are derived from the maximal element by reducing the stakeholder requirements to a subset of the stakeholder requirements associated with the maximal element.

This introduces a hierarchical ordering into the maximal set (and into the functional domain, as will be discussed in more detail later), in that an element that can be represented by combining two or more other elements can be said to be *larger* or more *complex* than those elements. The operation of adding one functional element, a, to another, b (e.g., adding an aspect to an existing element), will be formally denoted by $a \oplus b$; the result is again a functional element. The operator $\oplus$ may be called the *combination operator*, and it is defined as follows:

Definition 4.12: Let A and B be two subsets of the stakeholder requirements associated with a maximal element, and let a and b be the corresponding complete functional elements. Let c be the complete functional element corresponding to the set of user requirements $C = A \cup B$, then the combination operator, denoted by $\oplus$, is defined by $c = a \oplus b$.

The following theorem is self-evident:

Theorem 4.2: The maximal set is closed under the combination operator.

The result of applying the combination operator to two elements results in more than just adding the two elements together; it creates a new element through an *interaction* between the elements. But what interaction? Is there just one possible interaction? The answer to the latter question is "yes," and the reason is that we are dealing here not with any two functional elements, but only with elements belonging to the same maximal set. The interaction involved in combining two elements is the same interaction between the variables of the two elements as between these variables in the maximal element. If the two elements have variables in common, this duplication disappears in the combined element, and, in particular, $a \oplus a = a$. This is not true of interactions between elements from different maximal sets, as will be discussed in later chapters; for example, the interaction of identical elements from different maximal sets constitutes a very interesting class of systems whose members can have characteristics widely different from those of the individual elements. Both types of partitionings, within a maximal element and into maximal elements will be used in our quest to develop the representation of a complex functional element in terms of a number of less complex, but interacting functional elements first alluded to in section 1.3 and discussed further in section 3.1. Our aim is to develop a whole collection of standard, basic (i.e., of limited complexity) functional elements that gain wide acceptance within the engineering community because of their usefulness in the process of designing complex systems, but first we need to achieve a much better understanding of the nature of such elements.

Two comments need to be made with regard to the representation in Figure 4.1. First, the functions are not restricted to such between a dependency and a set of functional parameters, but include also functions between functional parameters only. (However, functions between dependencies only are not included; such functions would always be secondary in the sense of being derived from the functions determining the dependencies in terms of the functional parameters.) Second, interaction between two objects means that some property of the one object causes a change in the other object. This change can be caused either by a change in the first object (i.e., by the object doing something and actively driving the change in the other object) or by the first object having a certain property that by itself causes a change in the other object (i.e., the first object is a passive partner in the interaction). An example of the latter is when the color of an object is intended to have an effect on an observer (the observer being the other object). Relating this grouping of interactions to the stakeholder requirements an object is intended to satisfy, one obtains a picture as shown in Figure 4.2 — some of the stakeholder requirements will be satisfied by physical properties of the object, that is, by what the object is, rather than by active interactions, that is, by what the object does. Therefore, if functionality is to represent what an object does, it follows that in the definition of functionality, interaction must be restricted to mean active interaction, in the above sense.

The division of the set of variables that describe a functional element into three subsets, as shown in Figure 4.1, arose because of our desire to focus on the user requirements as the point of departure for the design process. In particular, it arose because we allowed a functional element to represent less than the full functionality of a physical object; the functional parameters describe only that part of the functionality of interest to a particular user group. As a corollary, it follows that the variables associated with a maximal element are all functional parameters; the other two subsets are empty. They only become populated in order to complement a limited set of functional parameters so that together they form a *self-consistent* description of functionality, in the sense that all the variables required by the set of

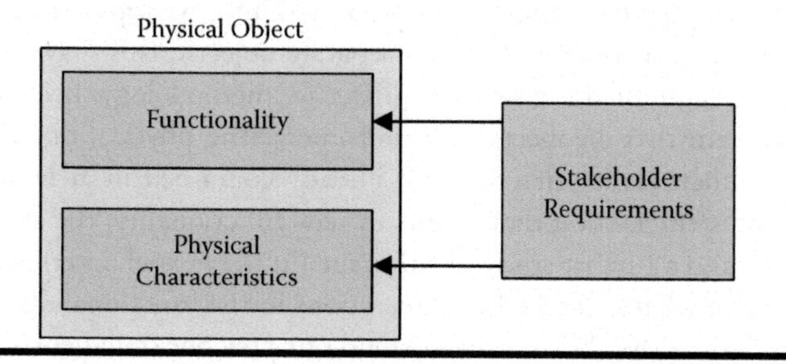

Figure 4.2 Meeting stakeholder requirements.

functions belonging to the element are present. The functional parameters take on a primary importance in the definition of a functional element; they describe the service offered to the users, and all the consequential variables and all the functions are secondary, in the sense of describing *how* that service is provided. Also, the concept of an aspect of the functionality can now be made precise; it is simply the description of the behavior of a single functional parameter.

4.2 The Functional Domain

The functional domain is the set of all functional elements. What can be said about this set; does it have any structure; can it be subdivided into significant subsets, etc.? We have already identified a class of subsets, the maximal sets, each of which contains all that can be known about the service provided by a given physical object, and within the maximal sets the combination operator introduced a hierarchical ordering. A first question is then whether such a maximal set contains more than one primary element, or, in other words, is it possible for the description of the functionality of one object to be identical to a part of the description of the functionality of another object? The answer is given by the following theorem:

Theorem 4.3: A maximal set contains one and only one maximal element, the one from which the set is derived.

Proof: Assume that the theorem is incorrect, and that a given maximal set contains a maximal element, say, x, in addition to the one from which the set is derived, say, y. Let Y be the set of functional parameters associated with y, and X the set of functional parameters associated with x. Then the functional parameters in X cannot have any linkage (functional relationship) to parameters in the complement of X in Y. But then the parameters in the complement of X in Y must also form a maximal set, so that the original physical object is really a collection of two independent (non-interacting) objects, which is contrary to the definition of a physical object. Thus, the assumption was incorrect, and x cannot be a maximal element.

Another way of stating this theorem is to say that if the description of the functionality of a physical object can be separated into two separate parts, then the physical object is also composed of two separate objects. This may seem obvious, but it lies at the core of the systems engineering methodology, because it follows from this theorem that a collection of non-interacting physical objects can never result in any functionality that was not already contained in at least one of the objects. It is only interaction that results in new functionality, the so-called *emergent properties,* and a further study of this issue (in the next chapter) will lead to an understanding of what is meant by "interactions in the functional domain."

As a corollary to the above we can extend our previous statement about the difference between interactions within a maximal set, as expressed by the combination operator, ⊕, and interactions between elements of different maximal sets, by the at-first paradoxical statement that combining elements within a maximal set increases the level of detail of the description of the functionality of the corresponding physical object, but does not change the functionality.

The intersection of maximal sets contains those aspects common to their maximal elements. In chapter 2 it was suggested that all maximal sets have one element in common, the functionality of return on investment. Any two maximal sets must differ in their maximal elements plus at least one other element; this is just saying that if two physical objects have the same functionality, in the sense that we cannot find a single aspect that is different, they are represented by the same maximal element in the functional domain.

It is not so simple to visualize the functional domain and its subsets, but one possible visualization is shown in Figure 4.3. The maximal elements are all points on a closed surface, shown here for convenience as a sphere. The complete elements are points in the interior of the sphere, and the irreducible element is close to the surface of the sphere so as to allow for elements of varying complexity. A maximal set is a "volume" (actually a collection of points) in the interior of the sphere, containing exactly one point on the surface (the maximal element) and at least two points in the interior, the irreducible element plus the element that distinguishes it from other maximal elements.

In the picture presented by Figure 4.3, the surface can be considered as the boundary between the functional domain and the physical domain. There is a

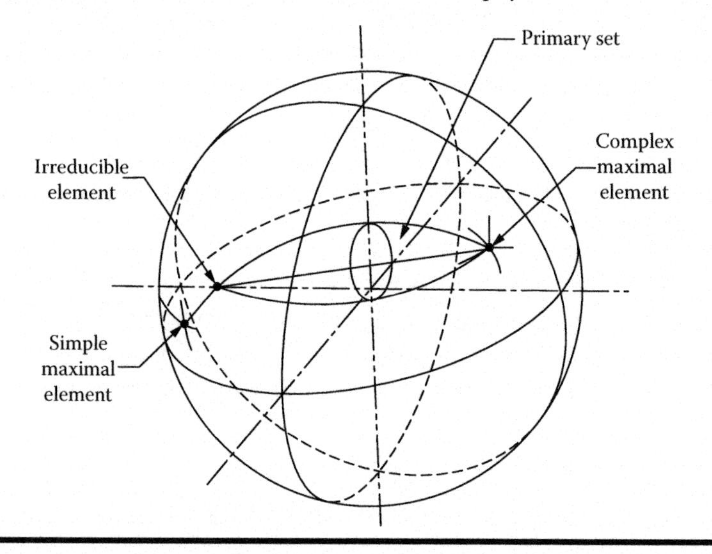

Figure 4.3 A graphical representation of the functional domain, with the maximal elements shown as points on the surface of a sphere, the complete elements as points in the interior of the sphere with the irreducible element close to the surface, and a primary set as a "volume" between the surface and the common element.

mapping between each maximal element and a set of physical objects (those with the same functionality), but there is no direct relationship between elements within the sphere and objects outside it. It is therefore reasonable to start the exploration of the characterization of the functional domain with the maximal elements and their variables, the functional parameters, and then to generalize to all functional elements.

4.3 The Functional Parameter Space

With each functional element there is associated a set of functional parameters. That is, if one considers the set of all possible functional parameters, the definition of a functional element defines a mapping between functional elements and subsets in this set, and, in particular, this is true of the maximal elements. Because maximal elements are complete descriptions of functionality (of physical elements), the following definition is relevant to both the physical and functional domains:

Definition 4.13: The *complexity* of a maximal functional element is its number of functional parameters.

This definition explains why the irreducible element was placed close to the surface (rather than in the center) of the sphere in Figure 4.3. Note, however, that Figure 4.3 is deceiving in that it implies that there is a limit to complexity (the size of the sphere). That is, of course, not true, and is easily overcome by realizing that the scale of the figure is undefined; the irreducible element can be placed only an infinitesimal distance away from the surface.

But the mapping between primary elements and subsets of functional parameters allows a much more interesting characterization of the structure of the set of maximal elements:

Definition 4.14: Let x and y be two maximal elements, and let X and Y be the corresponding sets of functional parameters. Then the *distance* between x and y, d(x,y), will be defined by $d(x,y) = c(X \cup Y) - c(X \cap Y)$, where c(X) is the cardinality of X (the number of parameters in X).

In practical terms the distance between two elements is a measure of their *dissimilarity*; the significance of the concept will become apparent in the next section.

It is then straightforward to prove the following theorem:

Theorem 4.4: The function d(x,y) defines a metric on the set of maximal functional elements.

Proof: In order for d(x,y) to be a proper measure of distance that defines a metric, it must satisfy the following four conditions:[1]

1. $d(x,x) = 0$
2. $d(x,z) \leq d(x,y) + d(y,z)$
3. $d(x,y) = d(y,x)$
4. If $x \neq y$, then $d(x,y) > 0$.

Of these, only (2) does not follow directly from the definition. First note that d(x,z) can also be written as $c(X) + c(Z) - 2c(X \cap Z)$. Then,

$$\Delta \equiv d(x,y) + d(y,z) - d(x,z) = 2[c(Y) - c(X \cap Y) - c(Y \cap Z) + c(X \cap Z)].$$

Let $Y = Y_A \cup Y_B$, such that $Y_A \subseteq X \cap Z$, and $Y_B \cap (X \cap Z) = \varnothing$, then

$$c(X \cap Y) + c(Y \cap Z) \leq 2c(Y_A) + c(Y_B), \text{ and}$$

$$\Delta \geq 2[c(Y_A) + c(Y_B) - 2c(Y_A) - c(Y_B) + c(X \cap Z)] = 2[c(X \cap Z) - c(Y_A)] \geq 0.$$

The above is valid for maximal elements, that is, points on the spherical surface in Figure 4.3. But maximal elements are of little practical use; useful functional elements with a limited number of functional parameters are points in the interior of the sphere. To see that the concepts developed for maximal elements can be generalized to all functional elements, consider any one maximal element and its associated set of variables (which are all functional parameters), $X = \{x\}$. According to the methodology in chapter 3, all the parameters in X arise from describing in more and more detail the functional parameters in the basic set (i.e., the two functional parameters S and L required to define the irreducible element), and with return on investment seen as the central purpose of any project, the partial derivative $\partial / \partial x(ROI)$ must be a measure of the importance of the variable x in achieving this purpose. This is illustrated in Figure 4.4, where X is shown as a roughly circular area in the set of all functional variables, with the basic set in its center, and with the importance of the variables decreasing with distance from the basic set. The largest possible area X is that of a maximal element. In this picture, functional elements arise by reducing the set of variables, often by discarding the less important ones or by combining variables, as shown by the set A, but not necessarily. If we are interested in a detailed description of a particular aspect, the set might look something like the one labeled B.

That is, while all functional variables are derived from the basic set, it is not necessary for any member of the basic set to be included in the set of functional

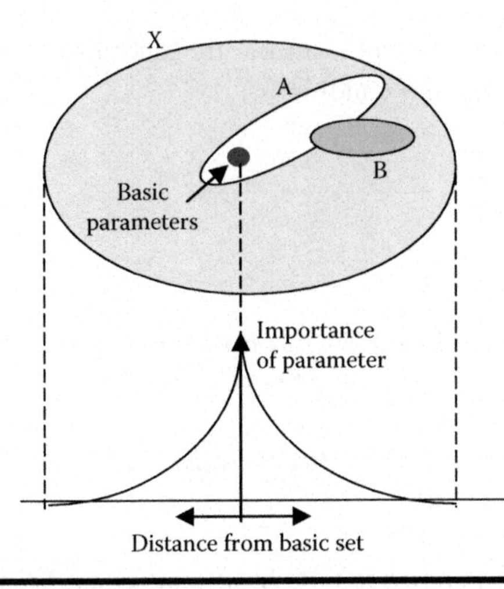

Figure 4.4 Declining importance of functional parameters away from the parameters of the basic set. Here X is the set of parameters associated with a maximal element, A are the parameters of a less-detailed description of the same functionality, and B are the parameters associated with a particular aspect of the functionality.

variables associated with a functional element. And while it is true that every complete functional element belongs to at least one maximal element, in the sense of being derived from that primary element by reducing the set of functional parameters, a functional element can also belong to two or more primary elements (the irreducible element belongs to all primary elements). So, we can now generalize the concept of a functional element beyond its being derived from a maximal element in the above manner, and thereby abstract it from its connection with a physical object, by redefining the concept of a functional element:

Definition 4.15: A *functional element* is a self-consistent description of the relationships between a set of functional variables.

This definition differs from definition 4.5 in two respects. First, instead of a reference to a physical object, the definition is wholly in terms of functional variables. That is, it defines a mapping between certain subsets of the set of all functional variables and the set of all functional elements.

Second, it introduces the notion of *self-consistency*, which was mentioned briefly at the end of section 4.1. This arises because all functional variables are derived from the basic set in a top-down process, so that any one variable is linked to one or more other variables through this process. However, it is important to recognize that this "linking" has two completely different sides to it. On the one hand, there is the linking that arises through the top-down development of a variable in more detail, such as splitting cost into acquisition cost, operating cost, and maintenance cost. Acquisition cost and operating cost are both "linked" to cost in a parent/child relationship that requires the sum of the "children" to equal the parent. The choice of "children" is usually based on normal usage (e.g., accounting practice for cost), and there may be more than one choice (e.g., cost could have been split into recurrent and nonrecurrent cost).

This process, subdividing variables into sets of variables that provide a greater level of detail in the description of functionality, will be called the *decomposition process*, with the converse process being called the *condensation process*, and we shall demand that the resulting parent/child relationships are uniquely defined by the condensation process. That is, to any one parameter there corresponds one and only one set of parameters that will allow a condensation to another parameter (e.g., for operating cost this set is maintenance cost and acquisition cost, which allows the condensation to cost). On the other hand, there are the relationships that are inherent in a functional element in order for it to be a description of functionality (or an aspect of functionality). Thus, as the set of variables is expanded from the basic set, relationships are created between the new (more detailed) variables. This is illustrated in Figure 4.5, where the rings represent successive levels of detail, and the fact that they are rings rather than disjoint "rays" radiating from the center is supposed to show that the variables within one level of detail are all related.

Therefore, if we have a stakeholder group that is only interested in a small number of parameters (or even just one), the corresponding functional element will need to include at least the variables to which these parameters are directly linked in order to be self-consistent; these additional variables are, of course, exactly what we have called dependencies in section 4.1. Then, if we want to go into more detail, we can expand the number of parameters considered, and thereby create a larger functional element. That is, the functional parameters express the stakeholders' view, and this subdivision into functional parameters and dependencies is what links a functional element to a stakeholder group.

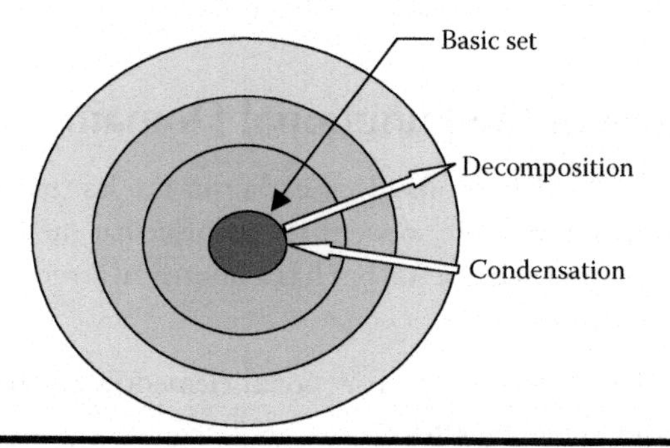

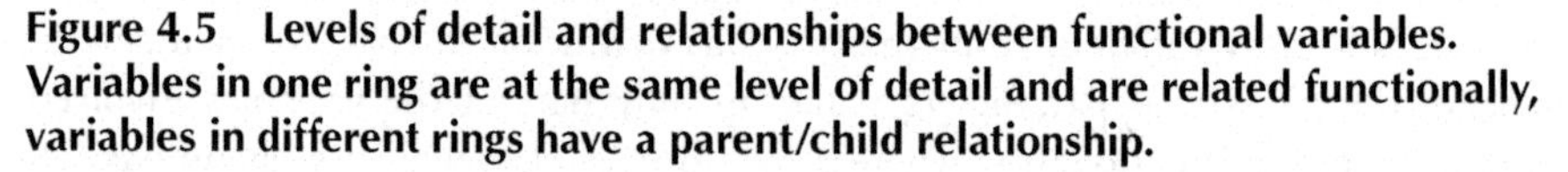

Figure 4.5 Levels of detail and relationships between functional variables. Variables in one ring are at the same level of detail and are related functionally, variables in different rings have a parent/child relationship.

At this point in the development it is appropriate to return to a statement made earlier (following the definition of the maximal set, definition 4.11) about the relationship between functional elements and the structure this introduces into the functional domain. A complex functional element can be said to include a number of less complex elements in the sense that these elements arise by reducing the number of functional parameters, either in order to achieve a simplification by condensing subsets of parameters or in order to concentrate on a particular aspect, as discussed above. The condensation process comes to a halt whenever a smallest self-consistent set of variables is reached, and how far inward toward the basic set it can progress depends on the size of the initial set of functional parameters. Using the image of Figure 4.5, the process will only reach the basic set if the initial set includes a complete "ring" of variables, and this leads us to the following definition:

Definition 4.16: The set of variables *associated* with a functional element consists of the variables of the element plus the variables that can be reached from them through the process of condensation.

The associated set is also subdivided into functional parameters and dependencies; the process of condensation does not mix the two types of variables.

Definition 4.17: The *included* set of a functional element consists of all those functional elements that are generated by forming self-consistent subsets of the set of variables associated with the element.

Recalling the definition of a functional element (definition 4.5) as a description of one or more aspects of the functionality of a physical object, the included set is the set of functional elements that describe those same aspects, but at higher levels (i.e., in less detail). As an example, if the aspect were reliability, the included set would be the descriptions corresponding to reliability block diagrams with less and less (larger and larger) blocks.

4.4 Structure of the Functional Domain

We can now extend the definitions given earlier in the last section to functional elements in general, and in doing so we shall discover that the functional domain is not a simple set of unrelated points, but has a nontrivial structure. The definition of complexity is straightforward:

Definition 4.18: The *complexity* of a functional element is equal to its number of functional parameters.

This is slightly different from the previous definition in that, for elements in general, the functional parameters make up only part of the variables involved in describing the functionality, whereas for primary elements they encompass all variables. The effect is to make the concept of complexity reflect the users' point of view; it reflects the number of variables they are exposed to (or are interested in).

The concept of distance cannot be generalized quite as easily, and an indication of why this is so can be gained by looking at Figure 4.3. As defined for maximal elements, the concept is confined to the distance between points on the spherical surface. When it comes to points in the interior, we must ask if the distance between two elements whose functional parameter sets are disjoint should be the same whether they belong to the same maximal set or not. Intuitively we would say no, because elements that belong to the same maximal set have something in common that elements belonging to different sets do not, and therefore the distance, which measures the difference between elements, should be less in the former case. Again, as with interactions between elements, there appears to be a basic difference between elements within the same maximal set and elements from different maximal sets.

The outline of an answer to this question emerges if we realize that functional elements fall into two groups; those for which the included set contains the irreducible element, and those for which it does not. The former group includes the maximal elements and all those elements that result from reducing the complexity of maximal elements by condensing the associated set of functional parameters, as described above. To see what the latter group (i.e., all other elements) contains, we recall that the starting point of our design methodology is a set of stakeholder requirements, and that the outcome of our design in the functional domain is a large functional element (or, rather, a system of smaller functional elements, to be further defined in the next chapter) such that when we make the transition into the physical domain, the resulting system will meet all the functional stakeholder requirements. That is, the functional element is an expression of the necessary and sufficient conditions for the physical system to meet the functional stakeholder requirements. Consequently, if the included set of the functional element does not include the irreducible element, there is at least one functional stakeholder requirement that is not determined by the element. So, according to definition 4.10, we can express this in the form of a theorem:

Theorem 4.5: The included set of a complete functional element contains the irreducible element.

Proof: A complete functional element includes all aspects of the functional stakeholder requirements, therefore also the requirement for optimizing the ROI, as expressed by the irreducible element.

The converse is, of course, not true; the included set of a functional element can contain the irreducible element without the element being complete. Completeness is not a property of the functional domain alone; it is defined only relative to a set of stakeholder requirements.

As was already discussed briefly after introducing definition 4.10, the issue of completeness is related to the issue of *physical realizability* if we expand the meaning of this concept beyond the immediate one of containing relations that are contrary to the laws of nature to include "realizable as a system that meets the stakeholder requirements." The situation here is the same as with a set of n variables and n linear equations between them; if we remove one of the equations we have a whole set of solutions (including the one obtained before the equation was removed). We often express this by saying that the $(n - 1)$ equations have no solution, meaning they have no one solution.

An example is reliability. On the one hand, a functional element expressing the behavior of reliability (i.e., its relationship to other variables) can be quite general and applicable to any system (or, at least, to a large class of systems), but it cannot, in itself, be realized as a physical object. There is no physical object that has as its purpose reliability; the purpose would always be to do something reliably. On the other hand, remove the aspect of reliability from a functional element, and there would be a large number of physical systems that would correspond to it, but only a small subset that would meet all the stakeholder requirements.

We now introduce the following two definitions:

Definition 4.19: A functional element is a *real* functional element if and only if its included set contains the irreducible element.

Definition 4.20: A functional element is an *imaginary* functional element if and only if its included set does not contain the irreducible element.

It follows that the functional domain consists of two disjoint parts, the real and the imaginary functional domain; the main significance of this will become apparent when we introduce the notion of systems of functional elements in the next chapter.

It is now clear that our earlier intuition about the distance concept was not quite correct; it is not so much a question of whether two elements belong to the same maximal set or not, as a question of whether they are both real elements. Between real elements the distance concept is a measure of what we would, in everyday language, consider to be the difference in functionality between the corresponding physical objects, whereas for imaginary elements we can only say that two elements either belong to the same (set of) concept(s) or to different ones, but it does not make sense to put a measure on the difference, as long as the elements are unrelated to any physical reality. Consequently, the earlier definition of distance and the accompanying theorem should be limited to real functional elements and modified as follows:

Definition 4.21: Let x and y be two real functional elements, and let X and Y be the corresponding sets of associated functional parameters. Then the *distance* between x and y, d(x,y), will be defined by d(x,y) = c(X∪Y) – c(X∩Y).

Theorem 4.6: The function d(x,y) defines a metric on the set of real functional elements.

The proof of the theorem remains the same as before.

However, despite the above, there is still a difference between real elements belonging to the same maximal set and elements in different maximal sets, and it may be helpful to visualize the difference in a modified, two-dimensional version of Figure 4.3. Consider the real element space to be spanned by a polar coordinate system, with the irreducible element at the origin, as shown in Figure 4.6. The distance from the origin measures the level of detail of an element as measured by the number of functional parameters, the angle (from some arbitrary reference direction) the purpose of the element. That is, the two elements *a* and *b* are describing the same purpose to different levels of detail, whereas the two elements *a* and *c* describe different purposes, but to the same level of detail. Clearly, all elements in any one maximal set lie on a radial line and, furthermore, because the set of parameters associated with *a* is a subset of the set of parameters associated with *b*, the distance between *a* and *b* is zero, whereas d(*a*,*c*) and d(*b*,*c*) take on values greater than zero.

It follows that the distance between real functional elements that have the same functional parameters is zero, and that leads to the following definition:

Definition 4.22: If the distance, d(*a*,*b*) between two real functional elements *a* and *b* is zero, the elements will be said to describe the same *type of service.*

The concept of a service type defines an equivalence relation on the real functional domain; two functional elements are equivalent if they belong to the same

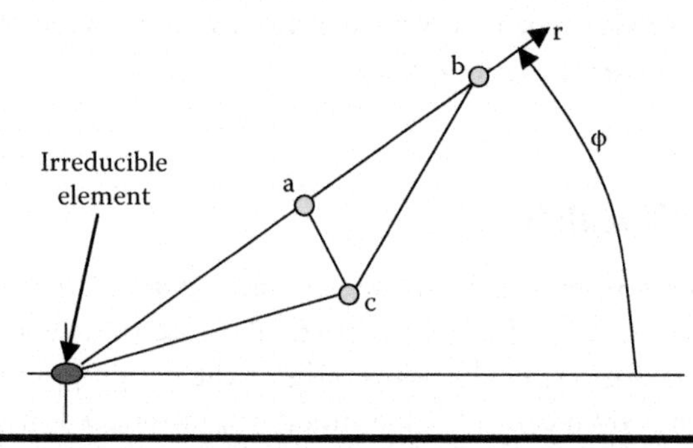

Figure 4.6 Visualization of the distance concept for real elements in a two-dimensional functional domain, with r representing the level of detail and ϕ the functionality.

service type. A service type is a subset of the real functional domain, and the different elements within such a subset represent different ways (or quality levels) of delivering the service (e.g., as in the difference in the service delivered by a Suzuki and a Lexus). The type of service is determined by the set of functional parameters; the actual level of service provided in a given case depends on the values these parameters take on in that case, and the detailed behavior of the object can also depend on the relationships between the parameters.

Let us summarize our development and understanding of functional elements so far. We developed the concept of a functional element from two points of view, on the one hand as representing the functionality of a physical object, on the other as representing the requirements of a user group, and it is this dual nature that makes functional elements central to our design methodology and its aim of assisting the designer to make the transition from user requirements to physical solution. The first point of view resulted in such concepts as maximal elements and physical realizability, the second point of view led to the concept of completeness. Both of these views are unified in the view of a functional element as an abstract entity representing functionality *per se*, consisting of a set of variables and a set of functions relating them to each other. This view would be devoid of any meaning and usefulness were it not for our axiomatic assertion that all variables are developed from the basic set. This process of decomposing the basic set, which is important in developing functional elements, and its converse, the condensation of variables, which is important in the design methodology, led to the concepts of included set, complexity, and real and imaginary elements.

It is apparent that the functional domain has a fairly complex structure, characterized first of all by the subdivision into two disjoint domains, the real and the imaginary domains. The concepts of complexity and combination operator apply in both domains; the combination of a real and an imaginary element is always a real element. But the concept of distance applies only in the real domain, as does the concept of service type. It makes sense to say that a sequence of real elements *converges* within a service type to a particular service, but a sequence of imaginary elements does not converge to anything.

4.5 Element States

In the previous section, we regarded the variables, and functional parameters in particular, as points in a set. But in any particular case these parameters take on specific values, and it is one of the main tasks of the design process to determine their optimal values. By means of value functions, introduced in section 2.5, the users specify their requirements of the service, and for any given representation of this service in terms of a functional element (and in the next section, a set of functional elements) the designer will endeavor to find an optimal set of values for the functional parameters of these elements under the constraint of the value functions.

Each parameter can take on values in a range of values that may be discrete, continuous, or a combination of both. The smallest range is that consisting of two discrete values (e.g., 0 and 1, or true and false), the largest is the real line.

However, the functional parameters of an element are not all independent; the functions between the variables of an element, and which form part of the definition of the element, place restrictions on the sets of values that can be delivered by the element, and this is what the users see as part of the *behavior* or *performance* of the element. In particular, as was already mentioned toward the end of section 4.1, there can be functions involving only functional parameters, and each such function lowers the number of independent functional parameters by one. However, the concept of independence is not particularly relevant to the set of functional parameters belonging to an element (so long as one parameter is not simply a fixed function of another, for example, multiplied by a factor, squared, etc.), and it is not possible to formulate a general rule as to which functional parameters should be called independent and which ones dependent. This becomes quite obvious if we consider the application of a functional element to a particular design, in which case one or more of the functional parameters might have prescribed, fixed values. If we recall that all functional parameters are derived from the parameters in the basic set (i.e., S and L) through the process of decomposition and, additionally, forming functions of the new parameters, we could perhaps at first think that decomposing S means describing the nature of the service, that is, *what* the element does, in more and more detail; the dependent parameters are, in a sense, additional measures of *how well* the element provides that service. However, a little further thought makes us realize that there is no actual difference between parameters arising from decomposition and those arising from subsequently forming functions of those parameters, because the decomposition process is completely free and only guided by what is useful. In some cases common usage makes certain variables more fundamental than others, but this is not relevant to our current investigations into the theoretical foundations of the top-down design methodology, and we shall simply consider all functional parameters as a set, but with *constraints* in the form of the stakeholder requirements and the functions defining the element.

However, speaking of decomposition, we note that decomposing L means subdividing the lifetime of the element into more and more segments, such as design, implementation, operation, refurbishment, and decommissioning, and as a consequence, each component of S is subdivided into a parameter for each segment of L. In the limit, L is subdivided so finely that the components of S simply become functions of time. Thus, a set of parameter values is a complete characterization of a functional element at a particular point in time, and this leads to the following definition:

Definition 4.23: The *state* of a functional element is a set of parameter values, one for each functional parameter of the element, and the space spanned by the functional parameters is the *state space* of the element.

As an element goes through its lifetime L, the state will pass along a *trajectory* in state space; the possible trajectories are limited by the constraints.

4.6 Functions on State Space

As we know, the definition of a functional element will also contain a number of functions of the functional parameters that express stakeholder requirements, and in light of definition 4.23, we would call them functions on state space. One such function is well known to us already — the return on investment (ROI) introduced in section 2.2, and we can now state the general applicability of this concept in different terms.

Theorem 4.7: For any real functional element, the ROI exists as a single-valued function on its state space.

Proof: By theorem 4.5, the included set of a real functional element contains the irreducible element, and by definition 4.16 the elements of the included set have variables that are related to the variables of the real element through the decomposition/condensation process. Consequently, the values of the variables of the basic set are determined by the values of the variables of the real element, and so is the value of the ROI. The single-valuedness follows from the definition of the irreducible element in section 2.2 and the requirement that the decomposition/condensation process results in unique mappings between parent and sets of child variables.

There is nothing profound about this theorem, as it follows directly from the fundamental axiom of our theory (i.e., that maximizing the ROI is the ultimate purpose of every project), but it expresses neatly the fact that for any real element there exists a set of parameter values that is optimal. There is never any doubt about what is the "best" solution.

We shall develop some further functions of this nature in later chapters, but let us just conclude this chapter by introducing a completely different type of function on state space. For a number of reasons, some of which were touched on briefly in section 1.4, the values of some or all of the variables belonging to a functional element will not be known to us, but we do know their probability density distributions. As a result, some or all of the functional parameters become random variables, s_i, and the state of the element is described not by a set of values of the independent parameters, but by a distribution density function, $\varphi(\mathbf{s},t)$, which we shall call the *service density function*. Here $\mathbf{s}$ is the set of functional parameters, and t represents whatever subdivision of L we have adopted. Such a state may be called a *superstate*, and the space of all such functions for an element (i.e., with the appropriate restrictions on the ranges of the variables s_i) may be called *superspace.*

Notes

1. Pervin, W. J., *Foundations of general topology*, Academic Press, New York, 1964, p. 99.

Chapter 5

Interactions and Systems

5.1 The System Concept

Let us start this chapter by recalling that our aim is to develop a design methodology that takes as its point of departure a set of stakeholder requirements and seeks to reduce the complexity of finding a physical object with a performance that will meet these requirements by a top-down process of subdivision of the requirements into smaller sets. The end result of this process of *design in the functional domain* is a set of interacting functional elements for which the transition into the physical domain is relatively simple and efficient. We now realize that this top-down process involves two basically different decompositions. On the one hand, there is the decomposition into functional elements that address one or more aspects of the functionality, and we know that these elements, which we called imaginary elements, all belong to the same maximal set and are related through the composition operator. The purpose of this decomposition is to be able to address different aspects of the functionality at different stages of the process; in particular, to be able to address global aspects (i.e., with a low level of detail) early in the process, when detailed design data are not yet available.

On the other hand, there is the partitioning of the complete element corresponding to the whole set of stakeholder requirements into a set of interacting elements that we called real elements, each one of which is complete with respect to some subset of the stakeholder requirements. (Stakeholder requirements and stakeholder group can be used interchangeably in this context.) Both types of decomposition play important roles in our design methodology, and the definition of a system, to be introduced shortly, does not distinguish between real and imaginary functional elements. It is perfectly acceptable to have functional systems made up of imaginary elements; such systems represent only certain aspects of a physical system's functionality. A well-known example is provided by reliability block diagrams. However, as our aim is to end up with a system of physically realizable elements, theorem 4.5 and definition 4.19 indicate that we should at first restrict our attention to real elements, and then, as required, extend any pertinent results to imaginary elements. Or, in other words, the decomposition into real elements may be considered to be the primary one in that it leads to the desired end result, whereas the decomposition into imaginary elements may be considered to be secondary in the sense that it supports the process of choosing the best primary decomposition.

The concept of a system was introduced briefly and in most general terms in section 1.5, and we recall that the basic idea behind the system concept was to be able to describe something complex, be it in the physical or functional domain, in terms of less complex entities, and thereby make it easier for the human mind to comprehend and manipulate. As such, the system concept is exactly what we require for our methodology, and this is the reason the methodology is part of what is called *systems engineering*, although this term is applied to a broad range of activities, including in particular the *management* of complex projects. In the latter application, which is an application in the physical domain, only the most rudimentary features of what might be called a general systems theory have so far been applied, and there has been no need for a rigorous theoretical treatment. But if we want to use this concept in an operational sense in our methodology, we need to make it much more precise and develop it in more detail.

There is no single, agreed definition of what a system is; there are a number of different definitions in use, and the choice of definition depends largely on what one wants to use it for or in conjunction with. In our case, the following definition, which is valid for any type of system, not just functional systems, will be adopted:[1]

Definition 5.1: A system consists of three related sets:

- a set of *elements;*
- a set of *internal interactions* between the elements; and
- a set of *external interactions* between the elements and the rest of the world.

It is important to note that the internal interactions do not represent anything (physical or functional) in addition to the elements; they express which of the possible interactions already inherent in the elements are realized in this system. They might be called *logical* interactions; they are either present or they are not, and there are no variables associated with the interactions themselves.

The external interactions are required, on the one hand, for the system to provide its service; on the other hand, for the system to maintain its operational state. All engineered systems are open systems.

In our case, the elements are, of course, functional elements, and the resulting system is itself a functional element. Thus, the forming of a system can be viewed as the result of applying an "operator" to a functional element. It is not a normal, simple operator, such as addition or multiplication, for a number of reasons. First, it is not defined by its action on any one functional element, only in terms of its action on all functional elements. That is, its action, which consists of identifying a set of elements and a set of interactions between them, depends on the particular element to which it is applied. Second, the operator is a composite operator, in the sense of a vector or a sequence of operators, such that each component in the sequence results in a system with more elements than that resulting from the preceding component but less than that resulting from the succeeding component. And third, it is not necessarily one-to-one, but may be one-to-many. That is, there will in general not be a unique representation of a functional element in terms of a system with a given number of elements. As a result, it will not be possible to give an explicit representation of the operator, but despite these difficulties, the existence of this operator is a useful concept.

Definition 5.2: The *partitioning operator* is that operator which, when applied to a functional element, results in a representation of the element as a system of elements. The number of elements in the system will be called the *order* of the operator.

The concept of a system introduces a hierarchy among elements; the elements making up a system can be considered to be at a lower (more basic) level than the element represented by the system. The design methodology in the functional domain is called *top-down* because it proceeds to partition a single, "large" element with complex functionality into successively more and more "smaller" elements with less complex functionalities, but with such interactions between the elements that the external functionality remains unchanged. This latter requirement on the partitioning process, the requirement of *traceability*, is a requirement on the partitioning operator, and it ensures that not only are the variables of the original element and of the resultant system the same, but the behavior of the two, as reflected in the functions between the variables, is also the same. Thus, an alternative view of the partitioning process is as a *representation* of the complex function between the external variables of the system in terms of a set of simpler functions, coupled together through additional variables internal to the system. This view should bring to mind the analogy we mentioned earlier (section 3.6) — the expansion of an arbitrary function in terms of an orthonormal set of functions. We are looking to develop "collections" of basic (i.e., relatively simple) functional elements such that any functionality can be expressed as a combination of such elements.

5.2 Interactions between Real Functional Elements

A functional element is defined by four sets; the set of functional parameters, the set of dependencies, the set of influences, and the set of functions between the elements of the first three sets. Both the first two of these sets could be involved in an interaction, but as the main purpose of a functional element is to represent what an object does, in particular as seen from the users' point of view, and the purpose of decomposing a functional element into a set of interacting elements is to partition the users' functional requirements and thereby reduce the complexity of the design task, it would be preferable to define interactions in terms of functional parameters. A starting point would therefore be the following definition:

Definition 5.3: Two real functional elements *interact* if the values of one or more functional parameters in one element depend on the values of one or more functional parameters in the other element.

This definition can also be formulated by saying that two real elements interact if the states of the two elements are related. Note that the definition does not introduce any direction of the interaction; functional interactions are always *between* two elements, even though the physical quantity involved in the interaction may flow from one to the other. Also, the dependency may be of any form — analytical, statistical, etc.

This definition says nothing about *how* it is possible for there to be a relationship between a functional parameter in element B and a functional parameter in element A. From the definition of a functional element, there are four different combinations of the two types of variables for each element, as shown in Figure 5.1. But in each case, the output of one element must be identical to the input in the other element. That is, the possibility of the interaction must already be inherent in the two elements; a system is created when one or more of these possibilities are realized in a particular case.

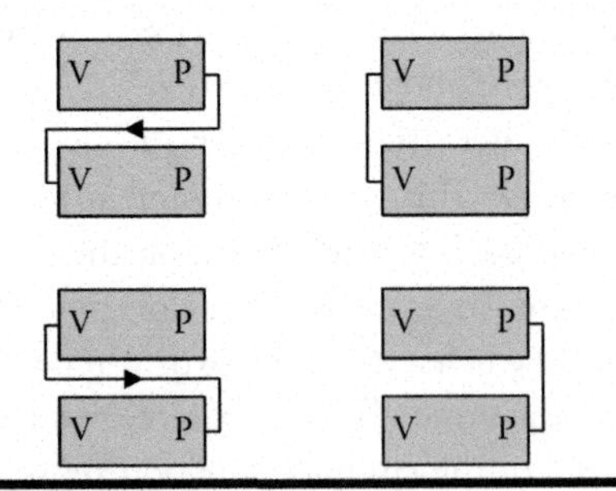

Figure 5.1 The four possible types of interactions. Here P denotes a functional parameter, and V a dependency. Parameters and dependencies not involved in the interaction are not shown.

From definition 5.3, it follows that the functional parameters involved in the description of the service provided by the system must be a subset of the functional parameters of the elements making up the system. In particular, if x and y are two elements with sets of functional parameters X and Y, respectively, then the set of functional parameters, Z, belonging to the system z resulting from the interaction of x and y satisfies the relation $Z \subseteq X \cup Y$. This is an important observation, because it means that any *emergent properties*, that is, properties of the system that were not present in the individual elements, can be completely described using the functional parameters of the elements; no new parameters are required. It may be much more *convenient* to introduce new functional parameters that directly characterize the emergent properties of the system, but these will always be related to, or expressible in terms of, the functional parameters of the elements. A little example from an important type of interaction, correlation, or coherence of identical elements, is given by considering a set of identical radiators being combined to provide the service of illuminating a small, remote spot with radiation. From the users' point of view, a most useful functional parameter is beam-width, and they might not even be aware of the relationship between beam-width and the parameters of the individual radiators (relative position and phase angle). Another example, from a different type of interaction, elastic collision between mass points, is the characterization of a gas in terms of pressure and temperature and, if the "service" of the gas is its ability to absorb and release heat, the functional parameter heat capacity, all of which are related (albeit in a statistical manner) to the parameters describing the individual mass points.

However, having said this, we note that there may be cases of practical importance where combining elements into systems creates emergent properties for which our (current) understanding of the elements provides no explanation. We believe that, with a more in-depth understanding of the elements and their ability to interact, we would eventually always be able to *explain* why the emergent properties occur; being able to *predict* the occurrence or details of the properties is a different matter. This is where the theme of this book series, complex systems and complexity theory, comes into play.[2] In systems that have the ability to maintain themselves by creating order (or negative entropy) to counteract the inevitable tendency of a closed system to decay into a disordered equilibrium state, this same ability can be used to change the system itself and, for example, result in the system *evolving* in response to changes in its operating environment. But it can also give rise to completely *unexpected* changes in the system behavior, often perceived as catastrophes, but sometimes providing the opportunities for drastic, beneficial changes.[3] The development of complexity theory now points in a direction that would allow us to explain or *understand* such behavior in terms of fundamental processes involved in the interactions between the system elements, but that would at the same time prove the inherent *unpredictability* of this behavior.[4]

The ability to create negative entropy and for a system to reorganize itself has been the distinguishing characteristic of what we would call living systems (although this distinction may now lose its absolute character and become a matter of degree), and in the case of the type of systems to which the methodology proposed in this book is applicable the living components are supplied by humans, mainly as operators and maintainers. Important as these components are for the viability of these systems, engineering has traditionally treated them as non-changing entities with fixed interactions with the non-living part of the systems, and even in this book, that is effectively the approach taken. The complexity we are aiming to handle with our methodology lies in the operational environment (technology and stakeholder requirements), not in the people within the system. With many of the systems now of interest to engineers being seen as *enterprises*, where the non-living part is seen as providing only a support to the human activities, the situation is in a way reversed, in that the behavior of the people within the system, treated as a subsystem of interacting elements, becomes the focus in understanding the performance of the overall, *complex system*, and many of the related issues are discussed in other books in this series.

5.3 Functional Systems

From definition 5.1, a system consists of three sets, and let us for the time being denote them by X (elements), Y (internal interactions), and Z (external interactions), so that the composition of a system A can be shown explicitly as A = (X(A),Y(A),Z(A)). With this notation, we can formulate the following definition:

Definition 5.4: Two systems A and B are *identical* if and only if X(A) = X(B), Y(A) = Y(B), and Z(X) = Z(B).

Or, conversely, two systems are different if any one of the three sets differs. This definition appears, at first sight, to be trivial; its significance only becomes clear once we realize that it would have been possible to define the identity relation differently. We could have said that two systems are identical if and only if X(A) = X(B) and Y(A) = Y(B), but not required Z(A) to equal Z(B). That is, we could have said that the external interactions are a matter of the users, that is, characterize the users and what they are interested in, or how the users perceive the system, thereby giving the system an existence independent of the users, as is true of a physical system. And, after all, a functional system tells us what a physical system does, so why should this version of the definition not be appropriate for the functional system also?

To see why this is not so, we need to recall our development of the concept of a functional element. It is true that we started out by recognizing that there are two aspects to describing a physical object — what it *is* and what it *does* — and this led us initially to the concept of a functional element. But then, in order to use this concept in our top-down design methodology, we abstracted from the physical origin of the concept and defined a functional element as "something" (that exists only in our minds, i.e., a thought element), which produces a service that satisfies a set of user requirements. That is, the concept of a functional element is now anchored in a set of user requirements rather than in physical reality, and this should be true of the concept of a functional system also. Taking the user requirements as the point of departure for our design, we would expect that systems resulting from different user requirements would be different, and this is reflected in definition 5.4. To illustrate this, consider two systems, A and B, such that X(A) = X(B) and

$Y(A) = Y(B)$, but $Z(A) \subset Z(B)$. As the elements in the two systems are identical, this means that some of the possible interactions between elements and the users (i.e., interactions inherent in the elements) are not used and remain undefined. In effect, user group A is saying "Yes, we want this service, but we are not interested in these aspects of it, we don't care what values these functional parameters take on." In particular, the parameters in the complement of Z(A) in Z(B) do not appear in the value function, and it would be a fluke if the optimization process with and without these parameters led to the same result, seeing that the parameters are coupled via the functions within the elements.

This brings up an important philosophical issue about who carries the responsibility for the outcome of the design process. The view advocated in this book is that the engineer is responsible for meeting the user requirements and that the values of parameters that are omitted from the user requirements may be chosen so as to minimize the cost. Of course, there are generally accepted requirements, such as statutory and other legal requirements, that the engineer is expected to be familiar with and comply with, but beyond that it is not appropriate for engineers to impose their own values on a design, nor should the users expect them to do so. Saying "we don't care" about some aspect of a design does not relieve the users from their responsibility for that aspect. But what about aspects that the users are not aware of, but that either are or become known to the engineers? This is where the importance of the requirements definition process comes in, a process that is very much a part of engineering and where the engineer has a leading role in both facilitating the process and ensuring that the result is unambiguous and complete (see, e.g., *The Changing Nature of Engineering*[5]). However, this is a process completely separate from and very different in nature from the design process, and getting the two confused can lead to unsatisfactory results and serious contractual difficulties.

Engineering is a creative activity, but it is a different creativeness from that of an artist. In a sense the two activities lie at opposite ends of a scale that measures the degree to which the creative activity is motivated by a desire to express oneself versus the desire to meet a challenge posed by someone else. The artist is creative for the sake of creativity itself; the engineer is creative as a means of solving a problem or meeting a challenge. Architecture lies somewhere between the two. In this connection, one could also consider that the two activities lie at opposite ends of a scale that measures the importance of previous work, with engineering heavily dependent on previous, successful designs (and the lessons learnt from the unsuccessful ones!).

Returning to the subject matter proper, we have noted that applying the partitioning operator to a functional element and thereby ending up with a representation of the element in the form of a system (of elements) must leave the external interactions intact. Consequently, the external interactions in the definition of a system are subdivided into the same three types as are the variables associated with an element: functional parameters, consequential variables, and influences. In definition 4.22 we defined the concept of a service type: two elements belong to the same service type if they have identical sets of functional parameters. Within the set of elements belonging to a particular system we can define another relation:

Definition 5.5: Within the set of elements defining a system, two elements a and b are *equivalent*, $a \leftrightarrow b$, if and only if interchanging them leaves the values of the system functional parameters unchanged.

Is there any difference between two elements being equivalent and them being identical? Yes, in a particular position within a system, the functionality inherent in an element may only be partially utilized, that is, some of its parameters are not used to interact with other elements, and this element is therefore equivalent to a simpler element that does not have this functionality in the first place. Note, however, that this relation is not transitive; if $a \leftrightarrow b$ and $b \leftrightarrow c$, this does not imply $a \leftrightarrow c$, and equivalent elements do not form an equivalence class (whereas elements providing the same type of service do). An easily visualized example of this is provided by people in an organization; a person on an intermediate level may have the capability (i.e., inherent functionality) to replace both a person on a lower level of the organization and one on a higher level, but the person on the lower level may not have the capability to replace the person on the higher level.

But why introduce this notion of equivalence? If our aim is to reduce the complexity of the design process, would we ever want to use unnecessarily complex elements with unused functionality instead of elements with just the required functionality? The answer is to be found in the concept of structure, which will be introduced and discussed in the next section, but to end this section, we observe that equivalence allows us to subdivide functional systems into two types, homogeneous and heterogeneous:

Definition 5.6: A system is ***homogeneous*** if and only if all its elements are equivalent, otherwise it is ***heterogeneous.***

Homogeneous systems are of particular interest in studying emergent properties; heterogeneous systems range from systems with only two groups of equivalent elements to completely heterogeneous systems, where no two elements are equivalent.

5.4 Structure of Systems

In discussing the definition of a system, we emphasized that the interactions between elements do not represent any functionality or have any parameters associated with them in their own right, they only indicate which possibilities for interactions between the elements are in fact realized in this particular system. That is, they are "logical" interactions; they are either present or not. But there could be more than one sort of interaction between two elements, depending on which variables in the two elements are linked by the interaction, as the definition of interaction allows more than one variable in each element to be linked, and so it would appear that an interaction will need some type of parameterization or indexation, that is, be characterized by a variable that can take on more than two values.

For our present purposes, we will consider the interactions to be purely logical, present or not present, in the sense that we will not differentiate between interactions between the same two elements. The situation is analogous to representing the level of water in a tank by a binary number; if the number has only one digit, we can only differentiate between full or empty (which might be defined in different ways, but typically more or less than half full), whereas with more digits we can represent the level in greater detail. In the case of functional elements, the equivalent of increasing the number of digits is to subdivide the elements into smaller elements, until finally there is only at most one possibility of an interaction between any two of them.

With this understanding, we can introduce the concept of an adjacency matrix,

Definition 5.7: In a system with n elements, x_i, with $i = 1, ..., n$, the *adjacency matrix*, **A**, with elements a_{ij}, with $i,j = 1, ..., n$, is defined by

$$a_{ij} = \begin{cases} 1, \text{if there is an interaction between elements i and j} \\ 0, \text{otherwise} \end{cases}$$

with the convention that $a_{ii} = 0$,

and the concept of structure follows immediately:

Definition 5.8: Two systems have the same *structure* if and only if their adjacency matrices are identical.

In this simplest definition of the adjacency matrix the interactions are between elements, not from one element to another, so the interactions have no direction, and the adjacency matrix is symmetric, that is, $a_{ij} = a_{ji}$.

The concept of structure in the functional domain is very similar to the concept of structure in the physical domain, and we are familiar with many examples of both. The structure of molecules arising from the interactions of the constituent atoms and the structure of crystals arising from the interactions of molecules are examples of structure in the physical domain that come immediately to mind, but there are many other, perhaps somewhat subtler ones, such as the structure of an organism or of a building. Similarly, in the functional domain (and, more generally, in the domain of ideas and concepts), the concepts in terms of which we understand Nature had been structured already by Aristotle; Kant developed a structure for his categories (four groups of three categories each); and grammar, as the structure of sentences, was developed by Chomsky. An organization can be either physical (i.e., with people as its elements) or functional (with functional elements, i.e., job descriptions, as elements), and in both cases we speak of a structure.

In both domains, the concept of structure is used in two different ways. The first relates to properties or features of systems that depend on structure alone, that is, properties that are independent of the nature of the elements. For example, crystals that belong to the same symmetry group have features, such as birefringence, in common. In the functional domain, a well-known example is the expression for the reliability of a system; it depends on the structure of the reliability block diagram (and reliability blocks are functional elements), but is independent of both the physical nature of the objects represented by the elements and the other aspects of their functionality.

Another feature of systems that depends on structure only is the connectivity of a system:

Definition 5.9: The *connectivity* of a system with adjacency matrix **A** is given by the value of the expression

$$\frac{1}{n(n-1)}\sum_{i,j} a_{ij}$$

This expression can, as it stands, take on values between 0 and 1. However, in order for a set of n elements to form a single system there must exist a path from any element to every other element, otherwise the set decomposes into two or more *disjoint* subsystems (i.e., separate systems), which means that the minimum value of the connectivity is $2/n$.

The connectivity is, on the one hand, a measure of the internal complexity of the system; on the other hand, it is a measure of the resilience of the system to the failure of individual interactions. But the exact interpretation of the connectivity, for example, as a measure of redundancy, is only possible once the details of the elements and of the interactions are known. Figure 5.2 shows four different systems consisting of four elements, each with a different structure, and with the connectivity of each indicated.

The second use of structure, and the one that has particular significance in the functional domain, is to consider some of the changes that take place in a system as a result of random failures and subsequent repair as changes to the structure, rather than changes to the elements themselves. Issues relating to maintained systems will occupy all of chapter 8; here we only want to develop the foundations for using the structure concept to describe such changes. The first step is to consider all the systems that arise from a given system by removing one or more of the interactions between the elements; this will be called the associated set of systems.

Definition 5.10: System A is *associated* with system B if the set of internal interactions of system A is a subset of the set of internal interactions of system B, while the other two sets are identical between the

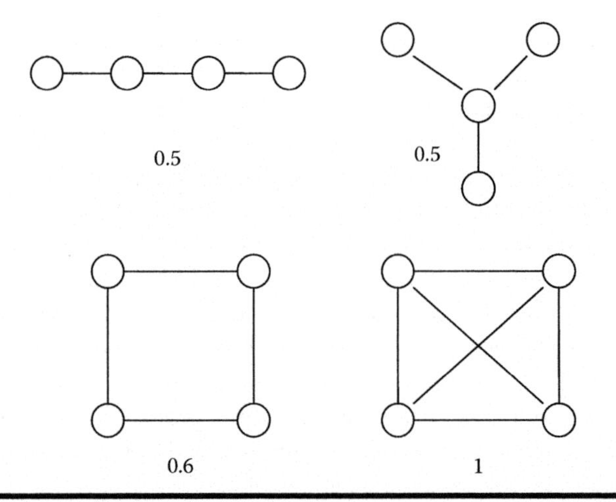

Figure 5.2 Four different structures of a four-element system. The number is the value of the connectivity for the structure.

systems, i.e., X(A) = X(B), Z(A) = Z(B), and Y(A) ⊂ Y(B). The set of all systems associated with system B is the *associated set* of system B.

It is straightforward to show that the associated set, which includes the system itself and the sets of disconnected elements (which are not systems, but only called so for convenience in the present context) has

$$\sum_{x=0}^{\xi} \frac{\xi\,!}{x!(\xi - x)!}$$

members, where ξ is the cardinality of the set Y of interactions, that is, the number of interactions between elements of the system. (This expression is just the number of ways we can select x elements out of a set of ξ elements, summed over all values of x.)

Consider a system and its associated set. If the elements of the system are independent of time (i.e., all their internal parameters are constants), the only changes to the system are that one or more of the interactions may be inactive. That is, the system is represented by one of the members of its associated set, and the member may change as a function of time. The service produced by the system is described by a set of parameter values, and each member of the associated set will produce this service with different values. Consequently, there is a one-to-one correspondence between the set of system states and the associated set, and a sum over states can be replaced by a sum over the associated set, with each member having a certain probability of occurring (and the sum of these probabilities over the whole set being equal to one, of course).

So far, nothing has been gained by introducing the associated set. However, in many systems there are groups of identical elements, and the structure of the system displays certain symmetries. The concept of *symmetry,* that is, the invariance of properties of a system under certain permutations of its elements, is as important in the functional domain as in the physical domain, but whereas in, for example, a crystal the permutations occur through actual rotation or reflection in space, in the functional domain the permutations are symbolic, as is the visualization of the symmetry, but the symmetry is very real in the expressions for the property in question. The members of the associated set can be ordered into groups of systems that produce the same service and have the same probability of occurring, and a sum over states is reduced to a sum over groups, with corresponding multiplicity factors.

We can illustrate this by a very simple example, a system consisting of four elements with the maximal connectivity, as shown in Figure 5.3. The associated set contains, according to the above formula, 64 members, ordered in Figure 5.3 according to the number of interactions that are removed from the original system. If two of the elements are identical, then the associated set is broken up into 24 groups of two associated systems each, plus 14 single associated systems, as follows (for the case when the two bottom elements are identical):

(1a,1f) (1b,1d) (2a,2j) (2b,2n) (2c,2i) (2d,2m) (2f,2o) (2h,2l) (3a,3t) (3b,3g) (3c,3r) (3d,3e) (3f,3s) (3h,3q) (3i,3p) (3j,3o) (4a,4j) (4b,4f) (4c,4i) (4d,4h) (4e,4o) (4q,4m) (5b,5f) (5d,5e)

If three of the elements are identical, the associated set is broken up into eight groups of three, four groups of nine, and four single associated systems each, as follows (for the case when the two bottom and the upper right-hand elements are identical):

(1a,1c,1f) (1b,1d,1e) (2b,2e,2n) (2g,2h,2l) (4b,4f,4i) (4c,4k,4l) (5a,5b,5f) (5c,5d,5e) (2a,2c,2d,2f,2i,2j,2k,2m,2o) (3a,3d,3e,3f,3i,3k,3p,3s,3t) (3b,3c,3g,3h,3j,3m,3o,3q,3r) (4a,4d,4e,4g,4h,4j,4m,4n,4o)

and if all four elements are identical, the associated set consists of six groups, with (1, 6, 15, 20, 15, 6, 1) members, respectively. Thus, the number of terms in a sum over states reduces as a function of the symmetry (identical elements), as shown in the following table:

Identical elements	0	2	3	4
Terms in sum	64	38	16	6

Clearly, this approach can result in a significant reduction in the work involved in calculating a sum over states (e.g., the expectation value of a functional parameter).

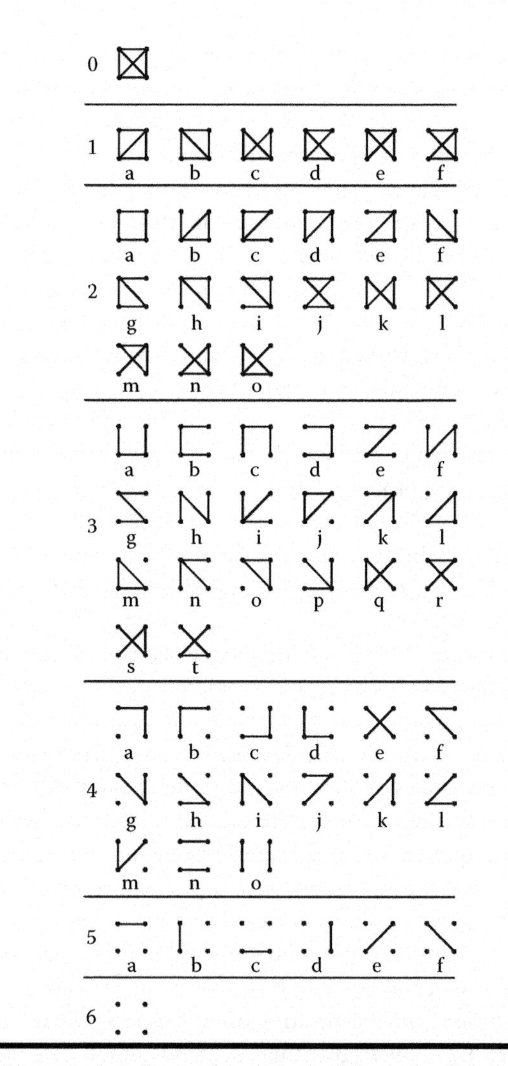

Figure 5.3 The associated set of a system of four maximally connected elements, ordered according to the number of inactive (or failed) interactions.

5.5 Systems of Imaginary Elements

The top-down design methodology relies on two different means of reducing complexity — the existence of simple functional elements with global parameters that still make significant statements about system properties, and the consideration of aspects of systems in isolation as a means of deciding between different options for further development of the design. The former will result in a hierarchy of real functional elements, starting with the irreducible element at the apex of the hierarchical structure, and with the branches of the structure arising from an increasing level of detail in the description of the functionality coupled with an increasing degree of specialization of the functionality itself. The latter will result in a similar hierarchy of imaginary functional elements, but when it comes to forming systems, the two types of elements show considerable differences.

A system of real functional elements arises from the partitioning of a complex element into a set of less complex, but interacting elements, and these elements may have very different functionalities in order to represent the complex behavior of the original element. With the original element there are associated a number of aspects, such as, for example, cost, reliability, productivity, etc., each of which may be represented by an imaginary functional element if we want to consider this aspect in isolation. However, while, as a result of the partitioning, the elements representing this aspect will be at a greater level of detail, they will still be representing the same aspect; they have only resulted from subdividing the original imaginary element, and the partitioning proper refers to real elements only. The "interaction" between imaginary elements is a simple one, such as addition (for cost elements) or logical combination (for reliability elements); the resulting "system" does not have a structure of its own, but mirrors that of the system of real elements, and there is no question of emergent properties.

Nevertheless, such sets of interacting imaginary elements meet all the requirements for being considered as systems, and in many ways they provide the best-known examples of functional systems. Models that represent particular aspects are commonplace: just think of cost models, reliability models, safety (or risk) models, distortion models (for transmission/amplification), delay models (in transport systems), and so on. However, their nature as functional models has been obscured by the way they have been applied, in modeling the behavior of a particular physical object, as a means of analyzing the behavior of a physical object to see if it meets given requirements. What is called a functional element in one case and a model in the other is exactly the same thing, in the sense of being a collection of variables and functions between them, and indeed, the reason for using them is also the same in both cases (simplification by concentrating on one aspect); what is very different is how and to what they are applied. By using the word "model" we imply that they represent a feature of a real physical object, and that we can verify the accuracy or otherwise of the model by making measurements on the object. A functional element, on the other hand, does not refer to any particular physical object; it was derived from a reference to a class of objects, but the application of them in our design methodology takes place *before* there is any consideration of physical realization. They are purely abstract.

The distinction between model and functional element may at first seem contrived and irrelevant, but that is mainly because we are so overwhelmingly used to the "model" interpretation that we find it difficult — for many almost impossible — not to automatically visualize the functionality in physical terms, in which case there is no distinction, of course. Our technology has been developed by studying and understanding Nature; the physical world has been the origin of everything we have developed so far. All our theories and representations, no matter how abstract they appear to be, are anchored in the physical world. What we are attempting to do with our top-down design methodology is to make functional user requirements the primary entities, with the eventual physical realization a secondary matter.

The above statements regarding the use of models in engineering need to be qualified in (at least) two regards. The first is in regard to software engineering, and in particular object-oriented software engineering. While individual objects themselves are often representations of physical objects, such as persons, accounts, etc., they are seen as instances of abstract *classes*. Such classes are very close in nature to our functional elements, and building models of software applications in terms of them is a cornerstone of modern software engineering.[6] The second is in regard to complex systems, mentioned earlier, where abstract models form the basis of studying types of behavior. Two very simple examples of such models are included in chapter 8, illustrating what is really a seamless interface between our approach and the current work in complex systems science.

Notes

1. Aslaksen, E.W. and W.R. Belcher, *Systems Engineering*, Prentice Hall, 1992, p.8.
2. Complex systems engineering and complexity theory have seen a huge upsurge in activity in the last decade, and this series of books is a testimony to that. For a discussion of recent developments and numerous literature references, a special issue of the INCOSE *Insight* newsletter, 11(1), January 2008, is a good start. For an in-depth discussion of the relationship between traditional and complex systems engineering, the reader is referred to another book in the present Taylor & Francis series, *Model oriented systems engineering: A unifying framework for traditional and complex systems*, by D. Hybertson. The relationship between complex systems and enterprise systems engineering is considered in Kuras, M.I., and White, B.E. Engineering enterprises using complex systems engineering, *Proceedings of the 15th Symposium of the International Council on Systems Engineering*, Rochester, NY, July 2005.
3. A very engaging discussion of unexpected happenings, our attitude to them, and how that attitude prevents us from adjusting our behavior appropriately can be found in *The black swan: The impact of the highly improbable*, by N.N. Taleb, Random House, New York, 2007. And with regard to abstraction, in the "Learning to learn" section of the Prologue, Taleb says "We don't learn rules, only facts.... We scorn the abstract, we scorn it with passion."
4. A very readable introduction to the implications of some of the developments in complex systems science is the book *The end of certainty*, by I. Prigogine, Free Press, New York, 1997.
5. The requirements definition process and the structure of requirements documentation have been major topics in systems engineering from its inception, and in INCOSE there is a separate working group dedicated to this subject matter; see www.incose.org/rwg. An extensive bibliography is available at www.ida.liu.se/labs/aslab/people/

joaka/re_bib.html, and an approach used on industrial projects is described in Aslaksen, E.W., *The changing nature of engineering*, McGraw-Hill, New York, 1996, Sec. 9.2.

6. Jakobson, I. *Object-oriented software engineering; a use case driven approach*, rev. print. ed., Addison-Wesley, Reading, MA, 1993.

Chapter 6

Properties of Systems

6.1 System States

Seen from the outside, a system is identical to a functional element; the representation of a functional element in the form of a system is just a means of reducing its complexity by introducing an internal structure. Consequently, the external interactions of a system are the same as the interactions of an element with the rest of the world, and can therefore be subdivided into the same three types of interactions as was done for elements in chapter 4. That is, we have functional parameters, dependencies, and influences, and, most importantly, these sets of (system) variables are subsets of the union of the corresponding element variables. As we know, the interactions themselves do not contain any functionality and do not introduce any new variables, and any so-called *emergent properties* of a system can be expressed in terms of the variables of the constituent elements (albeit that it may, as stated earlier, be more *convenient* to introduce some new variables that are functions of the element variables, much in the same sense that it is more convenient to transform to a new set of coordinates).

The issue of emergence and emergent properties of systems has been (and, to some extent, still is) the subject of discussion within the systems engineering community.[1] We shall take a very simple and straightforward view of this, consistent with our understanding of a system as a mode of description: Consider an engineered object and a description of it in terms of a system, that is, as a set of interacting elements. Then the emergent properties of the system are those that are not present if the interactions between the elements are inhibited. That is, emergent properties are only defined relative to a particular description of an object as a system. In the case of the description of an object as a single element there are no emergent properties; all the properties of the object are simply that.

As was the case with elements, it is the functional parameters that are of particular importance, and for the same reason; the point of departure for our design methodology is a set of user requirements expressed in terms of these functional parameters. And, as for elements, we will introduce the concept of a system state:

Definition 6.1: A system state, φ, is a particular combination of (system) parameter values. The set of all system states, $\Phi = \{\varphi\}$, is called *state space*.

The relationship between system states and element states can be developed further if we introduce the concept of basic system states:

Definition 6.2: Let a system consist of n elements, each with a set of states E_i, with $i = 1, ..., n$. Then the set of *basic system states*, Φ_0, is defined as the Cartesian product of the sets of element states $\Phi_0 = E_1 \times E_2 \times ... \times E_n$.

The sought-after relationship is expressed by the following theorem:

Theorem 6.1: There exists a one-to-one correspondence between the set of system states and a partitioning of Φ_0 into mutually disjoint subsets.

Proof: The proof proceeds in three steps:

1. Let U_i be the set of functional parameters characterizing the i-th element, and U the set of all element functional parameters $U = \cup U_i$. Let W be the set of functional parameters characterizing the system, then $W \subset U$. Assuming that this is incorrect, there then exists a system functional parameter that depends on at least one functional parameter that is not also a functional parameter of one of the elements. That implies that the interactions between the elements create functionality not already inherent in the elements, which in turn contradicts the definition of a system.
2. Let σ be the mapping $\sigma: \Phi_0 \rightarrow \Phi$. Then, for each $\varphi^0 \in \Phi_0$ there exists one and only one $\varphi \in \Phi$ such that $\sigma(\varphi^0) = \varphi$. That is, σ maps onto Φ (but not generally one-to-one). From (a) it follows that giving values to all the element functional parameters determines the values of the system functional parameters, which, by definition 6.1, uniquely defines a system state.
3. Let the equivalence relation ~ be defined by $\varphi_1{}^0 \sim \varphi_2{}^0$ if and only if $\sigma(\varphi_1{}^0) = \sigma(\varphi_2{}^0)$. This equivalence relation partitions Φ_0 into mutually disjoint subsets, and the mapping $\sigma(\varphi^0) \rightarrow \varphi^{0*}$, where φ^{0*} is the equivalence class containing φ^0, is a one-to-one correspondence according to (b).

The content of this theorem can perhaps best be illustrated by a simple example (which we will come back to in chapter 8). Consider a system consisting of *n* equal elements, each contributing an amount Q to the service being produced by the system, and with each element being characterized by a single parameter that can take on only one of two values, 1 if the element is operating, and 0 if the element has failed. Consequently, there are n element parameters, and 2^n basic system states (i.e., Φ_0 is a discrete set with 2^n members). Assume now that the users are only interested in whether the service level equals or exceeds mQ or not, with $m < n$, so that there is only one system parameter, s, which also takes on only two values, 0 and 1. The subset of Φ_0 that corresponds to the state $s = 0$ then has a number of members, depending on m, which equals

$$\sum_{x=0}^{m-1} \frac{n!}{x!(n-x)!}$$

and we could say that these basic states have *condensed* into the one system state.

As this example shows, and as is also implicit in both the theorem and in the system concept itself, there are always fewer system parameters than element parameters. But it is not possible to formulate a theorem similar to theorem 6.1 for the parameters; there is, in general, no correspondence between system parameters and disjoint subsets of the element parameters. If we visualize the content of theorem 6.1 by graphically overlaying Φ on Φ^0, as shown in Figure 6.1 for a simple case where the system has only six states, then it is tempting to do the same for the parameters by overlaying W on U (as defined in the proof), but this is obviously not correct. It is important not to confuse the partitioning operator with the decomposition process; the first applies to elements, the second to parameters.

This also explains why theorem 6.1 in no way excludes emergent properties; in general, the set of system functional parameters includes parameters from more than one element. However, and this is most important to note, even if the system parameters are those of a single element, the *behavior* (i.e., time dependence) of those parameters can be completely different to that of the isolated element, due to the interaction between elements.

The reduction in the number of parameters and in the number of states that occurs when a set of interacting elements (i.e., a system) is represented as a single element are both aspects of what, in the area of object-oriented software design, is called *information hiding.*

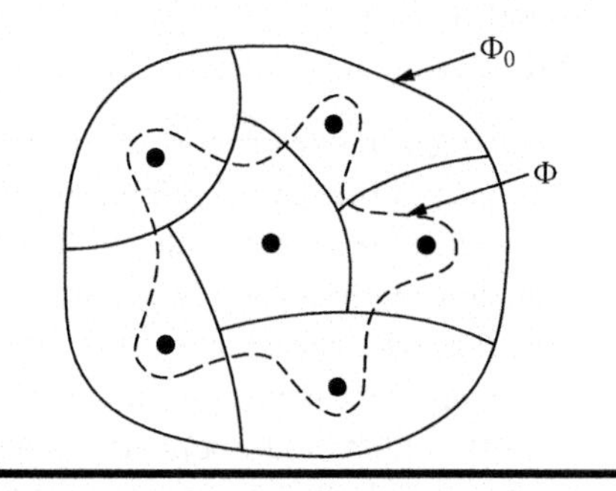

Figure 6.1 The set of system states, Φ, here shown as six dots, overlaid on the set of basic system states, Φ_0, subdivided into the six corresponding subsets.

6.2 Changes of State

If the concept of a system state is going to be useful in our design methodology, we have to understand what determines the state of a system or, conversely, what makes a system change its state. But before doing so, we need to return to an issue that has been mentioned briefly a couple of times in earlier chapters, and that is the role of people in our methodology and view of a physical object as a system. Already in section 1.5 we distinguished two groups of people: The ones within the system carrying out roles as parts of functional elements, such as operators and maintainers, and the ones outside the system, but who determine the success of the system in the sense of its meeting their requirements; they are the stakeholders. In our view, the people within the system are treated in a mechanistic fashion; their behavior is considered to be deterministic, albeit that it might require many parameters to describe it and that many of them might be in the form of distributions. In other words, they are treated as no different from other (non-living) components of the elements; the description of the behavior of such components may also include distributions, such as for the time between failures. Achieving the desired behavior is an important part of the engineering effort; operators and maintainers are not designed and manufactured, they are trained, and designing that training is just as much part of the system design as designing an electronic circuit.

This view, which is anathema to true complex systems engineers, is valid simply because we have limited the applicability of our methodology to systems where the purpose of the system is achieved mainly by means of the non-living components; the people play mainly a supporting and supervising role. In some of these systems it is even true that some operator functions are redundant and exist only due to stakeholder perception, such as drivers on trains. In true complex systems, mainly in the form of enterprises, it is the other way around, and the equipment plays only a supporting role.

We can now immediately discern three main causes of change:

1. Changes to one or more of the elements of the system during design, that is, as part of the optimization process. Each possible design option will result in a particular system state and therefore a particular value of the ROI, and our design methodology is a structured, systematic approach to identifying the option with the greatest ROI.
2. Changes occurring during "operation" of the system, for example, through wear, failure, and maintenance actions, but possibly also through changes in the operating environment, as described by the influences and dependencies. Of course, we cannot be talking about real operation here, as there is as yet no physical system; it is the reflection of the time-dependence of the functional elements. It is another example of the same type of abstraction that allowed us to define functionality, and operation in this context is the behavior of functionality under the passage of time (i.e., a model of the intended performance).

3. Changes that occur as a result of actions by the users (i.e., that subset of the stakeholders that use the service provided by the system). For example, the ROI depends on the revenue, which depends on the value function, which depends again on the attitude of the users (or a subset of them).

Let us examine this last type of change first, because it reveals an interesting characteristic of the interface between a system (i.e., service provider) and the users, or, more precisely, a statement about the consistency of the stakeholder requirements. We had previously noted that to each functional parameter there belong two types of values, the rated value and the actual value. We now note that some parameters have only the first type of value; typical examples are such functional requirements as "the total harmonic distortion will not exceed 5%," "the capacity will be adequate for a 1-in-100-year storm," and "the repetition frequency will lie in the range 1–4 per month." These values are not changed or influenced by the users during operation. However, a parameter such as power output has a rated value, say 500 MW, which is not changed or influenced by the users during operation, but it also has an actual value (i.e., the load) determined by the users, say, anywhere in the range 50–500 MW. There would be other stakeholder requirements that depend on the actual value of the power output; for example, the ROI would depend on the value over time, that is, the duty cycle, and it is not possible to determine the optimal system solution without knowledge of this duty cycle. If that information is lacking from the stakeholder requirements, they are not *consistent*; that is, it is not possible to demonstrate that all the requirements are met.

This check on the consistency of the stakeholder requirements as the system of functional elements is developed is a side benefit of the methodology. But it is more than that; it brings out an important aspect of the very early phase of a project, what in defense is called *capability development* and in other industries something like *business case development*. In principle, this activity starts with some of the future stakeholders that want to do something, to carry on a business. In analyzing what it takes to be able to do this, they find that they need certain capabilities, and that some or all of these will require the application of technology. They will also require the involvement of other parts of the community, in the form of suppliers, partners, distributors, etc., so that at the end of this early phase, the description of the required capability has evolved from "a good idea" to a complex model not only of the business activity itself, but, above all, of the environment in which it is to take place and the interaction between the two. Both of these will in general be complex in the sense of being described by many variables, being dynamic, and being unpredictable, and that complexity needs to be *reflected* in the stakeholder requirements for the systems that are to support the business activity. The enterprise carrying out the business is the complex system, in the sense of the other books in this series; it is the system that has to be agile and adapt to the variations in its environment. However, the designer of the supporting (or infrastructure) systems that our methodology targets needs to understand this complexity and the strategy the enterprise is intending to adopt to handle it, in order to optimize the design. The adaptability of these systems has to be designed in; they are not, in general, able to accommodate unforeseen changes in their operating environment.

The result of this is, then, that the requirements of the service itself are only part of the functional stakeholder requirements; the other part is made up of the numerous additional requirements arising from the conditions under which the service is to be provided, and it is the formulation of the implications of these additional requirements of the functionality of the system that involves the actual values of some of the parameters and opens up a whole new dimension of complexity as far as the design process is concerned. A different way of expressing this same change to what is expected of the design is that, whereas there used to be an information wall between the designer and the client, with the client specifying what he wanted, but not why he wanted it, today that wall is disappearing, and "client satisfaction" demands a much greater influence and reflection of the client's business in the design process.

The second type of changes of state, item (2) above, are intrinsic to the system, but they generally depend on external influences, which may range from physical variables, such as temperature, to such socio-economic factors as educational level or income level. They may also depend on the input variables, so that the changes of type (2) and type (3) are not independent. Changes of type (2) are always present, because only an ordered collection of elements can produce a service (complete disorder produces nothing), and any order is subject to deterioration (i.e., the entropy cannot increase) and must be restored by carrying out maintenance, but their importance may vary considerably from case to case. Chapter 8 is dedicated to studying the effects of changes of type (2).

From the point of view of our purpose, to develop the foundations of a top-down design methodology, changes of the first type, item (1) above, are of particular interest. They arise because, at every point in the design process, the designer has a number of options when it comes to fulfilling the requirements at that level, and each option must be characterized by having different values of some variables. But which variables are they? They cannot be the functional parameters, nor the influences, and the dependencies arise because they are demanded by the functional parameters. So that leaves only the functions that relate the variables to one another — in other words, the choice of elements used in the partitioning and the values assigned to the parameters involved in describing the functionality of them. The choice of elements, that is, the choice of partitioning, is, of course, the essence of the design process, and remains a creative activity dependent on the individual designer. The purpose of our methodology is to support this process by, on the one hand, offering a collection of functional elements that are useful in very many applications, and, on the other hand, giving a clear prescription for how to determine the optimal solution. Our approach to both of these objectives is further advanced in part C of the book.

6.3 Service Density Function and Superspace

Returning to our comments on the changes of type (a), we recognize that our descriptions of functionality so far, as exemplified by the irreducible element, contain a very major simplification: The descriptions are *static*, and the parameter values must, in some way, be averages over time. If we now want to take the time dependence into account, the previous discussions (starting already with section 1.4) show that this time dependence must be of two kinds: an explicit dependence, arising, for example, from a defined duty cycle, and a statistical dependence, arising from the stochastic nature of some of the parameters, such as the time of failure. In addition to this statistical time dependence, there is uncertainty attached to the dependencies, arising from the lack of detailed knowledge early in the design process and from the fact that design addresses the future. It is common practice in engineering to make a distinction between the two types of unpredictable variations, that is, stochastic and due to uncertainty, and to treat the two somewhat differently. The latter is usually treated by means of sensitivity analysis and some associated decision criterion, such as Minimax or Hurwicz' criterion,[2] the stochastic variations are handled by making the actual values of the relevant functional parameters (i.e., those that are not given as only fixed values, as discussed above) random variables and assigning a density distribution function $\varphi(x,t)$ to each output parameter x, such that the probability of x having a value between x' and x' + dx at time $t = t'$ is given by

$$P(x' \leq x \leq x' + dx; t = t') = \varphi(x',t')dx,$$

with the condition

$$\int \varphi(x,t)dx = 1, \text{ for all t.}$$

The introduction of this probability density distribution function needs some further consideration. What does "the probability of having a value ...." actually mean? It can only mean that if we had a very large number (in the limit, infinitely many) of identical systems operating under identical conditions (called an ensemble in statistical mechanics), and we measured the value the parameter x at time t = t' for all of them, then $\varphi(x',t')dx$ is the proportion of the measured values that would lie in the range $x' \leq x \leq x' + dx$. But, given that in most cases we have either a single system or a small number of identical systems, what is the practical use and interpretation of φ? The common interpretation is that the distribution over many systems at one point in time is identical to the distribution over many points in time for a single system. For this to be true, however, there can be no t-dependence in φ; the distribution must be *time invariant*. While this may be true for some systems, and may be a very convenient approximation in many cases, we know that it cannot be true in general; just think of all the systems that are subject to *ageing*. The approach to this issue comes through realizing that while the distribution $\varphi(x)$ provides information about the fluctuations in x, it is not an exhaustive characterization of these fluctuations; in particular, it gives no information about the frequency or time dependence of the fluctuations. This is best illustrated by a very simple example:

Let the parameter x be restricted to the range $0 \le x \le 1$, and assume that the fluctuations in the value of x are periodic, with period τ, and in the form of a square wave with $x = 1$ for a fraction a of the period and $x = 0$ for a fraction $(1 - a)$ of the period. In this case, $\varphi(x) = (1 - a)\delta(x) + a\delta(x - 1)$. Now, if τ is very much smaller than the lifetime of the system, say T, then it is justified to set the ensemble distribution equal to the temporal distribution. But if τ is of the order of T, then this is not justified, and the concept of a probability distribution is not useful as far as system design is concerned, even though the distribution function is independent of τ. Now let a be a function of time, and let T' be the time it takes for the value of a to change significantly. Then the above reasoning holds, but with T' substituted for T, so that the t-dependence of φ only makes sense if τ is much smaller than T', or, in other words, splitting the temporal variation of x into a random part, characterized by $\varphi(x)$, and a part characterized by the time dependence of φ, only makes sense if the fluctuations are rapid compared to the rate of change in the probability distribution. This is an implicit assumption whenever the notation $\varphi(x,t)$ is used, and we recognize it as essentially the same limitation as we are familiar with from equilibrium thermodynamics.

It goes almost without saying that as a result of functional parameters having distribution functions attached to them, so do the aspects defined in terms of them.

Focusing for the moment on the case of the irreducible element, the only element we have defined so far, there is only one functional parameter that could have uncertainty attached to it, the quality of service, S, and if we denote the corresponding random variable by s, the distribution $\varphi(s,t)$ will be called the *service density function* (SDF). The explicit time dependence will vary from system to system, but there is one step in expanding the level of detail used in describing the irreducible element we can take without impacting the universal applicability of the element, arising from the fact that the system must be created before it can go into operation and provide its service. The distinction between creation and operation is therefore fundamental to the concept of a life cycle, and while the relative lengths of the two time periods in the life cycle may vary from almost all of one to almost all of the other, in general they both have to be present. They will be denoted by l_1 and l_2, and are measured in *accounting periods*. An accounting period may be any time period, but is most often a year. Also, it will be most convenient to choose the origin of time to be the transition between the two time periods, that is, the time at which the system goes into operation.

With these definitions, the relationship between $\varphi(s,t)$ and S is given by

$$S = \frac{1}{l_2} \int_0^{l_2} \int_0^{\infty} s\varphi(s,t)dsdt. \tag{6.1}$$

This introduction of the stochastic nature of the service delivered by the system is of fundamental importance to the view of engineering presented here; ignoring it by using a deterministic variable is an unwarranted simplification that has led to numerous erroneous claims about what engineering can and cannot do. That the behavior of the service may be deterministic at a level of much greater detail, say at the level of individual atoms, is a different matter. This level of detail is not available to the engineer, and therefore, *as far as engineering is concerned,* the stochastic behavior of the service is inherent. Of course, to this must be added any uncertainty arising from *averaging* over the known behavior of a variable, or from neglecting the influence of certain variables. This uncertainty is not due to any lack of knowledge, but to a deliberate simplification in the sense of an approximation.

At any particular time, the function $\varphi(s,t)$ becomes a function of s only. The distribution $\varphi(s)$ may be called a *superstate*, and the space of all such distributions may be called *superspace.* As time evolves, $\varphi(s)$ for a particular system will describe a trajectory in superspace; if $\varphi(s,t)$ is time invariant, this trajectory degenerates to a point.

The domain of $\varphi(x)$ is the set of values x can take on; in general this set will consist of intervals on the real line plus a countable number of discrete points. At a discrete point, say $x = x'$, $\varphi(x)$ will be proportional to $\delta(x - x')$.

The above expression for S is an example of a function on superspace; another example (availability) is given in the next section. Such functions are called *functionals*, and a whole special area of mathematics is devoted to their analysis.

6.4 Availability

While the superstate of a system is given by the service density function φ, the quality of service seen by the users at any particular point in time is some particular value of s, and when s goes below some minimal acceptable value, s_1, the system is said to have *failed.* An important aspect of the operation of maintained systems, which would include the majority of systems, is the *availability* of the service, denoted by A. It is the proportion of time the system is in its operating state and is, still limited to the description in terms of the irreducible element, defined by

$$A = \frac{1}{l_2} \int_{t=0}^{l_2} \int_{s=s_1}^{\infty} \varphi(s,t)dsdt. \tag{6.2}$$

In the most common case of time invariance as far as the maintenance regime is concerned, the service density function $\varphi(s,t)$ is a function of of s only, and the availability is also time invariant. Otherwise, the above expression defines the availability of the service as an average over the operating time period, and any refinement is only possible once the lifecycle is further subdivided, so that A can become a function of time.

A time-invariant availability, as defined above, makes perfectly good sense. This is not so with the concept of *reliability*; a time-invariant reliability makes no sense. Reliability is defined as the probability that the system will operate, under defined operating conditions, for a time period T without failure; it is an inherently time-based concept, and as such, *it is impossible to determine the system reliability from a knowledge of* φ *alone.* It is necessary to know the time dependence of the fluctuations, such as the value of τ in the simple example at the end of the last section, and in the next chapter we develop a model that is of fairly wide applicability, although there will be situations to which it is not directly suited.

6.5 The Basic Design Process Revisited

With the understanding we have now gained of functional elements and systems of such elements, we can add some detail to the steps of the basic design process (BDP) introduced at the end of chapter 3. In the first step, the *analysis* step, a part of the system represented by a single functional element at that point in the design process is *partitioned* into a number of smaller, interacting functional elements. From the discussion in section 4.4, we can now add that the element being partitioned, as well as the elements into which it is being partitioned, must be real elements. The partitioning process relates to the user requirements, and the elements at each level in the step-wise process satisfy subsets of the user requirements. However, the union of all these subsets does not normally equal the set of user requirements — there is a subset of the user requirements that is satisfied by the *interactions* of the elements; they are the *emergent* properties.

With the real elements at any one level there will be associated a number of imaginary elements that describe such aspects as cost and reliability. In the analysis step of the BDP the elements at one level get *subdivided* (rather than partitioned) into the elements of the same type at the next lower level; it is a process that is sometimes referred to as *requirements flowdown* or *requirements allocation.*

But what we now also understand is how this analysis step is supported in our methodology. Every real element can be condensed into an irreducible element, and that becomes our starting point. The irreducible element is then *expanded* by increasing the number of variables used to describe the functionality; that is, the single functional parameter quality of service (QoS) is expressed in terms of a few of its component parameters, and each one of these becomes the QoS of a new (smaller, simpler) element. At the same time, the other variables are expressed in terms of more detailed variables, and new imaginary elements are formed to describe those aspects that are of interest at this particular level of the design process.

Three comments need to be made with regard to this step. First, when we speak of our design methodology as a "top-down" methodology, we realize that this relates to the expansion from the general (condensed) to the detailed and not only (or even primarily) to the partitioning process. Second, the expansion of the irreducible element is clearly a central activity, and the whole next chapter is dedicated to exploring it. Third, and perhaps most important, there is usually more than one possible partitioning of a given element, and we have not given any rule for how to choose a partitioning or, more correctly, a set of partitionings, as discussed below. And there probably is no one, universally applicable rule; the choice will always, to some extent, rely on the experience and judgment of the designer. However, the task will be very much easier and less time consuming if we have a catalog of elements to choose from, and we have noted in several places throughout the development of our methodology that the creation of such a catalog of standardized elements is a prerequisite to a widespread acceptance of the methodology. Chapter 9 makes a small start on this task by proposing a practical development framework.

The second step in the BDP is the *optimization* step. We have already defined the optimal design as the one that maximizes the ROI, and because the partitioning preserves the definition of the ROI, this is in principle straightforward. But there are at least three issues that complicate the process. First, the optimization takes place on two levels — the choice of the best partitioning, and then the optimization of the parameters of the elements involved in that partitioning. The latter is the well-known problem of finding the maximum of a function of several variables, and needs no further elaboration here. But how do we know that the optimal partitioning is among the set of partitionings we have chosen to optimize? And how do we make that set as small as possible, in order to increase the efficiency of the process? As noted above, the main tool here is the existence of a catalog of standardized elements; this will certainly allow us to choose a small set of partitionings. But optimizing a couple of partitionings may be unavoidable, and this leads us to the second issue, the accuracy of the information available. All design in the functional domain is based on *estimates*, and the higher up in the top-down process we are, the less accurate are the estimates available to us. It may therefore happen that the accuracy of the optimization is not good enough to discriminate between two possible partitions, in which case both will have to be pursued to the next level down in the process, with a corresponding increase in the design effort. But maximizing the ROI may not be the only decision criterion, and may well not be the appropriate one at a particular level. For example, if the various possible partitionings represent basically different system architectures, then reliability may be the appropriate criterion; only one of the architectures can (at any practical cost) meet the reliability requirement. This is the third complicating issue — finding those aspects of the functionality that are particularly relevant to the choice of partitioning at that point in the design process, and this means choosing the appropriate imaginary elements. Again, a catalog of such elements is the answer (or at least a very substantial part of the answer).

The final step in the BDP is verification through *synthesis*, that is, the elements are combined through their interactions to form a system, and the performance of that system is calculated or simulated in order to demonstrate that it corresponds to the performance of the single element from which the partitioning into a system arose, much as one does in verifying a design in the physical domain through computer-based emulators or through building a prototype and testing it. In the case of the BDP this step is, of course, potentially almost trivial if we already have the models of the individual elements, *and if these models are all compatible,* in the sense that they have all been derived through the same top-down process so that all elements on the same level fit seamlessly together.

Notes

1. Recent discussions of emergence in systems engineering can be found in Hitchins, D.K., *Systems engineering: 21st century systems methodology*, John Wiley & Sons, 2007; Ryan, A., Emergence in systems engineering, *Insight*, 11, 2008, 23–24. Emergence in complex systems and, in particular, organizations, is the subject of a separate journal, see http://emergence.org/.
2. Aslaksen, E.W., and Belcher, W. R. *Systems engineering*, Prentice Hall, New York, 1992, section 5.3.3.

Chapter 7

Expanding the Irreducible Element

7.1 Introduction

So far, we have defined the concept of a functional element and developed many of the characteristics of this concept, but we have really only defined a single functional element, the irreducible element. According to our theory, we should now be able to develop all functional elements from this single element, and the last three chapters of this book are devoted to making a start on that task. However, before we start, it might be good to once again state the scope and nature of this task, as a clear understanding of the objectives is required in order not to lose our way.

The purpose of functional elements is to act as the elements of design in the functional domain, just as standardized components (bolts, bearings, resistors, ICs, etc.) act as the elements of design in the physical domain. And the criterion for making something into an element is the same in both cases — usefulness. There is no law of Nature that proves the necessity of an element in either domain; elements exist because they have proven themselves to be useful.

The usefulness of an element has two sources. On the one hand, it arises because the features of the element are those required in many situations. On the other hand, it arises from the fact that the element has achieved wide acceptance, possibly having been incorporated into a national or international standard. There is therefore something self-fulfilling about the usefulness of a standard design element, and this is particularly true when it comes to the exact details of an element. Many design elements would be just as useful if they were slightly different (e.g., a brick if it were 10% larger than a standard one), but would not be useful if they were not standardized. This same issue is vividly illustrated when it comes to measures; the concept of a standard unit of length is clearly extremely useful, but the choice of unit, say, in terms of a particular wavelength of light or the circumference of the earth, is largely arbitrary. If the meter were 10% longer than it is today, it would be just as useful; its usefulness lies solely in its universal acceptance. A final example is the way in which we express a number as a sequence of digits with a particular significance to the position of a digit; this way of expressing it would be just as useful if we had chosen the base as, say, eight instead of ten. The usefulness lies in the universal acceptance of the notation.

With this in mind, most of what is presented in this chapter is not new; on the contrary, in order to make it as acceptable as possible it conforms largely to what has been standard practice over many years.[1] However, because the details of that practice vary from author to author and from application to application, we want to be as prescriptive as possible in order to provide a starting point for a standardization process.

7.2 The System Life Cycle

Let us start the expansion of the irreducible element by considering the temporal dependence of the aspect expressed by the element, the return on investment. The fact that the return must come after the investment implies that time must play a significant role, and in the irreducible element time is represented by a single parameter — the system lifetime, L. This global parameter will normally be hiding a great deal of detailed information about the system, and the lack of this information limits the use we can make of the element. For example, we already saw, in section 6.2, that it was not possible to define the relationship between the basic variable S and the stochastic variable s without some further information about the temporal behavior of the system, and the same was true of the definition of availability in section 6.3.

But the parameter L is useful even if it hides much detailed information — it can be identified for every system, and it allows a number of quantities to be defined, such as total cost and total revenue. Any parameters resulting from a subdivision of the lifetime should meet the same criteria of usefulness — it must be possible and convenient to identify them for (almost) all systems, and forming averages and totals over their respective time periods must be significant and useful. So, if we consider the three other basic variables, we should try to identify time periods within the system lifetime in which summing or averaging these variables makes sense; these periods will be called *phases* of the *system life cycle.*

First, both the performance (i.e., QoS) and the revenue are only different from zero during that part of the lifetime the system is operational, so this period is significant when it comes to defining total revenue and average performance.

Definition 7.1: The operational phase is the time period that starts when the system has been created and is available to provide its intended service and ends when the system will not be available to provide its intended service at any point in the future.

However, we recognize that the lifetime of a system does not necessarily end when operation ceases. We cannot generally just walk away from a system at the end of its operational phase; there are often significant activities and costs associated with decommissioning a system. It is a second time period that can be clearly identified, and it is useful in defining, for example, total decommissioning costs.

Definition 7.2: The decommissioning phase starts at the end of the operational lifetime and ends when no further cost and/or revenue related to the system will accrue to members of the system's stakeholder group.

In the wording of this definition, we have deliberately narrowed our focus to the stakeholder group, as otherwise the effects of a system propagating outward through society and through time would leave the decommissioning period of most systems completely open ended. However, even with this limitation the decommissioning phase will include the product liability period and any litigation that may result from such liability.

There now remains the time period required to create the system, that is, the time period from the beginning of the system lifetime to the beginning of the operational lifetime. Within this phase we can distinguish two sub-phases in which the activities and associated costs are of quite different nature; the design and development phase and the implementation and test phase.

Definition 7.3: The design phase starts with the first expenditure attributable to the system and ends when the system design is fully documented and approved.

Definition 7.4: The implementation phase starts with the end of the design phase and ends with the start of the operational phase.

While the activities from which these two phases arise are quite different, the time periods during which they are actually performed may in many case overlap. That is, design may have finished and construction started on some part of the system while another part is still being designed. In the above definition, any such overlap is minimized by restricting the design in the design phase to system design, leaving detailed design to be carried out in the implementation period. Defining the transition between the two phases at that particular point in the system life cycle is also in conformance with common practice, as the financial decisions and the project go-ahead are most often tied to the end of system design (sometimes called preliminary design), and detailed design may be contractually combined with construction in a design and construct package. Whatever overlap is still present is included in the design phase.

As a result of the above considerations, we have expanded the single parameter L into a life cycle vector **L** with components l_i, with i = 1,2,3,4, as follows:

Design phase	l_1
Implementation phase	l_2
Operational phase	l_3
Decommissioning phase	l_4

The relationship between these durations and the basic parameter L is simply

$$L = \sum_{i=1}^{4} l_i \tag{7.1}$$

7.3 Cost Components

In the basic set there is a single variable, C, which characterizes all costs associated with the system and, as such, it hides all details about different types of costs and when these costs occur; it is a total cost over the whole lifetime of the system. To bring out the details in C, we start by dividing up the cost according to the time period in which it occurs and to characterize it further by when it occurs within each time period. The total cost can then be considered to be a vector, **c**, with four components:

1. Design and development costs, c_1
2. Implementation and test costs, c_2
3. Operating (and maintenance) costs, c_3
4. Decommissioning costs, c_4

Where a project consists of several systems of the same *type*, design and development costs are assumed to be incurred only once, whereas the implementation and test costs as well as the decommissioning costs are assumed to be incurred equally for each system.

Within a period, costs occur at various times, but it is more convenient and useful to divide time up into discrete periods, called *accounting periods*, and account for all costs incurred during an accounting period at the *end* of that period. The accounting period may be anything, with something between a week and a year

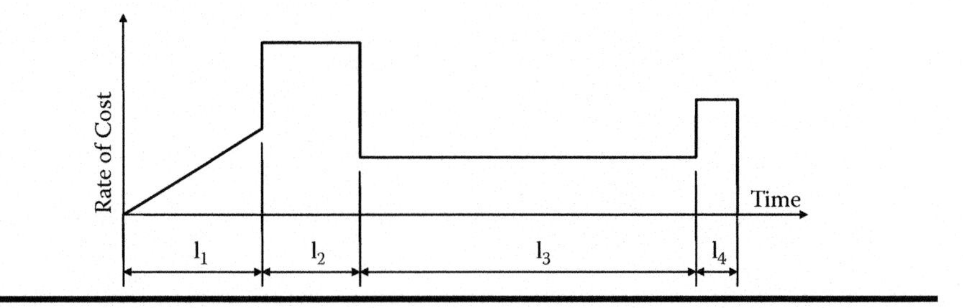

Figure 7.1 The assumed rates at which costs are incurred in the different phases of the system life cycle.

being the most common. The durations of the phases are then also measured in accounting periods.

Of the four cost components, c_3 is always given as a cost per accounting period, whereas the other three are lump sums for the corresponding phase; this is in accordance with common usage. The next step in detailing the costs is then to specify how the costs in the three phases are incurred, that is, at what rate. We shall assume that the implementation cost and the decommissioning cost are incurred at a constant rate, but that the development cost is incurred at a linearly increasing rate. This assumption is a good one for many projects, but even where it is clearly wrong, it can still be perfectly useful, because for any system feature that does not in itself depend on the manner in which the cost is incurred, just on the present value (see below), we can always define an *effective* cost that results in the same present value.

With the above assumptions, the rate of cost over the system life cycle will have the general form shown in Figure 7.1.

In order to operate with costs that are incurred at different times, we need to reference them to a common point in time. This point in time is often referred to as "the present," even though it can obviously not be the present in its true sense to the reader, and the result of translating a cost (or a revenue) to this reference point is called its *present value* (PV). We shall make the following definition:

Definition 7.5: The reference point in time for all PV calculations will be the start of the operational phase.

Finally, we need to give a name to the sum of the PVs of all costs:

Definition 7.6: The sum of the PVs of all costs associated with a system over its life cycle will be called its life cycle cost (LCC).

We are now in a position to transform the cost vector into the LCC by forming the *cost transformation vector,* $\boldsymbol{h}$, with components h_i, $i = 1,2,3,4$, given by

$$h_1 = \frac{2(1+p)^{l_1+l_2}}{l_1(l_1+1)} \sum_{n=1}^{l_1} n(1+p)^{-n}$$
$$= \frac{2(1+p)^{l_2}}{p^2 l_1(l_1+1)} [(1+p)^{(l_1+1)} - p(l_1+1) - 1] \tag{7.2}$$

$$h_2 = \frac{(1+p)^{l_2} - 1}{p l_2} \tag{7.3}$$

$$h_3 = \frac{(1+p)^{l_3} - 1}{p(1+p)^{l_3}} \tag{7.4}$$

$$h_4 = \frac{(1+p)^{l_4} - 1}{p(1+p)^{l_3+l_4}} \tag{7.5}$$

In these expressions, p is the *discount rate per accounting period.* As it is most common to express costs in constant dollars, that is, ignoring inflation, this discount rate is the actual interest rate (cost of money for the project) minus the inflation rate.

The LCC is then given by the scalar product of the two vectors,

$$\text{LCC} = \mathbf{c} \cdot \mathbf{h}\,. \tag{7.6}$$

As always, we need to demonstrate the relationship between these new, more detailed cost variables and the basic variable, C, and it is obviously given by

$$C = LCC(1+p)^{-(l_1+l_2)} \tag{7.7}$$

7.4 Subsystems and Cost Allocation

In the irreducible element, functionality is characterized by a single parameter, the quality of service, S, and as such, this parameter hides all the details of the processes taking place within the system in order to produce the service. These processes will, of course, vary from system to system, and the question is whether there is a more detailed description that is still of (almost) general validity and, most importantly, would serve a useful purpose in our design methodology.

We can approach this question by observing that any system represents an ordering of its parts; an unordered or random collection of parts is extremely unlikely to produce a specified service. And we also know that any order tends to decay (i.e., become disordered) due to random (thermal) interactions with its environment unless work is performed within the system using energy extracted from the environment (this is the process of generating negative entropy mentioned earlier); the ease and economy with which the order can be sustained is termed the *maintainability* of the system.

Consequently, most (but not all) systems contain within them not only the processes that produce the service, that is, what we have been calling the functionality of the system, but also processes that maintain the functionality and thereby allow the QoS to remain constant, or above a certain level, over the operational phase of the system's life cycle. The system can then be represented not by a single functional element, but by two interacting elements, and this representation will be pursued in detail in the next chapter. Staying with our current objective of expanding the variables of the irreducible element to obtain a more detailed description of the ROI while still representing it as a single element, we can go one step further by observing that in most (physical) systems both the service processes and the maintenance processes involve a combination of equipment and personnel, interacting through what in most general terms might be called person-machine interfaces (PMIs). As a result, the internal structure of a system may be represented by the block diagram shown in Figure 7.2 (this view of a system was proposed already in section 13.3.1 of *The Changing Nature of Engineering*[1]).

The *production subsystem* contains the equipment (i.e., hardware and software) needed to produce the service, whereas the *operations subsystem* encompasses the humans involved in producing the service. The *maintenance subsystem* is that part concerned with the direct maintenance of functional subsystems and, while it contains both human and equipment elements (e.g., a maintenance management system), the element of human involvement is predominant. The *support subsystem*

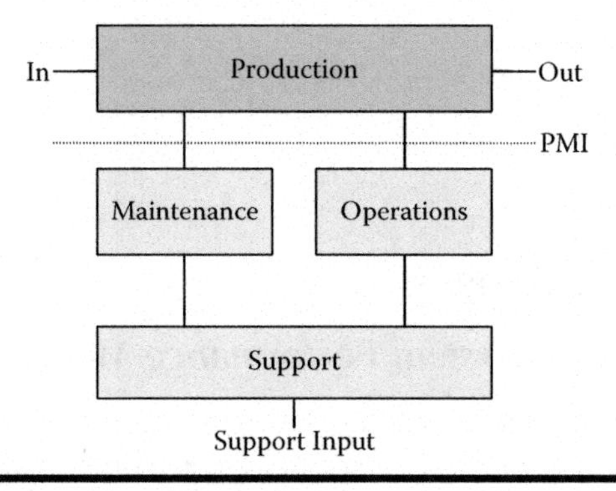

Figure 7.2 The four main subsystems of a system.

provides the support functions that allow the human involvement to be sustained and also provides the auxiliary (non-production) interface with the outside world (e.g., recruiting and spare parts provisioning).

Besides the fundamental reason given above, a reason for the partitioning into a production and a support process is to be found in the normal subdivision of costs into acquisition costs and operating/maintenance costs, or into non-recurring and recurring costs, and this is, of course, also the major reason for choosing the reference point in time for the PV calculations as the beginning of the operational phase. The reasons for subdividing each of these two processes further into a human and an equipment-based process are also partly to be found in the type of costs associated with these two subsystems, but partly because it leads to such characterizations as degree of automation and productivity.

As a result of introducing this internal structure for the purpose of cost allocation, the cost incurred in each phase of the life cycle can be subdivided into four *types* of cost:

1. Cost associated with the production subsystem
2. Cost associated with the operating subsystem
3. Cost associated with the maintenance subsystem
4. Cost associated with the support subsystem

and consequently we can define a *cost matrix,* **C**, with elements C_{ij}, where the first index indicates the subsystem, and the second index indicates the phase. Some typical contributors to these cost elements are indicated in Table 7.1.

The LCC can then also be correspondingly subdivided into four components, one for each type of cost,

$$LCC = \sum_{i=1}^{4} (lcc)_i , \tag{7.8}$$

with

$$(lcc)_i = h_i \sum_{j=1}^{4} C_{ij} .$$

7.5 Stochastic Aspects

7.5.1 A Stochastic System Performance Model

As we mentioned at the end of section 6.3, certain aspects of system behavior, such as reliability, cannot be treated in a static or averaged limit, that is, what in section 6.2 we compared to the thermodynamic limit. The random fluctuations in the

Table 7.1 Contributors to the Individual Cost Elements

Element	*Typical Contributors to the Cost Element*
11	Feasibility studies, system design, detailed design
12	Acquisition, construction, system integration, testing, V&V
13	Raw materials, services (including power), repair parts, consumables
14	Tear-down, demolition, site amelioration, disposal
21	Operations analysis, organization development
22	Not used
23	Operating staff salaries
24	Redundancy payments
31	Maintenance subsystem design
32	Test equipment, maintenance software, maintainability demonstration
33	Maintenance personnel salaries, support equipment maintenance
34	Not used (included in 14)
41	Logistic support analysis, support system design (stores, etc.), training course design
42	Personnel hiring, initial training, documentation
43	Personnel management, ongoing training
44	Disposal

service provided by the system cannot be described just by their statistics, that is, long-term averages; the temporal or spectral properties must be taken into account. To this end, we expand our description of the irreducible element by a model that reflects the most basic aspects of probabilistic system behavior with a minimum of assumptions.

The model is based on the view that the service is produced by the interaction of numerous elements, and that the fluctuations in QoS occur as a result of these elements and/or the interactions between them undergoing changes in the form of failures and repairs. The nature of these failures (and the subsequent repairs) does not need to concern us at this stage; we are only interested in the effect on the QoS, and we will call each decrease in s an *element failure* and each increase an *element repair*, even though we do not at any stage identify any elements. The word "element" simply refers to something that takes place on the element level, as opposed to the system level. In particular, as we do not identify any elements, we are really assuming that all the elements are equal, or that they can effectively be represented by an *average* element, defined by the effect it has on the system. As before, we will say that the system has *failed* whenever $s < s_l$.

Our model is based on the following four assumptions:

1. The element failures occur at random with a constant failure rate λ per unit time (i.e., they are the result of a stationary Poisson process). [Note that $1/\lambda$ is *not* the mean time between failures (MTBF) of a single element.]
2. The decrease in s resulting from an element failure, denoted by δ, is a random variable with a probability density distribution that decreases linearly from a value of 2Δ at $\delta = 0$ to zero at $\delta = \Delta$.
3. The time it takes for an element failure to be rectified, denoted by t_f, is a random variable with a triangular probability density distribution, rising linearly from zero at $t_f = 10$ to a value of 0.066667 at $t_f = 20$, then falling linearly to zero at $t_f = 40$. The mean time to repair a failed element (which is equal to $1/\mu$, with μ being the repair rate applicable to each element failure) is therefore equal to 23.333 units of time.
4. The system starts its operation at $t = 0$, in what we will call its *design state*, that is, $s = s^*$. The value of s^* is determined by the following consideration: The mean value of δ equals $\Delta/3$, so that as all the elements contribute equally to the QOS, there are $3/\Delta$ such elements, and the element failure rate is $\lambda\Delta/3$. As the mean time to repair a failed element is 23.333 units of time, the value of s^* equals $1/(1 + 7.77778\Delta\lambda)$.

In the following, we round off the above values and set the mean time to repair (MTTR) equal to 23, and s^* equal to $1/(1 + 7.8\Delta\lambda)$.

The choice of the two probability density distributions is to a certain extent arbitrary, but they have been chosen so as to be reasonable reflections of reality in a large number of existing systems. In particular, the triangular distribution of δ was chosen in preference to an exponential one, as we do not want a single element failure to result in a system failure.

Assumption (c) ties the repair rate to the unit of time; that is, our model operates with a unit of time that equals $0.043/\mu$ units of the time in which μ is measured (e.g., hours or accounting periods). Therefore, μ does not appear in the equations and λ is measured in this new unit of time, but at the end of the calculations, all time-dependent values have to be converted back to accounting periods.

The service density function, $\varphi(s)$, is a smooth, continuous function of s, except at the point $s = 1$, which is a singular point. To determine the value of the singularity, which will be denoted by $\chi(\Delta,\lambda)$, we use the fact that, within our simplification of identical elements, each element has a probability of being in the failed state equal to $7.8\Delta\lambda/(1 + 7.8\Delta\lambda)$. Allowing a Poisson distribution for the number of elements in the failed states, the probability of finding the system in the state where no element is in the failed state is given by

$$\chi(\Delta,\lambda) = \exp\frac{-23\lambda}{1+7.8\Delta\lambda} \tag{7.9}$$

The function $\chi(\Delta,\lambda)$ is shown in Figure 7.3.

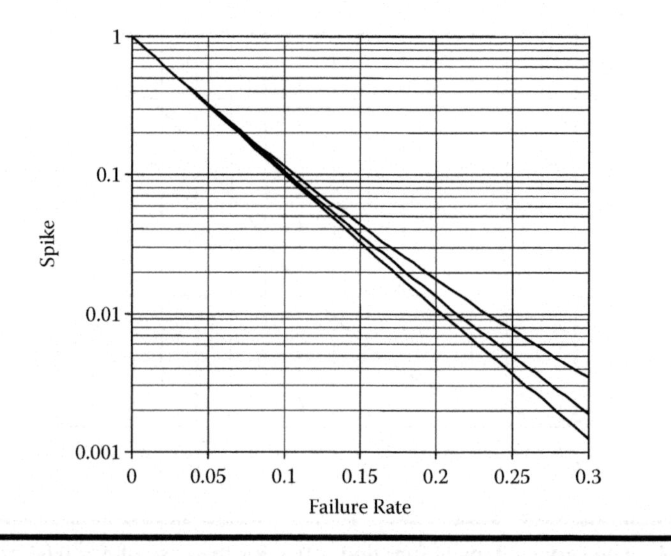

Figure 7.3 The probability of $s = 1$, as a function of the rate of element failures, λ, with the failure effect, Δ, as parameter, $\Delta = 0.1$, 0.05, and 0.02, from top to bottom.

Consequently, the service density function is made up of two components,

$$\varphi(s) = \varphi^*(s;\,\Delta,\lambda) + \chi(\Delta,\lambda)\cdot\delta(s-1) \tag{7.10}$$

and we will call χ the *spike*, so that when s = 1, the system is "in the spike." Also, let us call the event that s becomes 1 a *strike*, that is, the QoS "strikes the spike."

The distribution function $\varphi(s)$ is not a simple function that can be expressed in closed form, but it can be evaluated by simulating system operation, stepping through time unit by unit. At each step, element failures may or may not take place, and the calculation accounts for up to two element failures per unit time. (This puts an upper bound on the failure rate of about $\lambda = 0.3$, at which point the error in the failure rate arising from neglecting higher order terms is about 1.6%).[2] When an element failure occurs, random values of δ and t_f are determined according to their distributions, and δ is subtracted from s in this time step and added to s t_f steps later. At each step, a histogram vector for s is updated by having one of its components incremented by 1.

The calculation is started with $s = s^*$ and by assuming the previous $3(1 - s_0)/\Delta$ steps resulted in failures with $\delta = \Delta/3$. A step counter, n, is incremented after each step, and the value of s is compared with the value of s_1. If $s \geq s_1$ for $n = n' - 1$ and $s < s_1$ for $n = n'$, then the value of n' is entered into a failure register, and a failure counter incremented. Furthermore, if $s < 1$ for $n = n' - 1$ and $s = 1$ for $n = n'$, then the strike counter is incremented, and for each step with s = 1, the spike counter is incremented. Once the failure counter reaches a preset value, the simulation run is

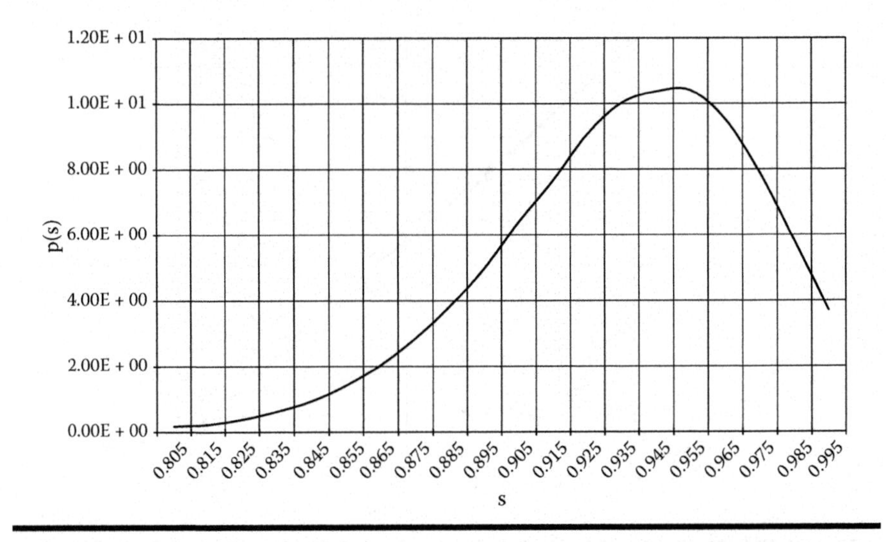

Figure 7.4 The service density function, $\varphi^*(s)$, for the case of $\Delta = 0.05$ and $\lambda = 0.2$.

terminated, and the function $\varphi^*(s)$, the histogram of time between failure, and the statistics, including MTBF, MTTR, MTBS (mean time between strikes), and χ, are generated and displayed.

We will first consider the function $\varphi^*(s)$, and a representative result, for the parameter values $s_1 = 0.8$, $\Delta = 0.05$, and $\lambda = 0.2$, is shown in Figure 7.4.

7.5.2 The Service Density Function $\varphi(s;\lambda,\Delta)$

The shape of the curve shown in Figure 7.4 leads us to suspect that the function $\varphi^*(s)$ can be approximated by a function of the form

$$\varphi^*(s;\lambda,\Delta) = \frac{G}{\sigma} e^{-\left|\frac{(s-s_0)}{2\sigma}\right|^{\alpha}}, \tag{7.11}$$

where the three parameters s_0, σ, and α are functions of λ and Δ. G is simply a normalization factor that ensures that the integral of the SDF over the range 0 to 1 equals 1. By comparing with the results of the simulation described in the previous section, it can be shown that a reasonable approximation can be achieved with a constant value of α, $\alpha = 1.6$. The errors encountered in choosing a constant value of α, rather than letting α be a function of λ and Δ, are not significant enough at this high level of the design process to warrant the additional complexity of varying α. Choosing this value of α, the values of s_0 and σ that provide the best fit for each set of (λ,Δ)-values can be determined, and a typical example is shown in Figure 7.5.

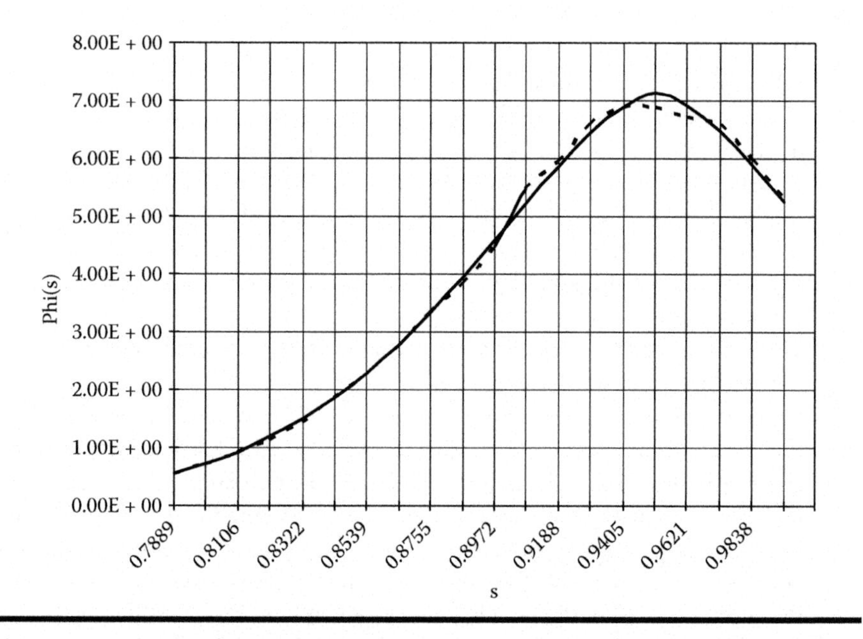

Figure 7.5 The fit of the simple exponential function to the result of the simulation, for Δ = 0.1 and λ = 0.1 (shown dotted).

Obviously, the lack of symmetry in the simulated function does not allow a perfect fit by using the simple, symmetric (about s*) exponential function, but again, at this high level of system design, it is adequate.

By running the simulation for a large number of values of λ and Δ, and each time determining the best fit of the function in equation 7.11, it is found that the functions $s_0(\lambda,\Delta)$ and $\sigma(\lambda,\Delta)$ can be approximated by the following expressions:

$$s_0 = 1 - 4.8\Delta\lambda\,, \tag{7.12}$$

and

$$\sigma = 0.765\lambda^{0.205}\Delta. \tag{7.13}$$

Both $(1 - s_0)$ and σ go to zero in the limits $\lambda \to 0$ and $\Delta \to 0$, as they must.

Before going any further, it is probably good to stop and review what we have done so far in this subsection, so as to clearly see its significance and limitations. At first glance the many simplifications and assumptions could lead one to dismiss it as simply an exercise in numerical analysis. But let us take a look at these simplifications and assumptions again:

a. *The system delivers a service, either directly or by means of a product.* It would be difficult to think of an engineered system where this was not true.

b. The service can be characterized by a single parameter, the quality of service (QoS), with value s, such that it is desirable for s to be as large as possible. Conceptually, there should be no problem with this; in practice there may not be an obvious choice. For example, take a mobile telephone service; there are a number of important features of this service, such as the normal QOS (probability of connection on the first try if the called subscriber is not busy), coverage, transmission rate, etc., and forming a combination of them that is representative of the overall performance of the system would be a matter of definition rather than a compelling necessity.
c. The range of s is 0 – 1. This is just a matter of convenience, and can always be achieved by a transformation of scale.
d. Element failures occur at random, with a constant system failure rate. That they occur at random is always true; if we knew when an element was going to fail, we would replace it before it failed. A constant failure rate is definitely not true of each type of element, but it becomes true for the system in the limit of a large number of elements once the effect of the initial condition (all elements in mint state) wears off, so that it is a reasonable assumption for the types of systems (large, with many elements) we are considering in this book.
e. The effect of failure, δ, has a density distribution function with a simple, universal form and characterized by a single parameter, Δ. It is true of all engineered systems (and many naturally occurring systems, too, such as storms, earthquakes, etc.) that the impact of failures is a decreasing function of their frequency. The exact form of the function is not important for our present, high-level development, but it must satisfy the condition that the maximum impact be (significantly) less than 1.
f. The time to repair a failure has a density distribution function with a simple, universal form and characterized by a single parameter, the mean time to repair. Between each type of element, both the MTTR and the shape of the density distribution function may vary greatly, but much of this variability will be averaged out, as seen from the system level. And it is generally true that the repair times will be clustered around some central value, determined by the criticality of the system and the corresponding design of the maintenance subsystem.
g. *All elements contribute equally to the QoS*. This assumption, which is obviously not true in general, is used to establish the connection between the element level and the system level by connecting the number of elements to the parameter Δ. That is, what we are really saying here is that if the stochastic fluctuations in *s* are fine-grained (as in, say, the telephone system), the system contains a large number of elements; if they are coarse-grained (as in, say, a power generating system with four generators), the system contains a small number of elements. The assumption is just the simplest possible expression of this observation.

So, while the degree to which our assumptions are fulfilled will vary from system to system, it would be rare for them to be *unreasonable*. What we have constructed is a model of the stochastic aspect of system behavior that is adequate for the purpose of providing a continuity of description of an engineered object from that of a single element to a system of elements. The model is neither unique nor of any other fundamental importance; its aim is simply to be useful in conceptualizing the behavior of complex systems. However, with this in mind, it is still of interest to compare the density distribution $\varphi(s; \lambda, \Delta)$ with a famous one, the Maxwell distribution for the momentum of particles, with mass m, in a gas at temperature T,

$$f(\mathrm{p};T)=(2\pi mkT)^{-3/2}e^{-p^2/2mkT}.$$

The derivation of this distribution involves some assumptions and approximations, in particular regarding randomness and the nature of the interaction between the particles, and it characterizes the system it refers to by only two parameters, the particle mass and the system temperature (i.e., neglecting any other characteristics). Despite that, it applies with a high degree of accuracy to a large number of situations and is both conceptually and practically of great value.

One further comment: The above description of system behavior related to random failures and their repair, that is, to corrective maintenance; it does not consider preventive maintenance in the form of scheduled shutdowns. That is, the model applies only to the time periods between such shutdowns.

In practice, then, whenever we need to operate on the function $\varphi(s)$ for a given set of $(\lambda,\Delta,)$-values, for example, integrate the product of this function with another function, such as the value function, we first call a small subroutine that generates the parameter values s_0 and σ and normalizes the resulting function (taking account of $\chi(\lambda,\Delta,)$, of course), and then use the exponential approximation.

7.5.3 Temporal Aspects

As previously noted, the service density function $\varphi(s)$, being a long-term average, contains no information about the temporal or spectral (frequency) characteristics of the stochastic behavior of the system. Of particular interest are the MTBF and the MTTR, which are related to the availability by

$$Availability=\frac{MTBF-MTTR}{MTBF}. \tag{7.14}$$

The distribution of the time between failures is an output of the simulation program, and a typical example is shown in Figure 7.6. Clearly, this is very far from

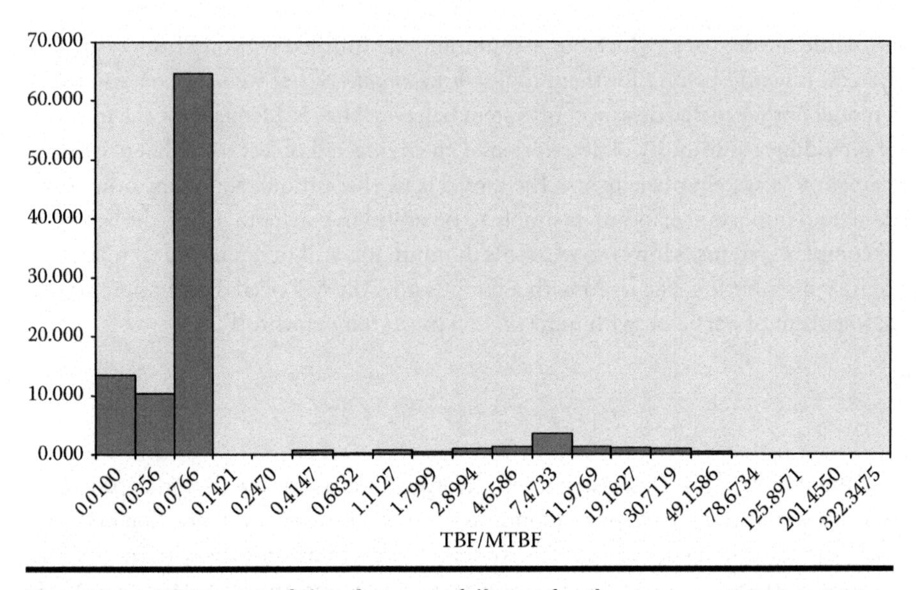

Figure 7.6 Histogram of time between failures, for the case $s_1 = 0.8$, $\Delta = 0.05$, and $\lambda = 0.2$, created by running the simulation until 5000 failures had occurred. The vertical scale is the percentage of times between failures falling within an interval; the numbers on the horizontal scale are the upper limit of the interval relative to the MTBF.

an exponential distribution, particularly when we consider the nonlinear interval scale.

Both MTBF and MTTR are functions of three parameters, λ, Δ, and s_1, and before we attempt to obtain useful approximate expressions for these functions, we need to consider the relationship between failures and the spike (introduced in section 7.5.1). Let us introduce two parameters, similar to MTBF and MTTR, for the spike, mean time between strikes (MTBS), and mean time in strike (MTIS). These two parameters are functions of λ and Δ, but, of course, do not depend on s_1. It is then straightforward to see that the following relationships must hold:

$$\lim(s_1 \to 1)\ \mathrm{MTTR} = \mathrm{MTBS} - \mathrm{MTIS}$$

and (7.15)

$$\lim(s_1 \to 1)\mathrm{MTBF} = 1/\lambda$$

This limiting value of the MTTR is shown in Figure 7.7, and it is, as we would expect from Figure 7.3, only slightly dependent on Δ.

Running the simulation program for a range of values of λ, Δ, and s_1, values for the MTBF are obtained, resulting in curves of the form shown in Figure 7.8.

For constant values of λ and s_1, the values of MTBF as a function of Δ produced by the simulator are reproduced very closely by a function of the following form

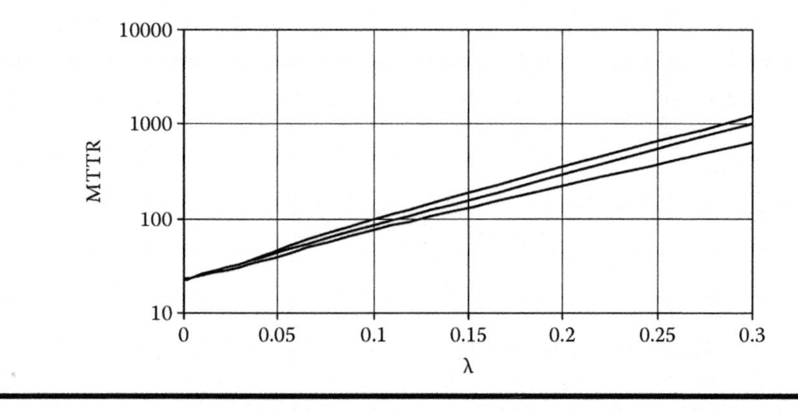

Figure 7.7 The MTTR in the limit of $s_1 \to 1$, for $\Delta = 0.1$, 0.03, and 0.01, from top to bottom. As $\lambda \to 0$, the value of MTTR equals the mean value of the element MTTR, i.e., 23 units of time.

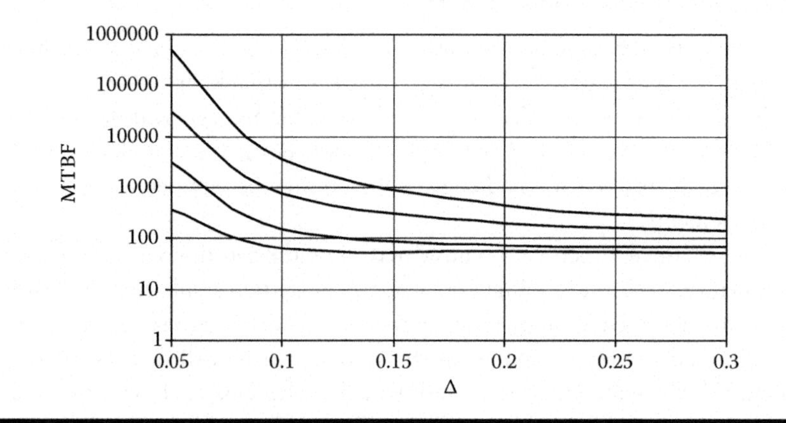

Figure 7.8 The MTBF as a function of Δ with λ as a parameter 0.01, 0.02, 0.05, and 0.1 from top to bottom, with $s_1 = 0.9$.

$$MTBF(\Delta) = ae^{\frac{b}{\Delta - c}}.$$

By considering the dependence of a, b, and c on λ, and taking account of the limiting value for MTBF in equation 7.15, a function that reproduces the MTBF values obtained from running the simulator is given by equation 7.16.

$$MTBF = \left[\frac{s_1^{3.11}}{\lambda} + 1065(1 - s_1)\lambda\right] \exp\left[\frac{(1 - s_1)e^{-17\lambda}}{0.0284(\Delta - 0.08\lambda^{1/2})s_1}\right]. \tag{7.16}$$

There is, of course, nothing unique about this expression for the MTBF; the purpose of developing it is simply to give a quantitative expression to the relationships that must exist between the expanded parameters of the irreducible element.

The availability is given by the expression

$$A(s_1,\lambda,\Delta)=\int_{s_1}^{1}\varphi(s,\lambda,\Delta)ds\,, \tag{7.17}$$

and the MTTR then follows from equation 7.14.

7.6 A Set of First-Level Elements

7.6.1 The First-Level System

As a result of the developments in the foregoing sections of this chapter, we have now arrived at an expansion of the basic set to the set of variables in Table 7.2.

The functional element that contains all these variables is a real element, because its included set contains the irreducible element, but it is a complex element and, in accordance with our methodology, it should now be represented by a system of simpler elements.

If the irreducible element is considered to be the zeroth level representation of functionality, in that it is a single element and no system at all, then the first breakdown into a representation by a system will be called a *first-level representation*, and the additional elements making up the system will be called *first-level elements*. In Figure 7.9, a representation in terms of five elements is shown, in which the

Table 7.2 An Expansion of the Basic Set

Basic Set Variables	*Expanded Variables*
Cost, C	4 cost types (related to subsystems), subdivided into
	16 cost elements, c_{ij}
	Cost transformation vector, ***h***
Revenue, R	Value function, W(s)
Quality of Service, S	Service distribution function, $\phi(s)$ (represented by s_0 and σ)
	System failure limit, s_1
	Availability, AVAY
	Reliability, MTBF
	Repair, MTTR
Lifetime, L	4 phases, with durations l_i

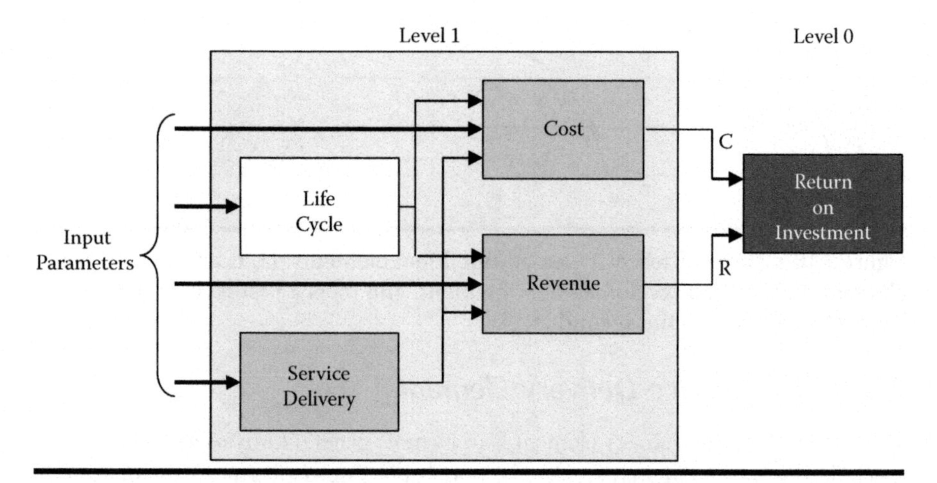

Figure 7.9 A representation of the ROI aspect of functionality in terms of a system of five elements, four on Level 1, expanding the Level 0 variables, and the irreducible element, as always, on Level 0. The life cycle element is a "virtual" element.

elements have been chosen so as to group the variables and the properties of the overall element in a natural manner, reflecting the way in which we would calculate the ROI using these variables. However, these five elements are not of the same type; only the ROI element is a real element. It is the same element as the irreducible element, except that the implicit dependence on S and L has been removed and made explicit by means of other elements.

Of the four other elements, which are imaginary elements, the life cycle element is a special case, in that it does not actually reflect any aspect of the functionality of the system, but contains the definitions of the four phases of the life cycle. In this sense it acts as a filter, and extracts the phase durations from the information contained in L. Perhaps the best analogy is the header or declaration part of a software module. However, it must still be considered a bona fide imaginary functional element, as it expresses the life cycle aspect of the system.

Figure 7.9 is an illustration of the understanding of functional elements we developed in chapter 4. The first-level functional element, represented as a system of five elements in Figure 7.9, is a real element, because if we condense the set of variables, we arrive at the irreducible element. The four imaginary first-level elements make up what would be the innermost ring in Figure 4.5. However, if we now imagine that one of these elements, say, the cost element, was expanded in terms of a set of second-level elements, then the condensation of this set does not end with the irreducible element; it ends with the (imaginary) first-level cost element. An illustration of this nomenclature is shown in Figure 7.10.

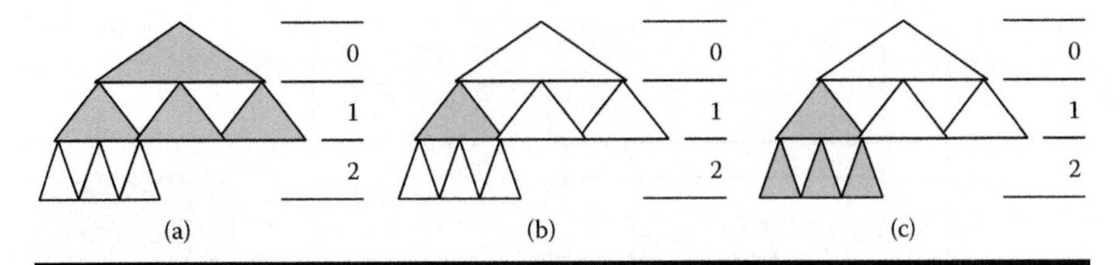

Figure 7.10 Three different types of functional elements. (a) is a first-level real element, (b) is a first-level imaginary element, and (c) is a first-level imaginary element expanded to the second level.

7.6.2 The Service Delivery Element

The first-level service delivery element is in its most general form, as we have not yet specialized to any particular service. It models the stochastic aspects of the system behavior; it models the effects of internal failures and repairs on the system performance, which is still characterized by the single parameter QoS. The functional parameters are:

Service density function parameters	s_0 and σ
Mean time between failures	MTBF
Mean time to repair	MTTR
Availability	AVAY

The dependencies are:

Performance limit	s_1
Rate of element failures	λ
Element repair rate	μ
Element failure effect parameter	Δ

There are no influences associated with this element.

The functionality, that is, the relationships between the parameters and the variables, is given by equations 7.9, 7.10, 7.11, 7.12, 7.13, 7.14, 7.16, and 7.17. Note that the element repair rate, μ, needs to be included as one of the dependencies in order to define the time scale (as discussed in section 7.5.1).

7.6.3 The Cost Element

The element representing the cost aspects of the ROI in Figure 7.9 relates cost to time and to the four functional subsystems, as expressed by equations 7.2–7.8. This relationship also contains a definition aspect, as was discussed in conjunction with the life cycle element, in that the information about the time dependence of the costs must be present in the input data; the definition of the life cycle phases simply allows this time-dependent cost to be allocated to the phases.

The element also takes into account the cost of failure. The cost of a single failure, COF, is a dependency, as is the MTBF, and they are combined by defining a total cost of failure as the COF times the ratio of l_3 to MTBF, and then spreading this cost evenly over l_3, as it is not known when the failures will actually occur.

This element has only one functional parameter:

System cost	C

The dependencies of this element are:

The fourteen elements of the cost matrix	C_{ij}
The four components of the life cycle vector	l_j
The cost of failure	COF
The mean time between failure	MTBF

There is one influence:

Discount rate	p

7.6.4 The Revenue Element

This element calculates the PV of the revenue as the expectation value of the value function $W(s)$ with respect to $\phi(s)$. The value function $W(s)$ is given in terms of a nominal value per accounting period, W_0 (i.e., the value produced if s were identically equal to 1 throughout the whole period), and a value of the QOS, s_2, such that the value function W decreases linearly from W_0 to zero as s decreases from s_2 to s_1, as shown in Figure 7.11.

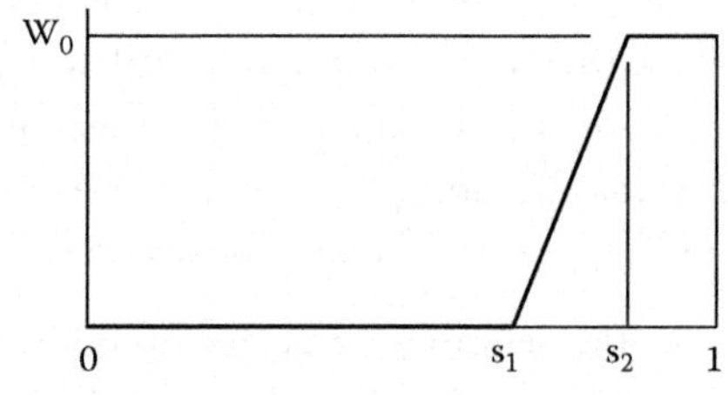

Figure 7.11 The value function W(*s*).

This element has only one functional parameter:

System revenue	R

The dependencies of this element are:

The service density function parameters	s_0 and σ
The nominal value per accounting period	W_0
The nominal value limit	s_2
The three components of the life cycle vector	l_1, l_2, and l_3

There is one influence:

Discount rate	p

7.6.5 The Return on Investment Element

This element, which is extremely simple, calculates the return on investment (ROI) as a function of the cost and the revenue, as defined in section 2.2. As the ROI is only a function of the ratio of the two, the reference point in time does not matter (as long as it is the same for both), and the values transferred from the cost and revenue elements are the present values.

7.6.6 Classifying Service Elements

So far, in our quest to develop functional elements, we have identified an element common to all systems, the irreducible element representing ROI, and it is implicit in this element that the ROI results from providing a service. But we have not made any attempt to differentiate between types of service. There is no single, unique way of subdividing functionality into a hierarchy of classes based on type of service and, as with most other aspects of this system design methodology, it will ultimately depend on what is perceived by the engineering community as being useful. That is also the reason why this subject matter is treated rather briefly; the development of classes cannot be done by one person in isolation, it requires interaction and discussion within a large user group. What is put forward here are *suggestions*, and must not in any way be interpreted as statement of facts. Having said that, some suggestions may be more plausible than others, and there are also two general criteria that would appear to be non-controversial:

1. The definition of a class should be simple and unambiguous, so that it is easy to determine if an element belongs to it or not.
2. At each level in the top-down development (and remember, the top-down development, starting with the irreducible element, is an axiom of our methodology), the union of all the classes must be able to contain every conceivable element.

The first-level subdivision of functionality is suggested by the fact that any activity is perceived by us to take place within a space-time framework (again, an acknowledgment of Kant's influence is appropriate at this point) and by the fact that functionality always involves doing something *to* something. That is, there is activity in the narrower, more abstract sense of doing *per se*, which then becomes concrete by specifying the *objects* to which it is applied.

The class of *transport elements*. A transport element moves a collection of objects from one set of positions in space to another, without altering anything about the objects themselves.

The class of *storage elements.* A storage element moves a collection of objects from one set of points in time to another, without altering anything about the objects themselves.

The class of *transformation elements.* A transformation element transforms a collection of objects into another, without any reference to space and time as far as the objects themselves are concerned. (However, the transformation process itself exists in space-time, so that we can, for example, speak of the MTBF of the process.)

An initial attempt at developing these three classes of functional elements was outlined in *The Changing Nature of Engineering*[3] and will not be repeated here. Further development of service elements will be most effective if it is done in conjunction with the development of real projects, as the front end of the conceptual design phase.

7.7 An Example: Underground Copper Mine

As was indicated in chapter 2, an early and very successful application of the top-down approach was the development of the design for the Northparkes E26 underground copper mine. Because in this case the ore body was in the shape of a cylinder of about 80 m diameter and 500 m depth downward from just under the surface and heavily veined with calcium, it was ideally suited for a mining method called *block caving.* This method consists of excavating a few parallel tunnels at the bottom of the ore body and interconnecting them with a large number of cross-passages. In the roof of each cross-passage a funnel-shaped shaft is excavated upward for about 20 m, and then a horizontal slice is excavated out of the ore body at this point. As the extension of this slice grows, the ore body above it will start to crumble under its own weight, and the resulting pieces of ore (rocks) fall down through the funnels and form heaps in the cross-passages, which are then known as *draw points.* The ore is extracted from the draw points using special front-end loaders known as load-haul-dump trucks, or LHDs.

Consequently, once the mine is developed (which involves, of course, a very significant initial investment), there is no further mining required in the traditional sense of drilling and blasting; the mine becomes what can be characterized as a *rock factory,* and the primary production activity is ore handling.

Before describing the development of the performance model, we need to understand the requirements on the output of the mine. The ore, which must have a rock size of 300 mm or less, is delivered to a concentrator plant that requires a steady input of about 800 tons of ore per hour and operates continuously except for a major shut-down each year. At the input end, the plant incorporates a stockpile that provides a buffer for up to 24 hours' interruption of the flow of ore from the mine; for any interruptions of longer duration, there is a significant penalty (e.g., $250,000 per day). Aside from the size limitation, there are no other quality requirements on the ore production, so it is not necessary to assign any value to the product; the service provided by the mine is characterized solely by availability (outside of the yearly shut-down) and cost.

In the initial, or top-level, model of the mine the service delivery model was a single element, characterized by the following parameters:

Production rate, the tons per hour that the mine was designed to deliver when operating at full capacity (this was the only operating mode considered, that is, all or nothing).
MTBF and *MTTR* of the ore delivery, which then determined the availability.
Non-recurring costs, consisting of a fixed part and a part depending on the production rate.
Recurring costs, which included personnel costs, spare parts and consumables, and power costs.

The production rate was determined by dividing the rate required by the concentrator by the availability, and the values of the other parameters were at first determined by comparison with recent similar operations. The output of the model was the LCC, determined by the same four elements as shown previously in Figure 7.9 (although these were all represented on the same worksheet). However, the stochastic aspects of the system performance was in this case handled somewhat differently to that outlined in section 7.5, in that instead of using a service density function, φ, the likely performance was determined using a Monte Carlo technique.

The application of this very simple model was in deciding between two different ore handling schemes; either trucking the extracted ore to the surface and crushing it there, or installing an automated ore handling system, which would require crushing the extracted ore underground. Even allowing for the great uncertainty of the results, the model was able to clearly differentiate between the LCC values of the two schemes, coming out clearly in favor of the automated system due to its lower recurring costs.

The service delivery model was then expanded to four elements — extraction, crushing, horizontal transport (meaning belt conveyors), and vertical transport (meaning hoisting) — each element characterized by a somewhat expanded (more detailed) set of the parameters of the single-element model, such that when summing up, the model produced the inputs to the single-element model and thereby provided continuity in the development and refinement of the LCC. The first use of this model was to decide on the split between conveyors and hoisting; hoisting turned out to be significantly more economical than conveying, so that the latter was confined to bridging the distance between the extraction point and the loading station of the hoisting system, as the hoisting shaft needed to be well away from the ore body.

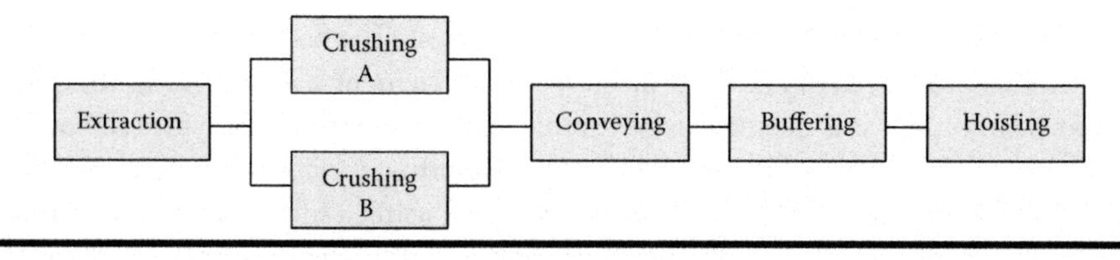

Figure 7.12 The service delivery model of the mine.

The model also showed that the hoisting was now the most expensive component, so that it became important to look at ways of minimizing the cost of it. That meant minimizing its capacity, which again meant ensuring that it would never be idle due to a lack of upstream ore supply. The most critical upstream element was the crushing, and the model demonstrated that having two crushing stations rather than one, with a degree of redundancy, would be more cost-effective. But even with this improvement in the availability upstream of the hoisting the model indicated a strong dependence of the hoisting cost on this availability, so consequently, a storage buffer was inserted in front of the loading station, and the size of this buffer was optimized, together with the degree of redundancy in the crushers, using the model (i.e., minimizing the LCC value).

The model now consisted of six elements, shown in Figure 7.12, represented by six worksheets, all linked to the worksheet that represented the single-element version and that produced the LCC values.

At this point in the design development, a number of important design decisions had been taken and rigorously justified without producing a single design drawing. In the further development of the design the model was continuously refined with more details as these became available from design calculations and from suppliers, and many choices between options, for example, for major equipment, were justified by showing that they minimized the LCC.

Of course, this was a relatively simple application, not least because it was possible to ignore what is usually the other half of the problem — modeling the value of the service. It demonstrated the usefulness of the approach, but it also demonstrated the problem of developing the model from scratch. It was quite time consuming, and under the time constraints of this project (i.e., a loss to the client in the order of $250,000 for each day delay) it was often difficult to maintain the model as a decision tool rather than as an after-the-fact means of justification.

7.8 Summary

In this chapter, we have suggested how the irreducible element could be described in more detail by representing the element as a system of smaller elements, each representing an aspect of the functionality. This subdivision into interacting elements was carried out by first expanding the basic set of variables through using more variables to describe the same concepts in more detail. The boundaries between the elements are, to a certain extent, arbitrary, and an element will survive in this particular form only if it is deemed to be useful. Each element now provides the point of departure for a further detailing of the corresponding functionality or aspect; for example, the cost element can be further developed using a subdivision of costs into various categories of labor costs and material costs, and the performance element can be developed into elements describing particular services.

More important than the exact form of the elements, which is likely to change as the result of any standardization program, is the methodology used to develop them and, in particular, the fact that it guarantees traceability. No new variable is introduced without its relationship to a variable on the level above being clearly defined.

Notes

1. Aslaksen, E.W., *The changing nature of engineering,* McGraw-Hill, New York, 1996, section 13.3.1.
2. As the failures occur at random, their occurrence is described by the Poisson distribution, $F(x;\lambda)$, this being the probability of having x or less failures per unit time. The error, ε, in the number of failures occurring over a (large) time interval is then given by

$$\varepsilon = \frac{1 - F(2;\lambda)}{F(2;\lambda) - F(0;\lambda)}$$

   and for $\lambda = 0.3$, the value of ε is 0.016, i.e., an error of 1.6%.
3. Aslaksen, E.W., *The changing nature of engineering,* McGraw-Hill, New York, 1996, chapter 11.

Chapter 8

Maintained Systems

8.1 Failure and Repair in the Two Domains

From our daily experience, we would conclude that maintenance seems to be an inescapable feature of systems. Cars need regular maintenance and occasional repairs, houses need to be maintained (painted, repaired), all appliances develop faults and are either repaired or replaced, organizations need constant management attention and training in order to remain efficient, and so on. To get a better understanding of what drives this need for maintenance, we must, for a moment, go back to the physical domain and consider some general characteristics of physical systems, as the reasons for maintenance in the systems we are considering are mainly physical in nature. However, that should not obscure the fact that there are many objects in engineering for which the need for maintenance is not driven by any physical deterioration, such as the methodologies that need updating in light of accumulated experience, data banks (as more accurate data becomes available), and even basic theories, such as the extending of the laws of physics to take into account relativity.

Physical systems are made up of interacting physical elements; it is these interactions that bind the elements together into a system. Each element is again a system of smaller interacting elements, and so on, until at some low level the elements are electrons and nuclei, and the systems are called atoms. The elements are bound together by a combination of short-range repulsive forces and long-range attractive forces, and the binding of each element can be expressed in terms of a potential energy that equals the work required to remove the element from the system. But, in addition to this potential binding energy, the elements also possess a kinetic energy; we may think of it as the elements rattling around in the "cage" formed by the potential energy. If this kinetic energy of an element becomes comparable to the potential energy, the binding may be overcome and the system (i.e., the atom) undergoes a change. To restore it to its original condition, we would have to supply a "spare" element (i.e., electron); this would be a maintenance action.

This picture of a system existing in a realm of two competing energies — the constitutive potential energy and the destructive kinetic energy — is very useful, even though it is highly simplified.

At the next level up in the hierarchy of systems, atoms are the elements in systems called molecules. The potential energy binding atoms together is called a (chemical) bond, and each atom has an amount of kinetic energy in the form of vibration within the confines of the bond. (It is sometimes useful to think of the bond as a spring or a rubber band between the atoms.) An atom of type A and an atom of type B may or may not be able to form a bond, and this *information* is carried in the *structures* of the atoms, that is, in how the elements are arranged within the system. But where two atoms can form a bond, there may be more than one type of bond possible, so that a set of atoms may result in one of several molecules. (Molecules with the same chemical formula but different *structures* are called *isomers.*) What emerges from this picture is that, in addition to the potential and kinetic energies of the elements, a system is characterized by an amount of *information*; the information carried in the structure of the system. The atom carries the information about how the electrons are arranged around the nucleus, that is, its structure (and which determines what bonds are possible); the molecule carries the information about which of the possible arrangements of atoms is realized in this particular molecule.

Continuing up to the next level, molecules combine to form materials, and again there is an amount of information that describes how, out of the many possible ways in which molecules can form materials, this particular material is constituted and structured. There are binding forces between the molecules (potential energy) and kinetic energy in the form of vibrations of the molecules, but because there are normally very many (say, on the order of 10^{22}) molecules in a piece of material used in any engineering design (except possibly in nanotechnology), it is neither practical nor necessary to look at the state of individual molecules, but rather to treat the collection of interacting molecules (i.e., the system, on this level) in a statistical fashion. This treatment, statistical mechanics, forms the link between atomic physics and the description of matter on a macroscopic level, called thermodynamics. For example, the kinetic energy is represented by the temperature, and as the temperature rises, the bonds between the molecules come under increasing strain, until at some point they begin to break and the material disintegrates or melts.

At this point in our journey upward in the hierarchy of systems a decisive change takes place. So far, the possible systems have been determined by the laws of Nature; the human influence has, at most, been in the *selection* of a desired system (e.g., molecule or material) out of the set of possible ones through the process used to create the system (e.g., pressure and temperature conditions in forming materials out of components). Or, to put it another way, the information contained in the system arises from the laws of Nature, not from any human input. But when we go to the next level up, the elements are components formed or machined out of materials in a manner determined by human *design,* and the way in which the elements are combined to form a system is also the result of human design. The system is a representation of the information contained in the design, originally contained in the designer's mind, then transferred to specifications, drawings, and other data forms.

As a consequence, designed or engineered systems have some very significant characteristics not found in the lower-level systems. First, every engineered system has a *purpose* — the purpose in the engineer's mind when the system was designed or, more precisely, in the terms of the earlier discussions in this book, the purpose expressed by the stakeholder requirements. A molecule or material, as such, has no purpose. Second, while the system is a (more or less good) representation of that purpose, knowing all the information contained in a system does not necessarily allow one to deduce the purpose (reverse engineering). Both of these characteristics have a decisive influence on our investigations into the, by far, most interesting of all systems, the human being. Does the human being have a purpose? If yes, this implies a designer (usually called God). On the other hand, even when we, as will soon be the case, know all the information contained in that system, it is highly unlikely that we will be able to reverse engineer it and so be able to determine if there is a purpose and what that purpose is.

Third, and closely related to the first characteristic, is the concept of a system's *functionality*, that is, that the system *does* something. A molecule or material does nothing, it only *is*; however, letting the molecules in a (suitable) set interact may result in an object (system) that does something, as in a living cell. This self-organization is one of the most fascinating issues in complex systems science. In the case of engineered systems ("engineered" taken here in the widest sense), the organization and the resulting functionality arise as a result of the work of the designer, and improving our ability to work in the functional domain is the aim of this book.

Finally, and of greatest importance with regard to the present subject matter, maintenance, is the manner in which statistics is used in the two cases (i.e., engineered and non-engineered systems) and the significance of the results. Take as an example of a non-engineered system one of the simplest, but also easily visualized systems, that of a fixed volume of a dilute gas. At any given point in time any particular gas molecule will have a definite position and momentum, or, equivalently, be represented by a point $(\mathbf{p},\mathbf{q})$ in a six-dimensional *phase space.* However, in the absence of an external potential, a molecule is equally likely to be found in any part of the volume, so that, as far as the state of the gas is concerned, we need only consider the three momentum variables. Through collisions with the walls of the container and with other molecules the values of these three momentum variables change frequently, but as there are very many (say, 10^{22}) identical molecules in the volume, the state of the gas can be described in terms of a probability distribution

$f(\mathbf{p})$, such that the probability of finding a molecule with momentum between $\mathbf{p}$ and $\mathbf{p} + d\mathbf{p}$ is given by $f(\mathbf{p})d\mathbf{p}$. As is well known, in this simple case the function $f(\mathbf{p})$ is the Maxwell distribution, from which can be derived the equation of state for a mole of a perfect gas, PV = RT (R being the gas constant), and other distribution functions apply to more general cases. The details of this case are not important, it is meant to illustrate one thing only; the manner in which statistics is used to connect a macroscopic quantity, such as the pressure P, to properties on a microscopic (molecular) level, and that this connection relies on the "most likely" or "equilibrium" distribution of the microscopic variables. Similarly, but using a variety of often complex distribution functions, changes on the microscopic level due to the "competition" between kinetic and potential energies can be expressed as changes to macroscopic parameters of the components that are the building blocks of engineered systems and, in particular, to the failure rates of such components. However, the "normal" state of an engineered system does *not* arise from the most likely distribution of the component parameters; it is a state that, in the sense of thermodynamics, is a highly non-equilibrium state, and statistics are used to determine deviations from this state.

If we now leave the physical domain and return to the functional domain, the concept of failure and the use of statistics to connect changes at the element level to changes at the system level need some thought. On the face of it, as we have endeavored to make functional elements independent of any physical realization, there is no direct connection between physical failure mechanisms and system behavior, with or without the use of statistics. This situation does not change when we subdivide the system into smaller and smaller functional elements; in principle, a functional element remains a functional element no matter how small it is, and so it would seem that the concept of failure and the need for maintenance do not apply to the functional domain. A thought-element can neither fail nor be maintained.

To resolve this apparent dilemma we only have to recall two features of the functional domain. First, that functional elements are not representations of physical elements; they are representations of what physical elements do. In order to represent failure and restoration (maintenance), a functional element does not itself have to "fail" or be "restored." Second, that whereas the physical domain is in its nature "bottom-up," progressing from the physics of the basic building blocks upward to more complex devices, equipment, and systems, the functional domain is "top-down," starting with the stakeholder requirements. Therefore, the relationship between element and system reliability (or availability or maintainability) is not one of cause and effect, but one of *allocation*. However, if a functional element is going to represent the failures of physical elements, it must also represent the fact that these failures occur at random, so that the treatment in the functional domain must also be a statistical one. That was the motivation for the statistical performance model introduced in section 7.5, and later in this chapter we will develop some models of the element-system relationship on this basis. But first we should explore the extension of a further concept from the physical domain into the functional domain — the concept of *entropy.*

8.2 Order, Information, and Entropy in the Two Domains

The statistical treatment of a gas introduced in the previous section is a reflection of our lack of detailed knowledge about individual molecules in the gas. Any one particular molecule will, of course, have a definite position and momentum at a given point in time, and so at any given point in time, the microstate of the gas is represented by a point in a 6N-dimensional phase space, if there are N molecules in the volume of gas. As time progresses and the individual molecules go through their trajectories and undergo collisions, the point representing the microstate of the system will also move around in phase space, and for a given thermodynamic state (i.e., macrostate) there is a definite probability of finding the system in a particular region (i.e., microscopic volume) of phase space. The central issue in statistical mechanics is therefore to find the relationship between thermodynamic states and their distribution functions.

The relationship turns out to hinge on the concept of entropy, S, as was first realized by Boltzmann. In thermodynamics, S is the extensive variable conjugate to the intensive variable T, the thermodynamic temperature, in the sense that the internal energy, U, can be expressed as a sum of products of intensive and extensive variables. For example, for a system with pressure, P, and volume, V, as the only mechanical variables, U = TS – PV, and the relationships between the extensive and intensive variables is given by differentials of U, such as $T = (\partial U/\partial S)_V$. If $f(q,p)$ is a distribution function, with q and p being 3N dimensional vectors, then the relationship between entropy and distribution function is given by

$$S = -k \int f \ln(f) dq dp \tag{8.1}$$

where k is the Boltzmann constant.

For our purposes, the interesting feature of this equation is that it shows entropy to be a measure of the *order* of the system, that is, of the degree to which we are certain that the system is in a particular state. The more disordered the system, the less information we have about what state it is in, and the greater the entropy of the system. If we know for certain exactly what microstate the system is in, then the distribution function is a 6N-dimensional δ-function, and the entropy is zero. If we know nothing about what microstate the system is in (total disorder), it is equally likely to be in any state, and f is a constant equal to the reciprocal of the allowable phase space volume (limited, e.g., by the internal energy of the system and the size of the container), and the entropy takes on its maximum value.

In the case of a functional system, there is also a lack of information, in the sense that we do not know what state the system is in at any given point in time; this was the reason for introducing the service density function, $\varphi(s;\lambda,\Delta)$, in section 7.5. The above relationship between disorder and entropy for a physical system then suggests that there might be an equivalent concept for functional systems, a functional entropy, E, defined by

$$E=-\int_0^1 \varphi(s;\lambda,\Delta)\ln(\varphi(s;\lambda,\Delta))ds. \tag{8.2}$$

The entropy E depends, within the accuracy of the numerical expressions introduced in the previous chapter, on λ alone and not on Δ, as we would expect as long as Δ remains much smaller than 1. However, before we display the result of evaluating equation 8.2, we will change the independent variable from the failure rate λ to the repair rate μ, as this will be more suitable for discussing maintenance in the rest of this chapter. That is, we will consider systems where λ is fixed (by the hardware and software design), but where μ can be varied by changing the maintenance regime. In the model developed in chapter 7, the variable λ is the element failure rate, but measured in units of $23.3\mu'$, where μ is the element repair rate, and the prime indicates that it is measured in fixed time units, for example, per hour. Consequently, $\lambda' = \lambda{\cdot}23.3{\cdot}\mu'$, or $\lambda = 0.0429(\lambda'/\mu')$. If we now measure the repair rate in units of λ', that is, $\mu' = \mu\lambda'$, then $\mu = 0.0429/\lambda$.

With this understanding, the integral in equation 8.2 can be evaluated as a function of μ, and the result is shown in Figure 8.1. As we would expect, the system entropy (i.e., the disorder) increases as we reduce the repair rate. However, remember that the results of the previous chapter were limited to $\lambda < 0.3$, or now $\mu > 0.15$; as μ becomes smaller than this value, the entropy must eventually start to decrease and become zero again for $\mu = 0$, as can be seen from considering the case of a functional system starting out in the state $\varphi(s) = \delta(s-1)$ and with entropy zero. If it is left to itself, that is, without maintenance, the service density function will broaden out, and the entropy increases, as shown above. But in the end, all interactions between the elements will have failed, and $\varphi(s)$ will equal $\delta(s)$, and the entropy will again be zero, which at first seems to make little sense within the analogy with the physical case.

This shows that, while the above similarity between the concepts of entropy in the two domains is intuitive, and the equations relating entropy to distribution function are identical, there is also a very significant difference between the two cases. In the physical case, for example, in the case of a perfect gas, the equilibrium state is the most likely state, the one in which the kinetic energy is distributed

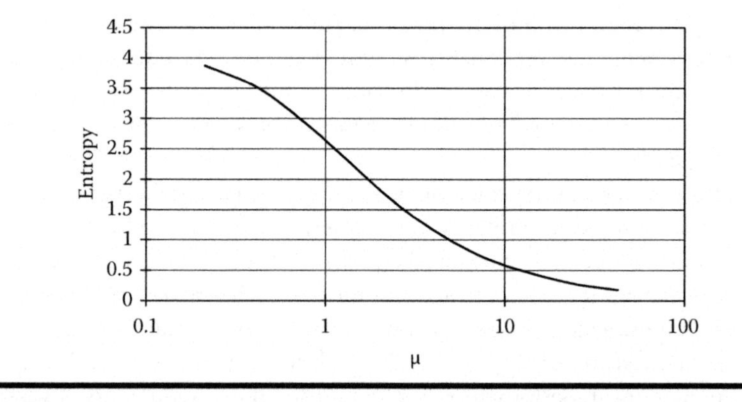

Figure 8.1 The system entropy, E, as a function of the element repair rate, μ.

equally over the molecules, and left to itself, the gas will attain this state, even if it starts out in the highly unlikely state where the whole kinetic energy is carried by one of the molecules. Correspondingly, the entropy goes from zero to its maximum value for this particular value of the total kinetic energy, as noted above. This process of entropy increase is *irreversible*; the entropy of an isolated volume of gas can never decrease of its own accord, whereas for a functional system, while the process is also irreversible, the entropy first increases, but then decreases to zero.

The explanation of this apparently paradoxical behavior of the system entropy is to be found in the definition of the two variables f and s in equations 8.1 and 8.2. The variable f indicates the state of the system; the possible values of f correspond to the possible states of the system, but it is not a measure of the energy of the system. On the contrary, in an isolated system the energy remains constant as the system moves toward equilibrium. The variable s is only indirectly a measure of the state of the system; it is a measure of the service provided by the system, which requires an expenditure of energy. To make the analogy with physics, we have to consider s to be a function not only of the configuration of the system (in the sense of the interactions between the elements), but also of the "energy" of the system (in the sense of being able to do something). As s goes toward zero, not only does the structure decay, but so, of course, does the capacity of the elements to do something (anything). The state $s = 0$ corresponds to a physical system (e.g., a gas) at a temperature of zero Kelvin, and it then also has an entropy of zero. Another view of this issue is discussed in section 8.5.

8.3 A Functional Element Representing Maintenance

Keeping in mind the purpose of this book, to develop functional elements that are useful in describing what systems do, that is, their capabilities, we now realize that most complex systems are capable of doing two things — producing a service, and maintaining this production over a period of time. Therefore, the next step in the process of describing the functionality in more detail is to introduce a separate element to account for the maintenance function, resulting in a subdivision of what in Figure 7.9 was represented as a single element, the service delivery element, as shown in Figure 8.2.

The output of the maintenance element, that is, what it delivers to the service element to maintain the performance of the latter, is characterized by the parameter μ, which can be viewed as a generalized repair rate. It was introduced in chapter 7, together with the failure rate, λ, to account for the stochastic nature of system performance in a most general fashion, without relating either failure or repair to any specific internal features of the system, although it was perhaps natural to assume that the fluctuations in performance resulted from failures and repairs of whatever elements the system was made up of. And indeed, in the definition of a system in section 5.1 we emphasized that the interactions do not in themselves represent any functionality; they simply indicate which of the possibilities for interactions inherent in the elements are utilized in a particular system. Thus, it would follow that interactions cannot fail or be repaired; it is always one of the elements involved in the interaction that fails or is repaired. However, in the following three sections we will develop three different models of the failure and repair process in which it is very convenient to use the interactions, rather than the elements, as the entities that fail and are repaired.

With regard to the two-element system representation in Figure 8.2, we see that while the maintenance parameter, μ, is an important input to the service element, and the value of this parameter is of great consequence to the service seen by the users, the manner in which this value is achieved is not directly important to the users. As a consequence, there is a tendency to concentrate on the design and optimization of the service element and to ignore, or leave as something to be considered later, the fact that the maintenance element may be a system of equal complexity as the service element, and that an optimized system design can only be achieved by optimizing the two elements together. And because the service

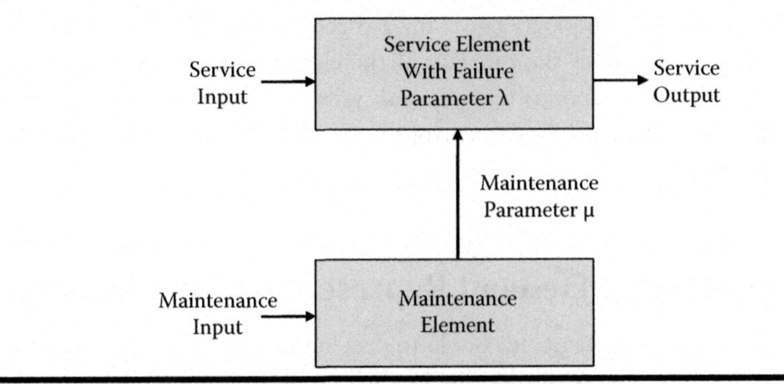

Figure 8.2 Separating out the maintenance function.

element is the one directly visible to the users, it is also quite common to speak of "the system" as being the service element only. We implicitly adopted this common usage in chapter 7 when we discussed classes of systems; these classes were classes of service elements only, but this is in no way meant to diminish the role of the maintenance element. A development of that element into classes could be the subject of a separate work.

8.4 A Model with Binary Interactions

Consider a system consisting of n functional elements, and let us assume that the system functionality has been so finely subdivided into elements (i.e., that n is so large) that the interactions between elements may be approximated by binary interactions. That is, an interaction is either operating or failed, and the system functionality is completely described by the functionalities of the elements plus the adjacency matrix, **A**, as already discussed in section 5.4.

The maximum number of interactions between the elements is $n(n-1)/2$; however, not every element will have the capability of interacting with every other element, and we shall represent the set of all possible interactions that are not already included in **A** by the matrix **H**, in which $h_{ij} = 1$ if an interaction between the elements i and j is possible and $a_{ij} = 0$, and 0 otherwise. As all interactions are bidirectional, the matrices **A** and **H** are both symmetrical, and if, as previously stipulated, the diagonal elements are defined as all equal to zero, the sets of interactions can be represented by upper triagonal matrices.

An interaction is characterized by a failure rate, and we shall, at least initially, assume that the failure rate is the same for each interaction, and denote it by λ. Each interaction is also characterized by a common repair rate, μ; however, when a failed interaction is repaired, there is a certain probability that the repair is done incorrectly, such that, instead of linking the two elements that were previously linked, it links one of them with another element, if another such link is allowed. At any point in time, the interactions actually operating are represented by the (upper triangular) matrix **X**. The repair process is modeled as follows:

Let the index pair of a failed interaction be (u,v), and define two associated sets, **U** and **V**,

$$\mathbf{U} = \{u_j = h_{uj};\ j = u + 1, \ldots, n\},$$

$$\mathbf{V} = \{v_i = h_{iv};\ i = 1, \ldots, v - 1\},$$

and let

$$U = \sum_j u_j ,$$

$$V = \sum_i v_{i.} .$$

If $U + V \neq 0$, there exists at least one allowed, incorrect interaction, and there is then a probability, to be denoted by ρ and satisfying the condition $\rho < 0.1$ (an arbitrary, but reasonable limit), that the repaired interaction will be one of the allowed, incorrect interactions rather than the original one, with each of the incorrect ones being equally likely. However, once an incorrect interaction has been activated, the corresponding correct interaction (i.e., the original one that failed) is blocked, and cannot be repaired until the incorrect one has failed.

The measure of the state of the system is the degree to which the current matrix, **X**, coincides with the design matrix, **A**, and the *correlation*, χ, is defined by

$$\chi = \frac{1}{\sum a_{ij}} \prod a_{ij} x_{ij} , \tag{8.3}$$

remembering that the matrices are triangular.

A small Visual Basic program was developed to simulate the system behavior; it steps through time in steps of a length that is the inverse of the units in which the failure rate λ is measured. For example, if λ is measured per hour, then the step length is one hour. The program starts with the initial condition $\mathbf{X} = \mathbf{A}$ and steps through a total of $1000 \cdot \mathit{Steplength}/\lambda$ steps, calculating the value of χ after each step, and forms the average value of χ from the last 90% of the steps (to eliminate the effect of the transient behavior before the steady state is reached). As an example, the following **A** and **H** matrices were chosen (a and h signify the elements that are equal to 1 in the respective matrices):

$$\begin{array}{cccccccc}
h & 0 & h & a & 0 & 0 & 0 & 0 \\
 & h & 0 & a & 0 & 0 & 0 & 0 \\
 & & 0 & a & h & 0 & 0 & 0 \\
 & & & a & 0 & h & 0 & 0 \\
 & & & & a & a & a & a \\
 & & & & & 0 & 0 & h \\
 & & & & & & h & 0 \\
 & & & & & & & h
\end{array}$$

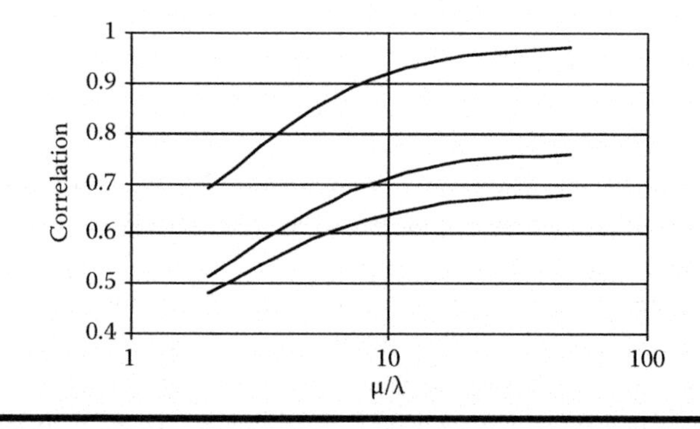

Figure 8.3 Correlation in the error-prone structure model as a function of the repair level, for three values of the error parameter, ρ = 0 (top), 0.01 (middle), and 0.1 (bottom).

and the results are shown in Figure 8.3 for three values of the error parameter ρ, 0, 0.01, and 0.1.

The upper curve in Figure 8.3 (with $\rho = 0$) reflects the normal behavior of a maintained system, in that the correlation goes rapidly toward 1 (i.e., the system is in its intact state most of the time) as $\mu/\lambda >> 1$. But for $\rho > 0$ the behavior is different; not only does the correlation decrease with increasing ρ (as we would expect), but the dependence on μ/λ shows a "flattening out." This is due to the special feature of the model that does not allow errors to be repaired.

8.5 Organizational Disorder

Following on from the model developed in the previous section, we shall continue to use the interactions to describe the changes taking place in the system, but consider that the changes are purely changes to the structure of the system and not caused by any failures as such. That is, an interaction, x_{ij}, may change by one or the other of the indices changing its value, but the number of interactions remains unchanged; such a change will be called a *mutation*, and the rate at which an interaction mutates will be denoted by λ. The state of such a system is completely described by its structure, and the role previously played by the space of basic system states is now played by the space of allowable structures, and an allowable structure will be denoted by $\mathbf{X}$; it is a matrix with elements x_{ij}, where i and j take on values from 1 to n, the number of elements in the system. The number of elements in the set of allowable structures is determined as follows: Using the upper triangular notation of the previous section, let the number of interactions in the adjacency matrix $\mathbf{A}$ be denoted by a and the number of additional allowed interactions in the matrix $\mathbf{H}$ by h, then the number of ways a set of a interactions can be selected from the total set of allowable interactions, say m, is given by

$$m = \frac{(a+h)!}{a!h!}.$$

Let us denote the set of allowable structures by **M**, then the *superstate* of the system is a probability distribution on **M**, which we may denote by f_i, $i = 1, ..., m$. Initially, of course, the superstate is a single basic state, the one given by the adjacency matrix **A**, and f_i takes on the value 1 for the corresponding value of i and is zero for all other values of i, so that the entropy, as defined by the equivalent of equation 8.2, is zero. But left to itself, such a system will change its structure in a random fashion, and after a while will be equally likely to have any one of the allowable structures. That is, the "equilibrium" superstate is a uniform distribution over **M**, and the entropy takes on its maximum value, $\ln(m)$.

In this model of system behavior, the concept of entropy is similar to the one we are used to from thermodynamics, and does not display the peculiar behavior we saw at the end of section 8.2. However, we have achieved this by defining entropy in terms of structure, a variable that is once removed from the performance (parameterized, e.g., by the quality of service), so that we now have the problem of relating structure to performance. But we can now also get another perspective on *why* the problem at the end of section 8.2 arose. In both the case of thermodynamics (i.e., the perfect gas) and that of system structure, the changes on a microscopic level, that is, the individual collisions of the gas molecules and the mutations of the individual interactions, are *reversible.* The irreversibility of the macroscopic process arises solely as a result of the statistics; the macroscopic process is irreversible because reversing it is so highly unlikely. But in the case where the macroscopic change is due to the failure of elements, the microscopic process, that is, the failure of an individual element, is itself irreversible, and no equilibrium state is reached, only a state of total destruction.

In most systems of practical interest, the structure is not allowed to change unchecked; maintenance is done in order to maintain the original structure or one as close to it as possible. This maintenance effort can be parameterized by a *restoration rate*, μ, defined as the probability per unit time that an interaction in **X** not in **A** will be changed back to an interaction in **A**, additional to any change due to the mutation rate, λ. As a result, the distribution function f_i will peak around the **A** structure, and the entropy will take on some value intermediate between zero and $\ln(m)$. Obtaining an explicit expression for this distribution as a function of μ is not practical nor would it be particularly useful; what we can obtain is, as in the previous section, the expectation value of the correlation between **A** and **X**, $\chi(\mu)$. This function must satisfy the boundary conditions $\chi(0) = a/h$ and $\chi(\infty) = 1$, and an example is shown in Figure 8.4 for the nine-element system used as an example in section 8.4. In this figure, the upper curve is for the **H**-matrix used in the previous section, that is, with $a/h = 0.5$, whereas the lower curve is for the case when the

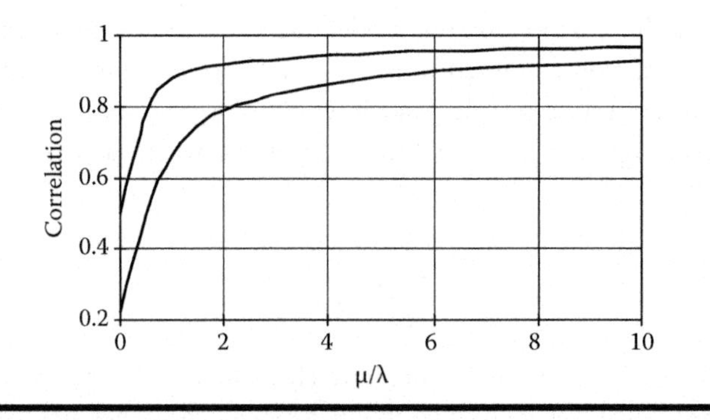

Figure 8.4 Correlation for the mutation model for two different H-matrices; the top one with *a/h* = 0.5, the lower one with *a/h* = 0.22.

allowable interactions equal the total number of possible interactions, that is, 36 interactions and $a/h = 0.22$.

The shape of the curves in Figure 8.4 is similar to that of the $\rho = 0$ curve in Figure 8.3 (but note the different μ/λ-scales); however, in that previous case, the a/h ratio was irrelevant. What this model demonstrates is the (not surprising) fact that the greater the freedom (i.e., the possibilities for change) the greater the effort required to maintain a particular structure (organization).

8.6 Coherence

In this last model of the process of failure and repair we shall consider a large group of systems in which the elements are either identical or similar, and all cooperate in producing the system output. Examples of such systems are all types of organizations (companies, military units, sports teams, political parties, etc.), markets (consumers), physical many-body systems (lasers), and engineered systems (phased arrays). In particular, we shall investigate a common feature of these systems called *coherence*, and to this end we consider a simplified, generalized system consisting of n identical elements, each producing an output characterized by two parameters, an *amplitude*, which, because the elements are identical, we may set equal to 1, and a *phase*, φ, which can take on any value in the range $-\pi$ to $+\pi$. That is, the element output can be represented by a two-dimensional vector in a polar coordinate system, with a fixed length of 1, and the system output will be the sum of these n vectors. This work was previously reported in *Systems Engineering*.[1]

Each element is subject to an influence that tends to change the phase in a random fashion; such changes take place at a constant *failure rate*, λ. But there is also an interaction between the elements that tends to align their outputs. Each element sees the combined output of the other $n - 1$ elements, called the *interaction*, and, at a constant *repair rate* μ, the phase of each element is aligned with the interaction phase. However, the model takes account of the fact that in many systems the interaction is limited to nearest neighbors or some other small group of elements (just think of people in society, in an organization, etc.) with, say, n_0 members, by multiplying the interaction amplitude by the factor

$$\frac{n_0}{n}(1-e^{-\frac{n}{n_0}}).$$

The behavior of this system is qualitatively as follows: If the system is initially in a state where all the element phases are identical, the coherence χ is initially 1, but under the influence of the failure process, the phases of the element outputs will start to differ from one another and the system output will start to decrease in a random manner. That is, both the system amplitude and the system phase become stochastic variables. For reasons that will become clear in a moment, it is preferable to use the amplitude as a measure of system performance, and the *expectation value of the relative amplitude* will be called the *coherence* of the system and be designated by χ; it takes on values in the range 0 – 1.

If there is no repair process, that is, $\mu = 0$, the coherence decays to its minimum value, which may be denoted by η and is a function of n. The function $\eta(n)$ was determined by the Monte Carlo method, and the result is shown in Figure 8.5. But if $\mu > 0$, then the coherence will stabilize at some steady-state value greater than η, as will be shown presently.

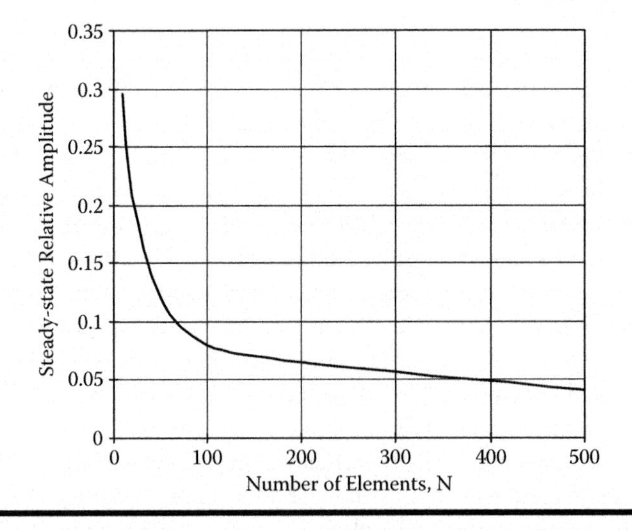

Figure 8.5 The value of the coherence, defined as the amplitude at t = ∞ divided by the number of elements in the system, *n*, for random distribution of the element phases, as a function of *n*.

If the system is initially in a chaotic state where each element phase is equally likely to be anywhere in the range $-\pi$ to $+\pi$, the coherence χ starts out being equal to η and then increases to its steady-state value, determined by μ/λ as above. This occurs because the element phases lock together and fluctuate around the phase of the system output, but the value of this system phase is arbitrary, and it also "drifts" in either direction. That is, the frequency spectrum of the system phase has significant components at both high and low frequencies, whereas the frequency spectrum of the amplitude decreases rapidly with decreasing frequencies (beyond a certain frequency, as we shall discuss shortly). This is the reason why the amplitude is a more convenient measure of system performance than the phase.

In order to study this behavior in a more quantitative way, a small Visual Basic program was developed that simulates the system behavior by stepping through time, again in steps with a duration equal to the unit of time in which λ is measured, with the element phase angles at the beginning of the step equal to $\vartheta 1(i)$, $i = 1, \ldots, n$, (called the "old" values), after the calculations equal to $\vartheta 2(i)$ (called the "new" values), and then, at the end of each step, $\vartheta 1(i) = \vartheta 2(i)$. The calculation starts from one of two initial conditions; either a uniform probability distribution of element phases (mode = 1), or all element phases equal to 0 (mode = 0). Then, for each element, the following sequence of calculations is carried out: First, if a random number is less than λ, the element phase is set to a random value within the range $-\pi$ to $+\pi$. Then the interaction amplitude A_i (i.e., the combined output from the $n - 1$ other elements) at the element is calculated from $\vartheta 1$, and if $\mu A_i/(n - 1)$ is greater than a random number, the element phase is set equal to the interaction phase. Once this has been completed for all n elements, the system amplitude and phase is calculated by converting each element vector to rectangular coordinates, adding the x- and y-components, and converting the result back to polar coordinates. Thirty of these steps are combined to form a GroupStep by taking the average of the amplitude and phase, and 20 such GroupSteps form a single Run. For each set of parameter values, ten Runs are completed, and the mean and standard deviation of amplitude (and phase, but that is not yet relevant, as was explained earlier) for each GroupStep calculated. The result of a calculation for a system with 10 elements is shown in Figure 8.6, and Figure 8.7 shows the results for the same parameter values, but for a system of 100 elements.

At this point, it may be appropriate to comment briefly on the statistics involved in this and the previous two models and, indeed, in any model that considers the dynamics of stochastic processes of this nature. At each time step, the measure of system performance (correlation, coherence) changes in a "random" fashion. The word "random" has been put in quotes because it needs to be qualified. If we understand "random" to mean "unpredictable," then it applies to our models. But if we understand "random" to also imply "equally likely to lie anywhere within its range," then this is not correct, and the reason is that there is a correlation between values of the stochastic variable that depends on the time interval between the values. In particular, if the rate of change of the variable (e.g., the failure rate λ) is comparable

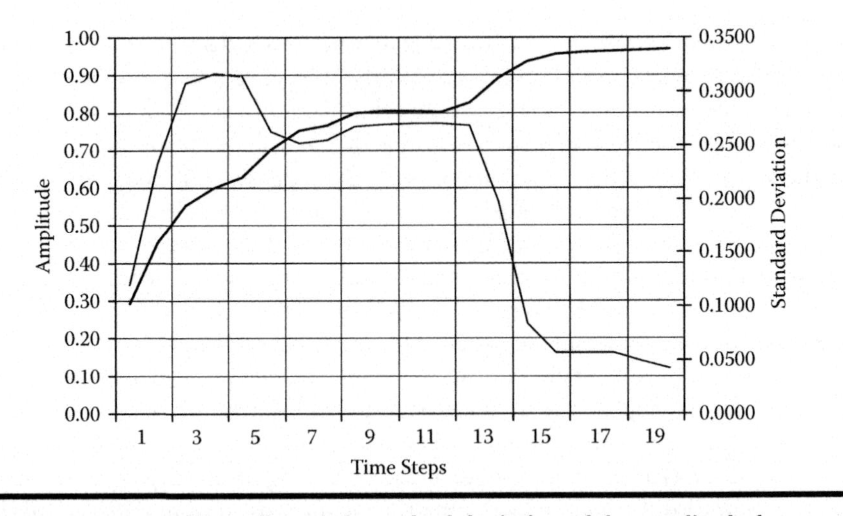

Figure 8.6 Mean (heavy line) and standard deviation of the amplitude for a system of 10 elements with λ = 0.0005 per unit time, μ = 0.02 per unit time, and $n_0 = n$. The time step equals 30 units of time.

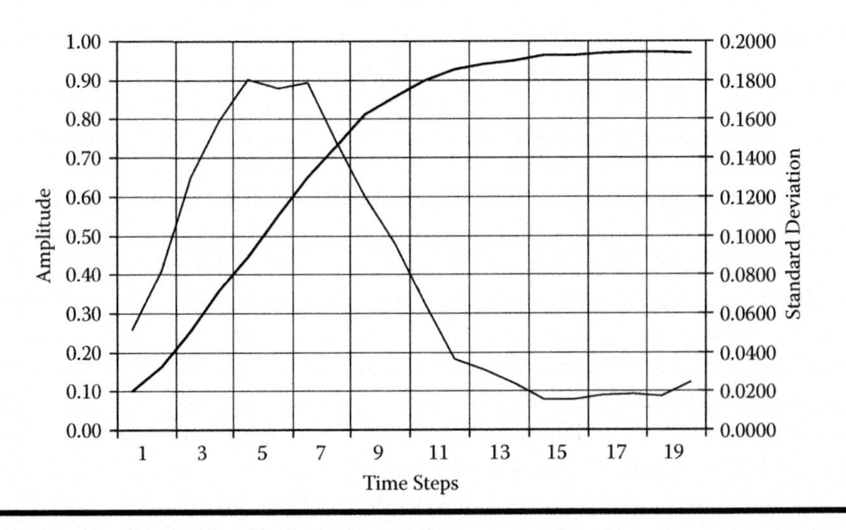

Figure 8.7 The same calculation as in Figure 8.6, using the same parameter values, but for a system with 100 elements.

to or less than the inverse of the interval between values, then two adjacent values are increasingly likely to be equal. The effect of this is best demonstrated by a very simple example: Let x be a stochastic variable that is equally likely to have a value anywhere in the range 0 – 1, and let the probability of x changing its value in a unit time interval equal λ. Start out with $x = 0.5$, and step through time in unit intervals, at each step generating a random number in the range 0–1 and checking if this number is less than λ. If yes, a new random number is generated and the value of x changed to it; if no, the value of x remains unchanged. For the moment, assume that $\lambda = 1$; then a very long sequence of (truly) random values of x is generated, say, 10,000 values, and the average value, which will be denoted by x_0, and the standard deviation, which will be denoted by σ_1, can be calculated. As we know, as the number of steps in the sequence increases, these values are increasingly likely to lie increasingly close to the values $x_0 = 0.5$ and $\sigma_1 = 0.2887$.

Now, group the values of x into groups of n consecutive values (i.e., in the language of statistics, each group forms a "sample" from the original "population" of values) and form the averages of the values in each group; we end up with a new sequence of random numbers, but with only $1/n$ the number of members as compared to that of the original sequence. The average of these new values obviously remains equal to x_0, but the standard deviation takes on a new value, σ_n, and from basic sampling theory we know that $\sigma_n = \sigma_1/\sqrt{n}$. So far so good; this is all very basic, but what happens as the value of λ starts to decrease from 1? The values of x within a group are no longer independent; indeed, if $\lambda \ll 1/n$, they are likely to all be equal, and we would expect the standard deviation σ_n to increase. This is exactly what happens, as shown in Figure 8.8, and the conclusion of this little digression is simply that we have to be careful when making statements about average or equilibrium behavior of a system from observations over a limited period of time or, conversely, when choosing what is an adequate period of time for determining equilibrium behavior.

Returning to the coherence model, Figures 8.6 and 8.7 illustrate how the element phases lock together over a short period of time (in the cases shown about 250 units of time), after which a "steady state" or "equilibrium" is achieved with the amplitude having a probability density distribution with a mean and a standard deviation that depend on the value of the ratio μ/λ (this will be shown shortly). The quotation marks are there to remind us that when dealing with stochastic systems,

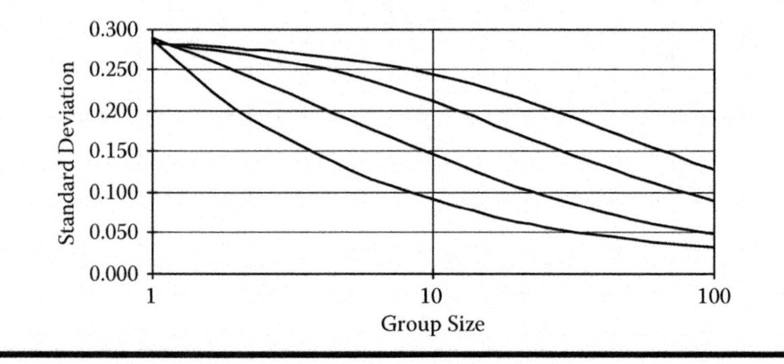

Figure 8.8 The standard deviation for a sequence of "random" numbers as a function of the group size, *n*, and the change probability, λ, showing curves for values of λ of (from top to bottom) 0.1, 0.2, 0.5, and 1.

words like "steady" have to be used with caution, and one aspect of this can be demonstrated in the coherence model.

If we run the model for a long time (i.e., many time steps), we can observe that the system phase angle, φ, "flips" to a new value from time to time, and such an event may be termed a *failure* of the system. What happens is, of course, that so many of the element phases take on random values at the same time that the system is momentarily back to its initial state of random phases, and then goes through the transient locking phase again, but to a new value of φ. An example of such a failure is shown in Figure 8.9; note, however, that the time scale is again in units of 30 units of time, so that an amount of averaging has already taken place. If a higher resolution had been used, for example, showing every time step, then the amplitude would have dipped down to its minimum value (about 0.28 for a system of 10 elements, as shown in Figure 8.5).

As an aside, such "flips" in the system phase can be observed in some everyday situations, one of them being a meeting to discuss and finalize a position on an issue. Documentation on the issue has been circulated previously, and the issue has been discussed one-on-one with all participants prior to the meeting, with everyone agreeing that a certain general direction would be the preferred one. However, during the meeting, discussing the details of the preferred direction, the opinions of the participants start swinging around in all directions, and suddenly lock on to a direction that is quite different to the one preferred prior to the meeting. No new facts were presented, no new aspects of the issue were raised that had not been raised before; the "flip" is solely a result of the system interaction. It must be that feelings,

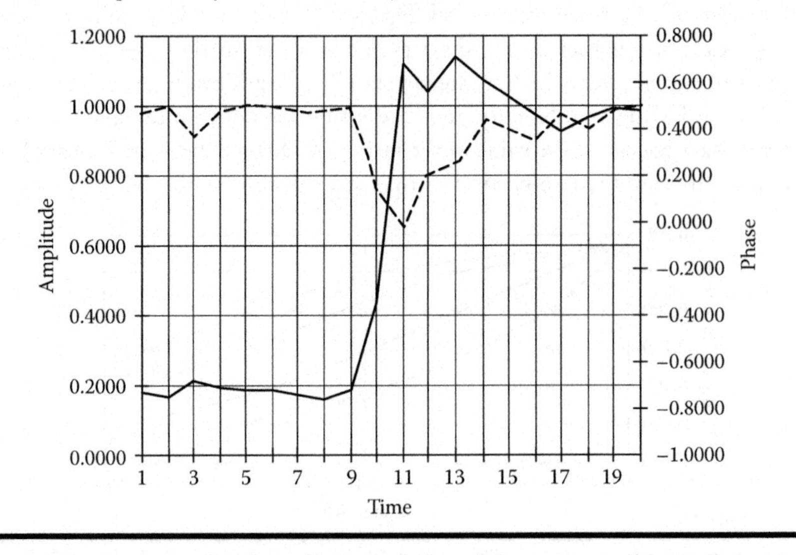

Figure 8.9 An example of a coherence failure for a system with 10 elements and $\lambda = 0.005$, $\mu = 0.1$, and $n_0 = 1$. The full line is the phase, the dashed line the amplitude. In this case, the MTBF was about 52,000 units of time, or $260/\lambda$.

opinions, attitudes, etc. that are repressed in a one-on-one situation because one wants to appear rational can blossom in a system setting; rationality is no longer the main criterion, it is the good feeling of being in agreement with everybody else.

A small routine was developed to detect coherence failures, and the dependence of the MTBF on the ratio μ/λ and the number of elements, n, and the results are shown in Table 8.1.

As we noted earlier, once the element phases have locked together, and disregarding the infrequent coherence failures, the system output amplitude is a stochastic variable with a probability density distribution characterized by a mean value and a standard deviation. These two statistics are, of course, related, as they both reflect the degree of coherence, but, first, the standard deviation depends on the number of elements in the system (decreases with increasing number) whereas the mean does not (at least to a first approximation) and, second, we have already defined coherence as the mean (or expectation value) of the amplitude. The results obtained for n in the range 10–50 elements are given in Table 8.2.

In the model so far, there has been no external influence, and therefore nothing that could fix the phase of the system output; it was arbitrary. We now amend the model such that whenever an element output undergoes a "failure" and the phase angle is set to a random value, this random value is multiplied by $(1 - \varepsilon)$. That has the effect of making $\varphi = 0$ the preferred value of the system output, and some qualitative observations are:

a. Even such a small external influence as $\varepsilon = 0.01$ is enough to ensure that the system phase aligns itself to $\varphi = 0$.
b. The duration of the transient "locking-in" behavior is now dependent on both μ/λ and ε, with an increase in ε from 0.01 to 0.1 decreasing the duration by at least a factor of three.
c. The MTBF increases dramatically with ε; above $\varepsilon = 0.05$ no failures could be detected.

Table 8.1 The MTBF for the Coherence Model, in Units of $1/\lambda$

$\mu/\lambda \backslash n$	*10*	*20*	*50*
5	40	300	650
10	54	450	1200
20	180	765	

Table 8.2 Coherence as a Function of the Maintenance Level

μ/λ	5	10	20	50
Coherence	0.82	0.90	0.95	0.98

The coherence model is particularly interesting in designing systems where some or all of the elements are persons; the influence represented by ε can then be interpreted as the management effort, but can also be a truly external influence, such as fashion. These issues require further development.

Notes

1. Aslaksen, E.W., A model of system coherence, *Systems Engineering*, 6, 2003, 19–27.

Chapter 9

The System Designer's Assistant

9.1 Introduction

Up until now, we have been concerned with developing a theoretical foundation for design in the functional domain, and have looked at a number of fairly general features of functionality. Hopefully this has led you to the conviction that there is such a thing as design in the functional domain, that it is possible to take design decisions, based on a set of user requirements, *before* there is any consideration of any particular physical system that might meet these requirements. But this is not enough; in order for design in the functional domain to be of any practical value and make a positive contribution to the overall design process, it must be *efficient,* and to this end we need to have tools that support the methodology. The outline of one such tool, called the System Designer's Assistant (SDA), is presented in this last chapter.

This outline has gone through several different versions since it was first developed as a tool to support the design of a particular system, the Mine, which we have encountered already. That version was really an expanded life cycle costing model that increased its level of detail as the project went from feasibility to detailed design, and it was built in the form of an Excel spreadsheet, as discussed in *The Changing Nature of Engineering.*[1] The idea of turning it into a tool for a variety of projects by making it a system of elements that could be combined in different ways and to which new elements could be added at any time was developed on and off over a number of years, and presented in various fora, for example, at the 4th International Symposium of the National Council of Systems Engineering.[2] This development resulted in a version programmed in Microsoft Visual Basic 6, with each element a separate form. However, this turned out not to be very practical; the framework within which a model for a particular project could be constructed was too rigid. It did not allow each element to be used to explore that particular aspect of the system in isolation, and the elements could not be easily modified by the user. The regular version was a compiled version, so no changes beyond changes to the parameters already included in the model were possible, and changing the source code version required a reasonable proficiency in VB programming.

The present version, which should be seen as a proposal, is again based on Excel, but with each element a separate spreadsheet. While this may not be very elegant as software tools go, it has a number of advantages. First, the VBA code, which represents the functionality of the element, is directly available to the user and can be followed step by step using the debugging tool, providing easy insight into exactly how the element models the functionality. Second, the user can create new elements by copying and modifying existing elements. Third, each element can be run in stand-alone mode, allowing a particular aspect to be studied in detail without running a large model. And, fourth, the spreadsheet user interface is very familiar to engineers and allows them to present the results in whatever way is most suitable in a particular case, for example, in the form of special graphics.

Of course, there is also a considerable downside to this approach; because it is so easy to make changes to the elements, it requires strict configuration management. A particular version of the SDA contains a set of *standard elements*, which should not be modified, but we shall return to this issue when we look at the structure of individual elements in section 9.4.

9.2 Structure of the SDA

The program is contained in an Excel workbook with a number of worksheets. The first sheet is used to define and run a model in the form of a system of functional elements; the following sheets each represent one functional element. A functional element is one or more aspects of a system's behavior, and this aspect is modeled in the VBA module attached to each spreadsheet. The calculations contained in such a module can be executed in two different ways: first, by being called from the module attached to the first, or model spreadsheet, as part of the program executing a complete model; second, from the element sheet itself, by clicking the Run button. In the first case, the module fetches its input parameter values from a list displayed on the model sheet and writes the values of its output parameters to the same list. This list can, of course, not make any distinction between input and output parameters, as what is an output parameter of one element is an input parameter of another element; this is how the elements form a system.

In the second case, the module fetches its input parameter values from a list of input parameters displayed on its associated spreadsheet and writes its output values to the list of output parameters on the sheet.

In addition to its input and output parameters, which are required in order to interact with other elements, an element may produce more detailed information on the particular aspect it is modeling, and this can be displayed in various ways using the graphics functions available in Excel. Such displays may be part of the standard elements, but may also be added by the user.

For complete documentation of a version of the SDA the workbook needs to be complemented by a *dictionary* of parameters and an *element definition document*, which defines each element in terms of its behavior, along the lines of the development of the service delivery element in chapter 7. In this regard, it should be recalled that while a functional element is defined in terms of the equations that transform the input parameter values into values of the output parameters, the module carrying out the numerical calculations is a *model* of the element. The behavior of the model should ideally mirror that of the element exactly; in practice it will only do so to an extent determined by the numerical accuracy of the variables and the algorithms used. For example, the accuracy of an integral will depend on the step size and the integration algorithm, the accuracy of statistical data on the sample size, and so on.

The documentation of the motivation for the SDA and of the fundamentals of functional elements is, of course, this book.

9.3 The Model Worksheet

The first sheet in the workbook, the model sheet, allows the user to build a model of system behavior or of aspects of that behavior by forming a system of interacting elements. As we discussed in several places earlier, functional elements, as they are defined in this book, fit into a hierarchical ordering, with the irreducible element forming level 0 (the top level). Elements on level 1 are either elements that have C or R (or both) as output parameters, or elements that complement these, such as the service delivery element in section 7.6. Elements on level 2 are those whose output parameters are input parameters for elements on level 1, and so on. In the model sheet, the elements contained in the SDA are listed on the left hand side, as shown in Figure 9.1, but while the return on investment (ROI) element is usually at the top of the list, the other elements do no have to be in any particular order, and an element at any level can be attached to the bottom of the list.

A model is a set of elements related in a particular way, and the relationship is that the outputs of elements on one level are linked to the inputs of element on the next level up. Consequently, the elements on the lowest level have to be executed first, and the number in the column to the right of the element name shows the order in which the elements are executed, with the element with the highest num-

System Designer's Assistant, Version 1.0, dated 27-Jan-2008

Run Model

Number of Elements in Library: 4

Model Size: 4

Model Composition:

Sheet	Library List	Include
Sheet2	Return on Investment	1
Sheet3	Basic Cost	2
Sheet4	Basic Revenue	3
Sheet5	Service Delivery	4

Parameter	Value
Return on Investment, ROI [%]	25
Cost, C [k$]	302,529
Revenue, R [k$]	378,432
Prod/D&D [k$]	8000
Prod/I&T [k$]	200000
Prod/O&M[k$/y]	10000
Prod/Decom [k$]	40000
Ops/D&D [k$]	1000
Ops/O&M [k$/y]	6000
Ops/Decom [k$]	1000
Maint/D&D [k$]	1000
Maint/I&T [k$]	2000
MaintO&M [k$/y]	6000
Support/D&D [k$]	100
Support/I&T [k$]	500
Support/O&M [k$/y]	800
Support/Decom [k$]	10
L1 [y]	2
L2 [y]	3
L3 [y]	15
L4 [y]	1
COF [k$]	1000
MTBF [hours]	14,439
Discount rate [%]	8

Parameter	Value
DF Mean, s0	0.9757
SDF Deviation, σ	0.0349
The Spike, Chi	0.3800
Nominal Value [k$/y]	70,000
Performance limit, s1	0.8
Value limit, s2	0.95
Failures, λ [per h]	0.01
Repair rate, μ [per h]	0.01
Failure effect, Δ	0.1

Figure 9.1 The model worksheet.

ber being executed first. This order has to be decided by the user of the SDA; it is not generated automatically.

The user also has to type in the values of the input parameters of the elements that have no element below them. In the example shown in Figure 9.1, which contains only the elements developed in chapter 7, the input parameters are all the ones listed except ROI, cost, revenue, MTBF, s_0, σ, and χ.

It is not necessary that these "input" elements all be on the same level; for example, it is perfectly good to build a model in which the cost is provided by the basic (level 1) cost element and the revenue is provided by a set of several elements occupying more than one level.

When a model is running, the word "Running" appears in red font to the right of the command button. However, depending on the model, this time may be so short that the message is not visible.

9.4 Element Format

A functional element is represented by a worksheet and an associated VBA module. The worksheet displays the name, version, and date (when last modified) of the element at the top, and below that lists the input and output parameters, as shown in Figure 9.2.

However, below this standard section, element worksheets may display calculated values in various forms, using standard Excel graphics. As an example, the basic cost element displays various views of the cost matrix, as shown in Figure 9.3.

The VBA module associated with each element worksheet has a fixed format, consisting of the following eight components:

1. The *Declaration section*, where the options and module variables are declared.
2. The *cmdRun_Click subroutine*, which executes when the Run button on the element worksheet is clicked. It calls the three subroutines Read_Input, Compute, and Write_Output.

Element Name:	**Service Delivery**		
Version:	**1.0**		**Run**
Date:	**27-Jan-08**		

Inputs:		**Outputs:**	
Performance limit, s1	0.8	SDF mean, s0	0.9757
Failures, λ [per h]	0.01	SDF deviation, σ	0.0349
Repair rate, μ [per h]	0.01	The Spike, Chi	0.3800
Failure effect, Δ	0.1	MTBF	14,439
		MTTR	30
		AVAY	0.9979

Figure 9.2 The worksheet for the element service delivery.

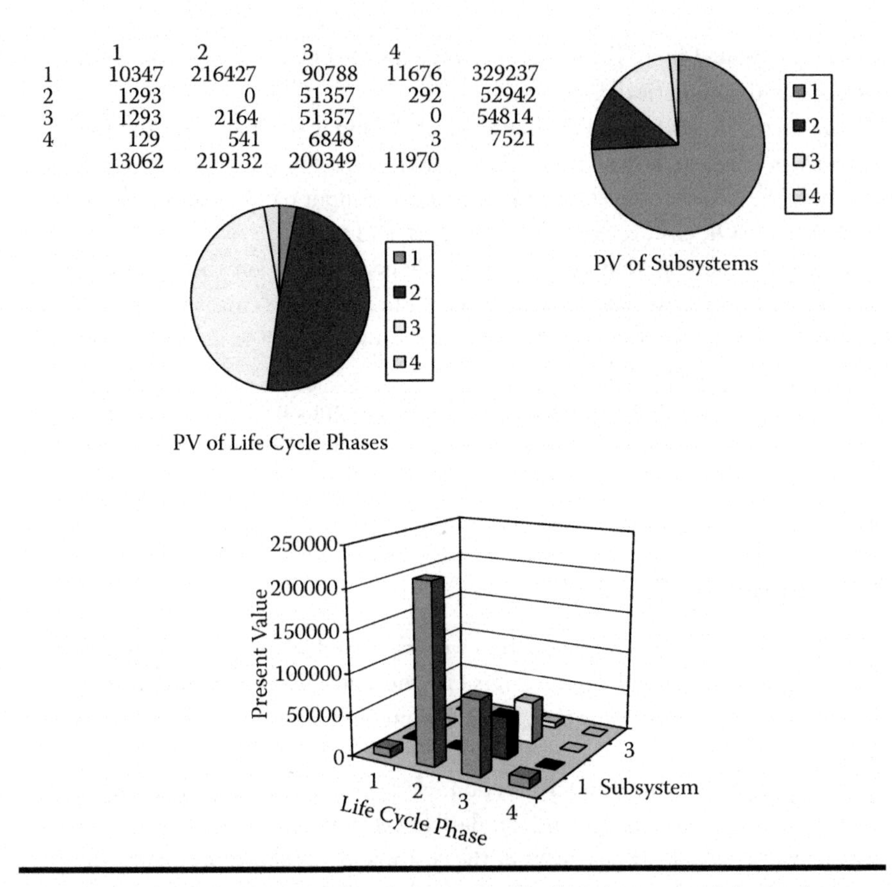

	1	2	3	4	
1	10347	216427	90788	11676	329237
2	1293	0	51357	292	52942
3	1293	2164	51357	0	54814
4	129	541	6848	3	7521
	13062	219132	200349	11970	

Figure 9.3 Various views of the cost matrix, displayed as part of the basic cost element worksheet.

3. The *Execute subroutine*, which executes when called from the Model module. It calls the three subroutines Read_Master_Input, Compute, and Write_Master_Output.
4. The *Read_Input subroutine*, which reads the input values from the worksheet.
5. The *Compute subroutine*, which carries out the calculations modeling the element functionality.
6. The *Write_Output subroutine*, which writes the results of the Compute subroutine to the worksheet.
7. The *Read_Master_Input subroutine*, which reads the input values from the Model worksheet.
8. The *Write_Master_Output subroutine*, which writes the results from the Compute subroutine to the Model worksheet.

As an example, the code for the basic cost element is shown below:

```
'Basic Cost Element
'Version 1.0, 28-Jan-2008
'This element transforms the basic 4 x 4 Cost Matrix into the
'single parameter Cost (as input to the expression for the ROI),
'taking account of the cost of failure.

Option Explicit
Option Base 1
'Input parameters
Private sngCost(4, 4) As Single 'Cost matrix [k$]
Private intL(4) As Integer 'Life cycle length vector [y]
Private sngCof As Single 'Cost of failure [k$]
Private lngMtbf As Long 'MTBF [h]
Private sngDiscount As Single 'Discount factor
'Output parameters
Private sngC As Single 'The cost C entering into the ROI [k$]
'Local output variables
Private sngCpv(4, 4) As Single 'Present value of cost elements
'Internal variables
Private sngH(4) As Single 'The transformation vector

Private Sub cmdRun_Click()
 Read_Input
 Compute
 Write_Output
End Sub

Sub Execute()
 Read_Master_Input
 Compute
 Write_Master_Output
End Sub

Sub Read_Input()
 With ActiveSheet
  sngCost(1, 1) = Cells(6, 3).Value
  sngCost(1, 2) = Cells(7, 3).Value
  sngCost(1, 3) = Cells(8, 3).Value
  sngCost(1, 4) = Cells(9, 3).Value
  sngCost(2, 1) = Cells(10, 3).Value
  sngCost(2, 3) = Cells(11, 3).Value
  sngCost(2, 4) = Cells(12, 3).Value
  sngCost(3, 1) = Cells(13, 3).Value
  sngCost(3, 2) = Cells(14, 3).Value
  sngCost(3, 3) = Cells(15, 3).Value
  sngCost(4, 1) = Cells(16, 3).Value
  sngCost(4, 2) = Cells(17, 3).Value
```

```
  sngCost(4, 3) = Cells(18, 3).Value
  sngCost(4, 4) = Cells(19, 3).Value
  intL(1) = Cells(20, 3).Value
  intL(2) = Cells(21, 3).Value
  intL(3) = Cells(22, 3).Value
  intL(4) = Cells(23, 3).Value
  sngCof = Cells(24, 3).Value
  lngMtbf = Cells(25, 3).Value
  sngDiscount = Cells(26, 3).Value
 End With
End Sub

Sub Compute()
 Dim sngD As Single 'The yearly multiplication factor
 Dim I As Integer
 Dim J As Integer
 'Intialise zero value elements
 sngCost(2, 2) = 0
 sngCost(3, 4) = 0
 'Add the yearly cost of failure to C13
 sngCost(1, 3) = sngCost(1, 3) + sngCof * 8760 / lngMtbf
 'Determine the transformation vector
 sngDiscount = sngDiscount / 100
 sngD = 1 + sngDiscount
 sngH(1) = 2 * sngD ^ intL(2) * (sngD ^ (intL(1) + 1) -
sngDiscount * (intL(1) + 1) - 1) / (sngDiscount ^ 2 * intL(1) *
(intL(1) + 1))
 sngH(2) = (sngD ^ intL(2) - 1) / (sngDiscount * intL(2))
 sngH(3) = (sngD ^ intL(3) - 1) / (sngDiscount * sngD ^ intL(3))
 sngH(4) = (sngD ^ intL(4) - 1) / (sngDiscount * sngD ^
(intL(3) + intL(4)))
 'Transform and sum phase costs
 sngC = 0
 For I = 1 To 4
  For J = 1 To 4
   sngCpv(J, I) = sngCost(J, I) * sngH(I)
   sngC = sngC + sngCpv(J, I)
  Next J
 Next I
 'Change the reference point for C
 sngC = sngC * sngD ^ (-(intL(1) + intL(2)))
End Sub
Sub Write_Output()
 Dim I As Integer
 Dim J As Integer
 'Display the local variables
 For I = 1 To 4
  For J = 1 To 4
```

```
    Sheet3.Cells(30 + J, 1 + I).Value = sngCpv(J, I)
   Next J
  Next I
  'Display the output parameters
  ActiveSheet.Cells(6, 7).Value = sngC
End Sub

Sub Read_Master_Input()
  sngCost(1, 1) = Sheet1.Cells(9, 5).Value
  sngCost(1, 2) = Sheet1.Cells(10, 5).Value
  sngCost(1, 3) = Sheet1.Cells(11, 5).Value
  sngCost(1, 4) = Sheet1.Cells(12, 5).Value
  sngCost(2, 1) = Sheet1.Cells(13, 5).Value
  sngCost(2, 3) = Sheet1.Cells(14, 5).Value
  sngCost(2, 4) = Sheet1.Cells(15, 5).Value
  sngCost(3, 1) = Sheet1.Cells(16, 5).Value
  sngCost(3, 2) = Sheet1.Cells(17, 5).Value
  sngCost(3, 3) = Sheet1.Cells(18, 5).Value
  sngCost(4, 1) = Sheet1.Cells(19, 5).Value
  sngCost(4, 2) = Sheet1.Cells(20, 5).Value
  sngCost(4, 3) = Sheet1.Cells(21, 5).Value
  sngCost(4, 4) = Sheet1.Cells(22, 5).Value
  intL(1) = Sheet1.Cells(23, 5).Value
  intL(2) = Sheet1.Cells(24, 5).Value
  intL(3) = Sheet1.Cells(25, 5).Value
  intL(4) = Sheet1.Cells(26, 5).Value
  sngCof = Sheet1.Cells(27, 5).Value
  lngMtbf = Sheet1.Cells(28, 5).Value
  sngDiscount = Sheet1.Cells(29, 5).Value
End Sub

Sub Write_Master_Output()
  Sheet1.Cells(7, 5).Value = sngC
End Sub
```

As already mentioned, a feature of this particular format of the model is that it requires only a modest knowledge of Excel and VBA in order to generate new elements describing detailed aspects of real systems and, in particular, to do this by modifying existing elements. However, the downside of this is that it could lead to pure anarchy; a mass of poorly documented elements understandable and useful only to their creators, thereby undermining the aim of increasing the efficiency of the design process by having a set of standard elements. One approach that would preserve both the standardization and the flexibility would be to form a user community, for example, in the form of a Wiki, where users could post and explain elements they have developed, and where other users could try them out and comment on them, and then have a central database, in the form of a single workbook, which contains all the elements that have been shown to have a certain degree of general applicability. This workbook would be under version control, and with each added element meeting the requirements for a standard element with regard to format, annotation, etc.

9.5 Application and Further Development of the Methodology

Functional models, for example, performance models or reliability models, are nothing new, and most of us have probably both developed and used a number of such models in our work. But in most, if not all cases, they have been modeling the performance or reliability of an existing system; that is, the system design had already been completed and the modeling was used to either verify the design or optimize the values of some of the design parameters. In this book we have argued that the functional domain, that is, the "world" of functional models, can be viewed as having an existence independent of any specific physical object, and that in the design process, which starts out with a set of user requirements, the functional domain should be *prior* to the physical domain. We have developed the fundamental features of the functional domain, and this necessitated defining a number of concepts.

We have also argued that the main reason the functional domain is generally not used in the design process is that the physical domain is so well developed and so familiar to us; we have millions of standard building blocks at our disposal, whereas the functional domain is comparatively unpopulated. But we also recognize that despite the existence of all these building blocks, the process of designing a physical system that meets all the user requirements, including minimal cost, becomes increasingly difficult and inefficient as the complexity of the system increases. The purpose of carrying out design in the functional domain is to reduce this complexity before making the transition from requirements to physical realization. The issue is how to make this additional step in the design process cost-effective, that is, that the effort expended on the top-down process is less than the resultant reduction in effort in the bottom-up process, and we have concluded that the only way to achieve this is to develop standard functional elements.

Consequently, when we consider "application of the methodology," we do not mean building another one-off functional model, but approaching the design of a system in a top-down manner, using functional elements that meet the requirements we have developed in this book. In particular, it means applying the methodology from the very inception of a project and starting with the ROI as the overarching purpose of the system. And this brings us to perhaps the most significant barrier to the application of the methodology — very many projects are not developed in a continuous manner, but in two phases that view the project from such disparate points of view that they are almost disjoint. In the first phase, the project is (correctly) viewed as an investment opportunity, but without involving anyone with an understanding of the design process. The persons involved are business managers with a law or commerce background, bankers, and investors, and the only way they can define the project is in terms of physical entities; a factory, a mine, a warehouse, a fleet of trucks, etc., usually by reference to existing entities. Engineers are only involved in the second phase — the realization of the venture — and the design is restricted to the design of already defined physical entities. If the functional design process is to be effective (or even possible), systems engineers need to be involved from the very beginning of the project definition.

With regard to the further development of the methodology, we can see that there are two aspects that need development. The first is a practical, efficient, Excel-based framework in which specific functional models can be constructed, and an attempt at this was outlined in the earlier part of this chapter. However, there is still some way to go before this particular framework can be presented as a commercial tool, and it would be beneficial if some of the readers of this book would take up that development.

The second aspect is, of course, that many, many more standard functional elements need to be developed and made available on an open market. The ideal start would be a user group within an existing association, such as IEEE or INCOSE, with an Internet site where elements can be posted and from where elements can be downloaded, either for free or as shareware, but that has not been achieved so far. [3]

Notes

1. Aslaksen, E.W., *The changing nature of engineering*, McGraw-Hill, New York, 1996, chapter 13.
2. Aslaksen, E.W., A leadership role for INCOSE, *Proceedings of the 4th International Symposium of the National Council of Systems Engineering*, San Jose, August 1994.
3. With regard to sharing composable elements of valuable functionality and the benefits this would bring, see Norman, D.O. and White, B.E., Asks the Chief Engineer: "So what do I go do?", *Insight*, 11, 2008, 25–27.

Index

D

E

F

G

H

I

L

M

T

U

V

W